Impact of Climate Change on Water Resources

Impact of Climate Change on Water Resources

Edited by Erick West

Larsen & Keller
www.larsen-keller.com

Impact of Climate Change on Water Resources
Edited by Erick West
ISBN: 979-8-88836-081-1 (Hardback)

Larsen & Keller
Published by Larsen and Keller Education,
5 Penn Plaza,
19th Floor,
New York, NY 10001, USA

Cataloging-in-Publication Data

Impact of climate change on water resources / edited by Erick West.
p. cm.
Includes bibliographical references and index.
ISBN 979-8-88836-081-1
1. Water-supply--Climatic factors. 2. Water-supply--Effect of global warming on.
3. Water-supply--Environmental aspects. 4. Climatic changes. 5. Climatology.
I. West, Erick.
TH6521 .I47 2023
696.1--dc23

For more information regarding Larsen and Keller Education and its products, please visit the publisher's website www.larsen-keller.com

Table of Contents

Preface

This book has been an outcome of determined endeavour from a group of educationists in the field. The primary objective was to involve a broad spectrum of professionals from diverse cultural background involved in the field for developing new researches. The book not only targets students but also scholars pursuing higher research for further enhancement of the theoretical and practical applications of the subject.

Water resources are natural sources of water that are beneficial for human beings. Water is used for verious purposes such as irrigation and drinking. Frozen water, surface water, groundwater and under river flow are examples of natural sources of fresh water. Climate change, water scarcity, water conflict, and water pollution, all pose a threat to the available water resources. Daily water security of the society can be affected by various climate change factors, such as melting of glaciers, severe droughts, heavy precipitation, and the magnitude and frequency of floods. The availability of subsurface and surficial water is also affected by factors such as climate change, and regional and global water cycles. This book contains some path-breaking studies on the impact of climate change on water resources. It will serve as a reference to a broad spectrum of readers.

It was an honour to edit such a profound book and also a challenging task to compile and examine all the relevant data for accuracy and originality. I wish to acknowledge the efforts of the contributors for submitting such brilliant and diverse chapters in the field and for endlessly working for the completion of the book. Last, but not the least; I thank my family for being a constant source of support in all my research endeavours.

Editor

1

Climate Change Impacts on Streamflow Drought: A Case Study in Tseng-Wen Reservoir Catchment in Southern Taiwan

Pao-Shan Yu [1,*], Tao-Chang Yang [1], Chen-Min Kuo [1], Hung-Wei Tseng [1] and Shien-Tsung Chen [2]

[1] Department of Hydraulic and Ocean Engineering, National Cheng Kung University, No. 1, University Rd., Tainan 701, Taiwan; E-Mails: tcyang58@hotmail.com (T.-C.Y.); jemkuo@mail.ncku.edu.tw (C.-M.K.); n8897108@mail.ncku.edu.tw (H.-W.T.)

[2] Department of Water Resources Engineering and Conservation, Feng Chia University, No. 100, Wenhwa Rd., Seatwen, Taichung 407, Taiwan; E-Mail: stchen@fcu.edu.tw

* Author to whom correspondence should be addressed; E-Mail: yups@mail.ncku.edu.tw

Academic Editor: Monica Ionita-Scholz

Abstract: The study aims to assess climate change impacts on streamflow drought in a catchment upstream of Tseng-Wen Reservoir which is the main water supplier in southern Taiwan. A singular-value-decomposition statistical downscaling method and a stochastic weather generator were applied to six different general circulation models for rainfall and temperature downscaling. Two emission climate change scenarios (A1B and B1) were used for rainfall and temperature projections during the two periods, 2010–2045 and 2081–2100. After rainfall and temperature projections, the HBV-based hydrological model was used to simulate future streamflows. By using the threshold level method, the characteristics of streamflow drought during the baseline period (1975–2000) and the future periods were calculated and compared. Important findings include: (1) the start and end times of scenario droughts occur earlier than those of baseline droughts; (2) streamflow under future scenarios during the dry period tends to decrease in January and February, but to increase in March and April; and (3) most of general circulation models' results support that the drought duration and magnitude tend to decrease for a given return period in the future, and the occurrence probability of severer drought tends to decrease.

Keywords: climate change; downscaling methods; streamflow drought

1. Introduction

A drought is a prolonged period of water deficit, and typically occurs when an area receives precipitation below usual levels for months. Droughts may develop during or following periods of low accumulated precipitation relative to normal conditions and is exacerbated by high temperatures. As a drought develops, the lack of rainfall leads to persistently dry conditions, low soil moisture, low river flows, reduced storage in reservoirs and less groundwater recharge [1]. Although Taiwan has abundant precipitation (annual average of about 2510 mm), it is a region prone to droughts due to uneven temporal distribution of precipitation and small storage capacity of man-made reservoirs. Yu *et al.* [2] found that annual rainfall in southern Taiwan has decreased significantly during the past century. Chen *et al.* [3] investigated historical trends of meteorological drought by using long-term precipitation records and found that the incidence of meteorological drought has increased in southern Taiwan since around 1960 due to the decreasing trend of rainfall. The study aims to investigate streamflow droughts, defined by the threshold level approach [4], in the Tseng-Wen Reservoir catchment in southern Taiwan to understand how the likelihood of droughts will change under climate change scenarios.

General circulation models (GCMs) are used to assess the possible changes of climate in the future by modeling the effect of particular future emission scenarios of greenhouse gases on the climate [5]. Generally, GCMs are of low resolutions. Therefore, GCMs' outputs need to be combined with a downscaling technique to provide spatially detailed information to resolve regional details. Fowler *et al.* [6] reviewed the current downscaling literature, examining developments in the downscaling field specifically for hydrological impacts. There are two kinds of downscaling methods, *i.e.*, statistical downscaling and dynamic downscaling. Statistical downscaling methods construct a statistical relationship between large-scale GCM outputs and local weather variables, and dynamic downscaling methods use high-resolution regional climate models with boundary conditions provided by a GCM to obtain local weather variables. Wilby and Wigley [7] made a comparative summary of the relative merits of statistical and dynamical downscaling techniques. Although both statistical and dynamic downscaling methods have their own advantages, statistical downscaling models were more commonly used in hydrological studies due to their less computational demand [8]. In this study, a statistical downscaling approach was applied by using the singular value decomposition scheme [9] to downscale the monthly precipitation from GCMs under climate change scenarios. The downscaling of monthly temperature under different scenarios is retrieved from the Global Change Research Center, National Taiwan University [10]. The projected changes of monthly precipitation and monthly temperature were then used in a weather generator to project the daily precipitation and daily temperature. After generation of daily precipitation and temperature under climate change scenarios, a conceptual hydrological model was used to simulate the streamflow under future scenarios for drought analysis.

In this work, an HBV-based model [11,12] was employed to simulate daily streamflow under future scenarios. The HBV model was designed by the Swedish Meteorological and Hydrological Institute, and has been successfully applied in over 30 countries [13–16]. This model was originally developed for use in Scandinavian catchments but has also been proven to work well in tropical and subtropical areas. For example, this model performs well in simulating daily discharge in Taiwan [11]. The HBV model has also been successfully employed to investigate the impact of climate change on water resources [12,17–20].

The rest of this paper is organized as follows. Section 2 introduces the study site and the data that is available for use in this research. Section 3 describes the techniques of spatial and temporal statistical downscaling, the hydrological model (*i.e.*, the HBV-based model), and the threshold level approach for drought event definition. Section 4 represents the results of daily rainfall and temperature downscaling, calibration and verification of the HBV-based model, the changes in drought characteristics, and drought frequency analysis. Conclusions and future work are finally drawn in Section 5.

2. Study Area and Data Set

The catchment of Tseng-Wen Reservoir, as shown in Figure 1a, is the study area of this work. With a storage capacity of about 7.8×10^8 m^3, Tseng-Wen Reservoir is the largest reservoir in Taiwan and has multiple functions of meeting the water demands for agriculture, domestic use, flood control, and hydropower generation. Tseng-Wen Reservoir catchment encloses an area of 481 km^2, and has a relief of 157 to 3514 m above the sea level. The mean annual precipitation is about 2744 mm and the mean annual runoff is about 2182 mm during 1975–2000. The study area receives temporally uneven precipitation. Around 85% of annual precipitation occurs during the period from May to October (wet period) but only 15% of annual precipitation occurs from November to April (dry period). The mean monthly rainy days, rainfall, and inflow for each month during 1975–2000 are shown in Figure 1b–d, respectively, which reveal the obvious difference between the wet and dry seasons. Recently, the shortage of water resource during the dry period has often caused the paddy field to lie fallow. Therefore, it is of great concern whether the distribution of water resources during this period is influenced by climate change.

Climate and hydrological data used in this work contain local-scale and large-scale data. Local-scale data on a daily basis, including precipitation, streamflow (*i.e.*, naturalized reservoir inflow) and temperature continuously from 1975 to 2008, are provided by Water Resources Agency in Taiwan. The daily precipitations are available from eight raingauges as shown in Figure 1a, from which areal precipitations on the Tseng-Wen Reservoir catchment were computed using the Thiessen polygon method.

Large-scale data, *i.e.*, GCMs data, were downloaded from the Data Distribution Center of the United Nations Intergovernmental Panel on Climate Change. The climate scenarios describe the emission conditions of greenhouse gases and more details are given in the report of IPCC [21]. Different scenarios used were 20C3M for historical climate, and A1B and B1 for future climate. Generally speaking, 20C3M is the scenario to represent past climate, A1B is currently regarded as the most likely climate scenario in the future and B1 describes a convergent world in which the impact on climate change is less significant than A1B.

Rainfall during typhoon and Meiyu seasons is the major source for seasonal precipitation in Taiwan. A coupled GCM (CGCM) that can simulate tropical cyclone index or large-scale circulation reasonably would be the candidate for downscaling study over Taiwan. Tropical cyclone (TC) activity in nine CGCMs submitted to IPCC AR4 is evaluated by means of yearly genesis parameter (YGP) and convective-YGP index [22]. On the other hand, the ability of CGCMs in simulating large-scale circulation pattern in East Asia is given by Min *et al.* [23] and Kitoh and Uchiyama [24]. As a result, six models mentioned in these literatures with reasonable performance are taken into account in the present study. Table 1 lists the summary of six used GCMs.

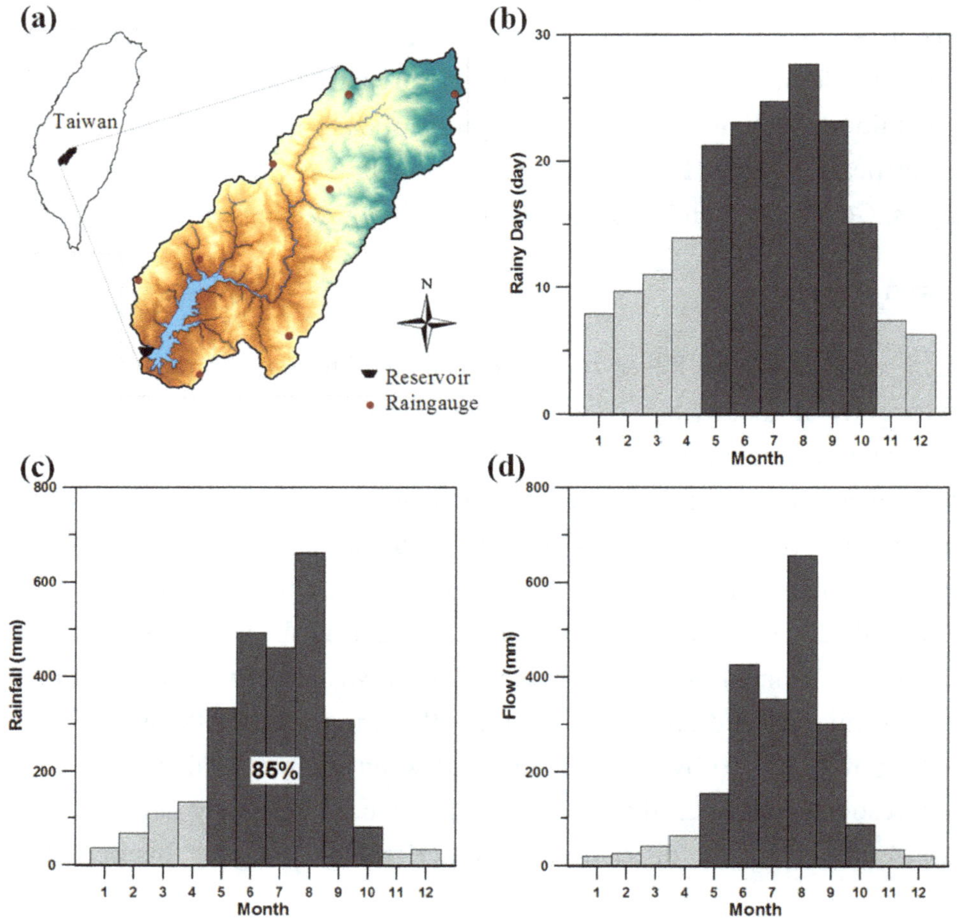

Figure 1. The basic information of the study area: (**a**) Tseng-Wen Reservoir catchment; (**b**) rainy days; (**c**) mean monthly rainfall; and (**d**) mean monthly flow.

Table 1. Summary of the GCMs used in the study.

Model Code	Model Name	Country	Resolution
GCM-1	CGCM3.1(T63)	Canada	T63, L31
GCM-2	CSIRO-Mk3.0	Australia	T63, L18
GCM-3	ECHAM5/MPI-OM	Germany	T63, L31
GCM-4	GFDL-CM2.0	USA	2° × 2.5°, L24
GCM-5	GFDL-CM2.1	USA	2° × 2.5°, L24
GCM-6	MIROC3.2(hires)	Japan	T106, L56

A time period with a length of 20–30 years is usually used for presenting the average climate condition during the period. Based on the common decision of the project teams in Taiwan who executed the studies on assessing the potential impacts of climate change on hydrology, three periods of 1975–2000, 2010–2045 and 2081–2100 were defined as the baseline, near-term and long-term periods, respectively.

3. Methodology

3.1. Spatial and Temporal Statistical Downscaling

GCMs are the main tool to project climate changes through the prescribed emission scenarios. However, due to the coarse resolution, GCMs are not able to represent regional topography and land-sea contrast properly, making local climate projection a big challenge. Thus, a two-stage statistical downscaling method was applied to generate future daily precipitation data from climate outputs run by six GCMs. In the first stage, spatial statistical downscaling was applied by using the singular value decomposition (SVD) scheme [9] to downscale the monthly precipitation form GCM to catchment scale. Monthly temperature downscaling results were obtained from Global Change Research Center, National Taiwan University [10]. In the second stage, temporal statistical downscaling was applied. In the temporal statistical downscaling, the projected changes of monthly precipitation and monthly temperature were further used in a weather generator to produce the daily precipitation and daily temperature. Daily data are more practical for hydrological purpose. After the daily precipitation and daily temperature are generated, the hydrological model uses them as inputs to simulate daily streamflows.

Spatial statistical downscaling provides relationships between local and large-scale variables to overcome the drawback of GCMs' coarse resolution. The first step of spatial statistical downscaling is data reconstruction by the empirical orthogonal functions to filter off noises. This method is the same as applying principal components analysis. Then the SVD is applied to extract coupled patterns between local precipitation and large-scale variables, which can be expressed in the following equations.

$$Z_{predictor}(t,x) = \sum_{i=1}^{m} U_i(x) \cdot S_i(t) \tag{1}$$

$$Z_{predictand}(t,x) = \sum_{i=1}^{m} R_i(x) \cdot K_i(t) \tag{2}$$

where $Z_{predictor}(t,x)$ and $Z_{predictand}(t,x)$ are normalized anomaly field pertaining to the large-scale circulation and the observed station rainfall, respectively; m is the total number of SVD modes; $U_i(x)$ and $R_i(x)$ denote respectively the singular vector of the predictor and the singular vector of the predictand in the i-th mode; and $S_i(t)$ and $K_i(t)$ indicate the time series of expansion coefficient of the i-th SVD mode regarding the predictor and the predictand.

The downscaling results can then be obtained by applying GCMs projection data to the following transfer functions.

$$S_i(t_p) = \sum_{j=1}^{x} Z_{predictor}(t_p, j) \cdot U_i(j) \tag{3}$$

$$PRJ(t_p,x) = \sum_{i=1}^{n} S_i(t_p) \cdot R_i(x) \tag{4}$$

where $PRJ(t_p,x)$ represents the projected predictand; t_p stands for the projection time horizon; and n denotes the total number of the SVD modes retained. In this study, the leading ten modes are retained.

The details of the method of downscaling can be found in Kim *et al.* [25] and Feddersen and Andersen [26].

In the second stage of temporal statistical downscaling, a stochastic weather generator was applied to downscale the monthly precipitation, derived in the first stage, to the daily precipitation. The daily precipitation generation is based on procedures proposed by Richardson [27]. The generator uses a Markov chain to model the occurrence of wet or dry days, and then uses a probability distribution to generate the precipitation amount conditional on a wet day modeled by the Markov chain. A first-order two-state Markov chain was used in this work. The occurrence of a dry or wet day is modeled by a transition probability matrix consisting of conditional probabilities, given a previous dry or wet day.

Many probability distributions were applied to generate daily precipitation amount, such as the exponential distribution [28,29], Weibull distribution [12], two-parameter gamma distribution [27,30–33], and mixed exponential distribution [31,34,35]. Among the probability distributions, the Weibull distribution most appropriately approximates daily rainfall in Taiwan [12]; consequently, this work used the Weibull distribution to generate daily rainfall.

Regarding the daily temperature generation, a first-order autoregressive model was utilized to generate the daily temperature sequences in each month. This daily temperature generation model is expressed as follows:

$$T_k = \mu_T + \rho_{1T}\left(T_{k-1} - \mu_T\right) + \sqrt{1-\rho_{1T}^2}\,\sigma_T \nu_k + \Delta\mu \tag{5}$$

where T_k is the temperature on day k; μ_T is the mean temperature in a certain month; σ_T is the standard deviation of daily temperature in the month; ρ_{1T} is the lag-one autocorrelation coefficient of daily temperature in the month; ν_k is the random standard normal variate, and $\Delta\mu$ is the mean temperature change in the month under a future scenario. Given the parameters, μ_T, σ_T, σ_{1T}, and $\Delta\mu$, a daily temperature sequence in a month can be generated by this model.

3.2. Hydrological Model

A continuous hydrologic model was used to simulate future projected streamflow, after the daily precipitation and temperature were obtained in the previous section by the downscaling method. This work used a continuous hydrologic model based on the structure of Hydrologiska Byrans Vattenbalansavdelning (HBV) model [13,14], which was initially designed for use in Scandinavian catchments by the Swedish meteorological and hydrological institute. Yu and Yang [11] adapted the HBV model structure to suit catchments in Taiwan. The HBV-based model uses both an upper and lower tank to model the rainfall-runoff behavior. Model structure mainly consists of three parts: (1) soil moisture module; (2) runoff response mechanism; and (3) water balance functions. Detail description of the HBV-based model, as well as its procedures for calibration and validation in this work, can be found in Yu and Yang [11] and Yu *et al.* [12].

In the HBV-based model, Hamon's temperature-dependent equation [36] was used to transform the daily temperature series into the daily potential evapotranspirartion series. The Hamon's temperature-dependent equation is as:

$$Ep(t) = 0.21 H(t)^2 e_o(t) / [T(t) + 273] \tag{6}$$

where $E_p(t)$ is the potential evapotranspiration (mm/day) on day t; $H(t)$ is the sunshine duration (hour) on day t; $e_o(t)$ is the saturated vapor pressure (milibar) on day t; $T(t)$ is the mean temperature (°C) on day t. Data of $H(t)$ are available at the Ali-Shan meteorological station near the study area, and $e_o(t)$ can be estimated by the following empirical equation.

$$e_o(t) = 33.8639 \times ((0.00738 \times T(t) + 0.8072)^8 - 0.000019 \times |1.8 \cdot T(t) + 48| + 0.001316) \quad (7)$$

3.3. Threshold Level Method for Drought Event Definition

The categories of droughts are frequently used as follows [37]: climatological drought (deficit in precipitation), agro-meteorological drought (deficit in soil water), river flow drought (deficit in river discharge), groundwater drought (deficit in groundwater storage), and operational drought (conflict of water shortage and water management demands). Within the present study the concept of river flow (or say streamflow) drought is adopted.

The threshold level method proposed by Yevjevich [4] is the most commonly applied approach to drought studies, such as [38–43]. The choice of the threshold level is important for the results of a drought study. Besides the option for a constant threshold value for all data over time or for a variable threshold (e.g., fluctuating on a monthly or seasonal basis), the value of the threshold level is highly significant. A threshold level which is too low might lead to a high number of no-drought years making the few identified drought events statistically uncertain to evaluate. On the other hand, with a high threshold level the likelihood for a series of small single drought events being combined into one severe multi-year drought (drought lasting longer than a year) increases [44]. To prevent the aforementioned conditions from occurring, a proper choice of threshold level should be made.

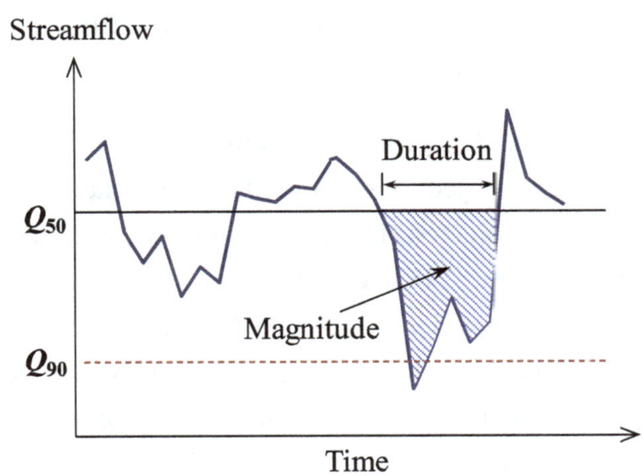

Figure 2. Definition of streamflow drought.

In order to study the features of major streamflow drought, this work defines a streamflow drought based on the annual flow duration curve from which the 50-percentile flow (Q_{50}) and 90-percentile flow (Q_{90}) are selected. A streamflow drought is a low flow event when streamflow series is continuously below the 50-percentile flow (Q_{50}) and the minimum flow during the period is less than the 90-percentile flow (Q_{90}) (Figure 2). The Q_{90} is required here to avoid minor drought events. Once a drought event is specified, drought characteristic can be quantified. Drought frequency (time/year) is the times of drought occurrence per year. Drought duration (day) is the time period between the start and the end of a drought.

Drought magnitude (mm) is the total amount of streamflow deficit, expressed in depth. During a streamflow drought, the daily streamflow deficit is calculated by Q_{50} minus the daily flow for each day. Then, the drought magnitude can be obtained by summing up the entire daily streamflow deficits during a drought.

4. Results

4.1. Precipitation and Temperature Downscaling

The analysis from Chu and Yu [45] demonstrated that the sea level pressure (SLP) and the meridional wind field at 850 hPa (V850) are appropriate predictors for downscaling precipitation over Taiwan. Thus, the two large-scale variables, SLP and V850, both of which are taken from six GCMs involving 20C3M, A1B, and B1 scenarios, are used as predictors for downscaling in this work. The 20C3M scenario is used to reproduce historical climate condition. GCMs' performance can be evaluated by comparing downscaling results and observation data during 1975–2000. In Figure 3, the left panel presents observation data and downscaling results (under 20C3M scenario) during the period of 1975–2000, indicating that the downscaling results can capture the rainfall pattern of observation data. The Gerrity skill score (GSS), proposed by Gerrity [46], is applied to evaluate the performance of GCM. GSS is easy to construct and, what is noteworthy, equitable, giving rewards and penalties reasonably to the forecast skill. The above mentioned points make GSS a skill score for categorical deterministic forecast recommended by the World Meteorological Organization (WMO) in the standardized verification system for long-range forecasts [47] (WMO 2002). The GSS used herein is only for inspecting monthly-rainfall downscaling performance of the selected GCMs during the baseline period under 20C3M scenario. The GSS value above zero stands for skillful downscaling performance. In Figure 3, the right panel shows that almost all GCMs have a reasonable simulation.

After spatial downscaling process, change rates of monthly precipitation from 20C3M to A1B and B1 scenarios can be obtained. Future projections of monthly precipitation under A1B and B1 scenarios can be derived by multiplying these projected change rates and historical monthly rainfall data together. The changes of monthly temperature from 20C3M to A1B and B1 scenarios during the periods of 2010–2045 and 2081–2100, respectively, were derived by the Global Change Research Center at National Taiwan University [10]. Then, future monthly temperatures can be derived by adding the projected changes to the historical monthly temperature. Figure 4 displays the projected monthly rainfall and mean temperature in each month under A1B and B1 scenarios during the periods of 2010–2045 and 2081–2100, respectively. In the figure, the ensemble results (*i.e.*, the mean results) of the six GCMs under A1B and B1 scenarios are presented, and the baseline indicates the projected values during 1975–2000 under 20C3M. The results reveal that the rainfalls during the wet season (May to September) tend to obviously increase. During the dry season (October to April), the rainfalls has no obvious change, except for March in which there exists an increase. The projected mean monthly temperature tends to increase and the increased change in summer is larger than that in winter. Moreover, the increased change under A1B scenario is larger than that under B1 scenario.

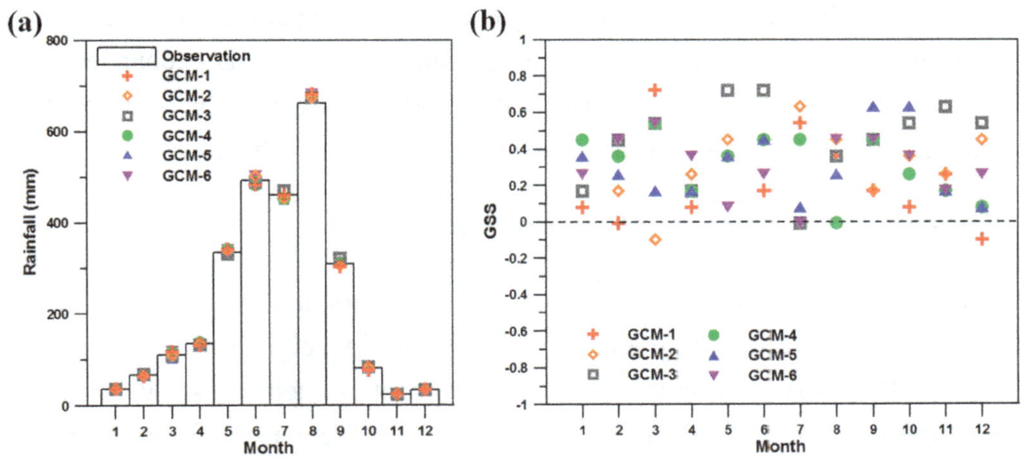

Figure 3. (**a**) Observation and scenario data during the period between 1975 and 2000 and (**b**) the Gerrity skill scores of each GCM.

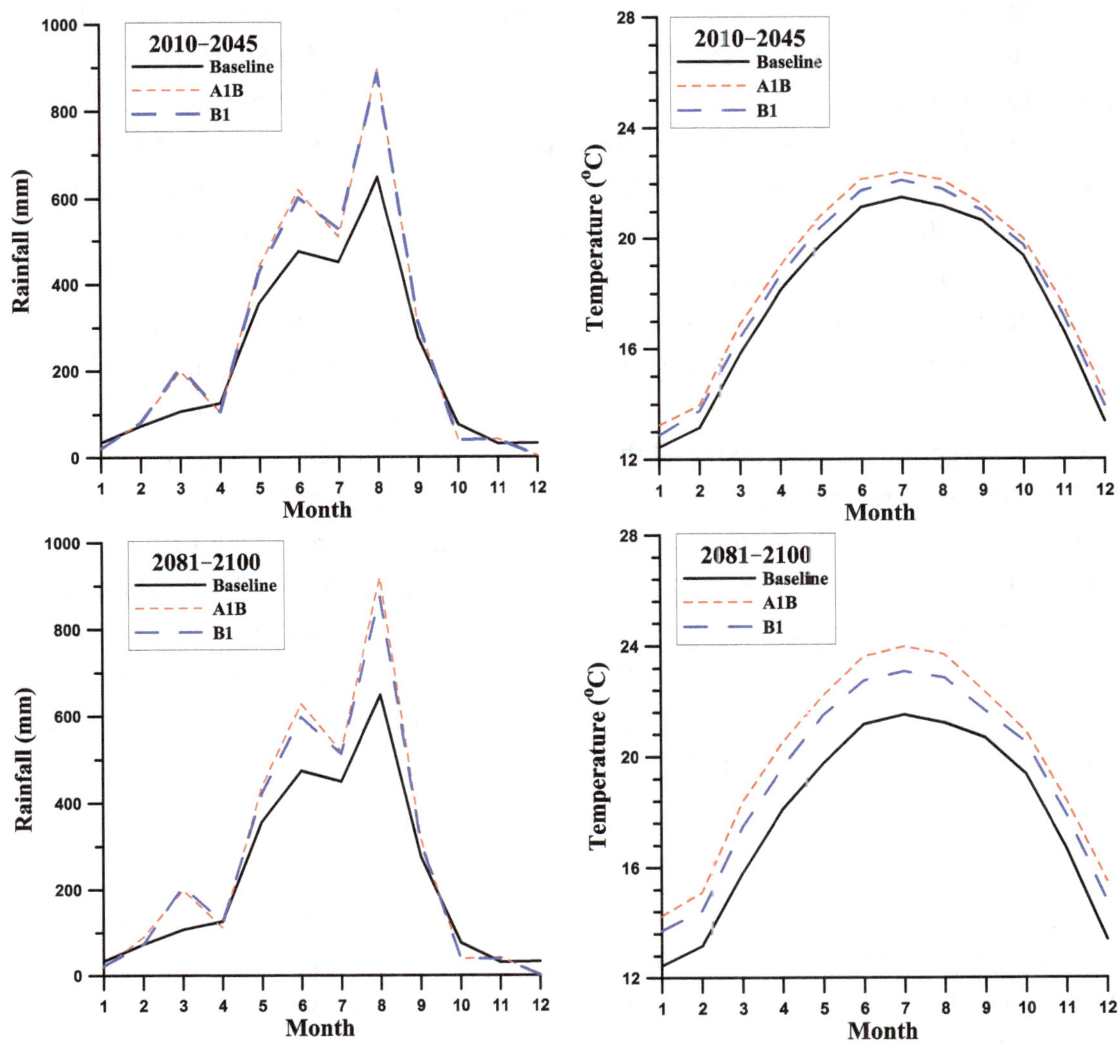

Figure 4. Projected monthly rainfall and mean temperature under A1B and B1 scenarios during the periods of 2010–2045 and 2081–2100, respectively.

Then, the projected changes of monthly precipitation and temperature were used in the weather generator to produce daily precipitation and temperature under climate change scenarios. Daily precipitation and temperature of a 200-year length were generated by the temporal statistical downscaling methods for baseline and future climate scenarios, respectively.

4.2. Calibration and Verification of the HBV-Based Model

During model calibration, the HBV-based model was applied on the Tseng-Wen Reservoir catchment to explore its ability of streamflow simulation. The fuzzy multiple objective functions, proposed by Yu and Yang [11], and the shuffled complex evolution optimization method [48] were used in the study. Historical daily rainfall and flow data from 1975 to 1998 were used for the HBV-based model calibration. The calibrated HBV-based model was further verified by historical daily data from 1999 to 2008. To assess the model performance, three criteria, including the ratio of simulated to observed runoff volumes (Vs/Vo), the root mean squared error (RMSE), and the coefficient of correlation (CC), were calculated pertaining to the calibration and verification periods, respectively. During the calibration period, the values of Vs/Vo, RMSE and CC are 0.86, 9.98 (mm) and 0.86, respectively. During the validation period, the values of Vs/Vo, RMSE and CC are 0.88, 9.90 (mm) and 0.96, respectively. Table 2 lists the values of Vs/Vo, RMSE, and CC for each calibration and validation year. From this table, most of years have good simulation performances. Figure 5 shows the calibration and verification results in 1992 and 2005, respectively. These results reveal the HBV-based model is able to simulate the rainfall-runoff behavior over the study area.

The HBV-based model was then used to simulate future projected streamflow, after the daily precipitation and temperature under the A1B and B1 scenarios were obtained in the previous section by the temporal downscaling. The flow duration curves of the dry season and the wet season, respectively, during the baseline period (1975–2000) and the future period (2010–2045) were shown in Figure 6 to display the influence of projected changes in climate on the hydrological regime. In the figure, the flow duration curve during the future period for each season is the ensemble result of the six GCMs. The future period (2081–2100) has the similar influence with the period (2010–2045) and its figure was not included here. From the average viewpoint, the flows during the wet season increase in future due to the increasing precipitation during the wet season (found in Section 4.1) and the flows during the dry season increase due to the increasing lower tank storage in the HBV-based model during the wet season. These figures implicate the magnitude and duration of the defined drought "may" reduce in future.

4.3 Changes in Drought Characteristics

The historical streamflow data during the baseline period of 1975–2000 were used to derive the annual flow duration curve from which the Q_{50} and Q_{90} were calculated (*i.e.*, 1.24 and 0.16 mm, respectively). Since Q_{50} and Q_{90} were used to define a streamflow drought in the present study, the two values for the future scenarios are the same as those for the baseline period. Based on the threshold level method described in the Section 3.3, the drought frequency (0.77 times per year), the mean drought duration (174 days), and the mean drought magnitude (148 mm) can be calculated. These three drought characteristics during the period of 1975–2000 are used as the baseline against which change is compared.

Table 2. Values of three criteria for each year during calibration and validation periods.

Calibration Period				Validation Period			
Year	Vs/Vo	RMSE (mm)	CC	Year	Vs/Vo	RMSE (mm)	CC
1975	0.93	23.89	0.46	1999	0.97	5.37	0.91
1976	0.97	20.43	0.74	2000	0.95	3.36	0.94
1977	1.00	25.55	0.60	2001	0.91	10.61	0.93
1978	0.86	15.87	0.70	2002	0.98	2.86	0.94
1979	0.95	8.36	0.90	2003	1.08	3.29	0.94
1980	0.65	6.91	0.84	2004	0.92	6.41	0.99
1981	0.92	9.11	0.97	2005	0.80	15.50	0.96
1982	0.88	4.61	0.97	2006	0.89	11.88	0.96
1983	0.99	5.20	0.94	2007	0.86	11.92	0.97
1984	0.58	5.98	0.96	2008	0.84	13.22	0.97
1985	0.54	10.63	0.92	Total Period	0.88	9.90	0.96
1986	0.59	9.06	0.64				
1987	0.93	14.92	0.72				
1988	0.60	7.30	0.98				
1989	0.77	15.18	0.92				
1990	0.72	7.97	0.98				
1991	0.87	3.71	0.96				
1992	0.96	6.92	0.97				
1993	0.98	2.87	0.96				
1994	0.94	5.73	0.94				
1995	0.96	2.48	0.97				
1996	0.82	15.09	0.98				
1997	0.97	4.06	0.95				
1998	1.01	5.96	0.92				
Total Period	0.86	9.98	0.86				

The change of streamflow drought can be assessed by calculating the differences between historical data (baseline) and projected data under different scenarios (A1B and B1) and during different future periods (2010–2045 and 2081–2100). Table 3 shows drought characteristics, including drought frequency, duration, and magnitude, based on historical data and projected data (results from six GCMs under A1B and B1 scenarios). The scenario droughts become more frequent (0.87 to 1.04 times per year), compared to the frequency of baseline droughts (0.77 times per year). From the ensemble average, the drought frequencies range from 0.94 to 0.97 times per year for different scenarios and different periods, revealing that the future drought frequency tends to increase.

The durations of projected droughts for different GCMs and different future periods are between 128.7 days and 205.9 days. The ensemble averages of projected drought durations are about 160 days for different scenarios and different future periods. As compared to the duration of baseline drought (174 days), the future drought duration decrease about a half month.

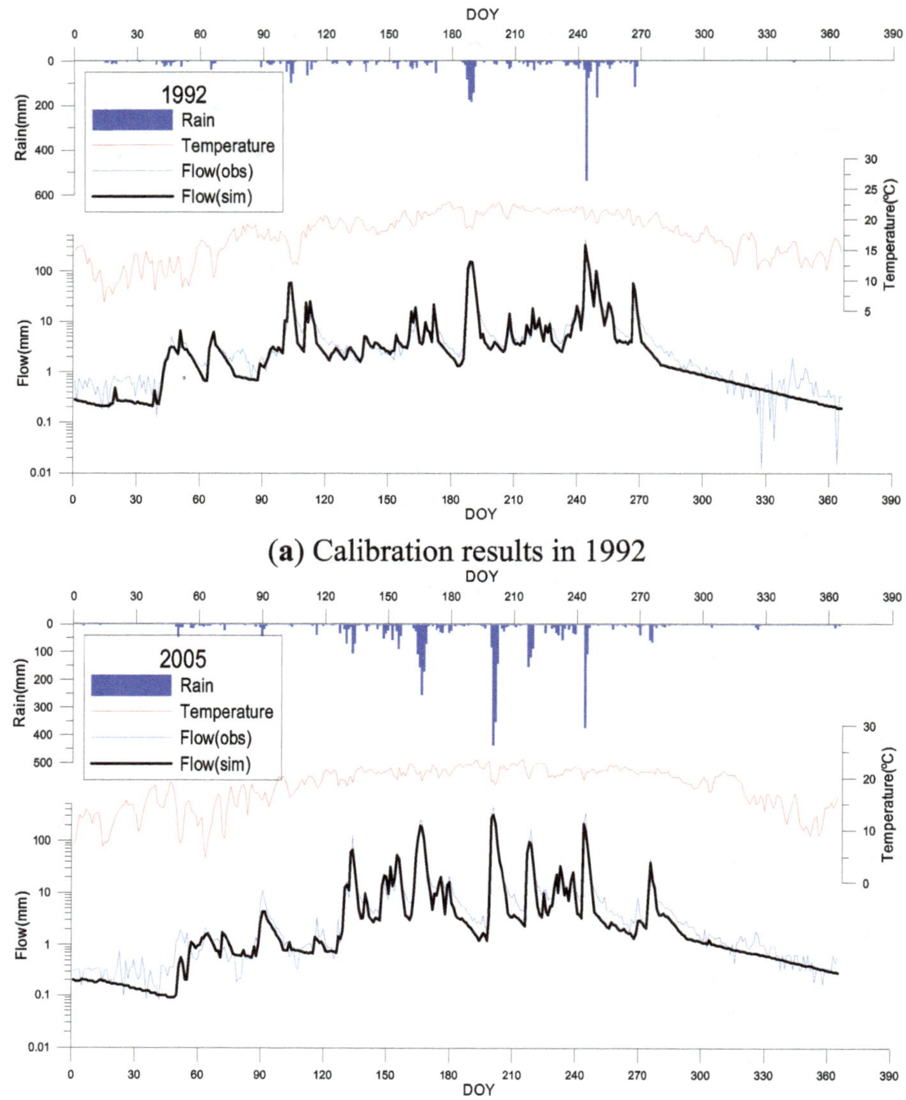

(**a**) Calibration results in 1992

(**b**) Verification results in 2005

Figure 5. Calibration and verification results for the HBV-based model in (**a**) 1992 and (**b**) 2005, respectively (DOY: Day of Year).

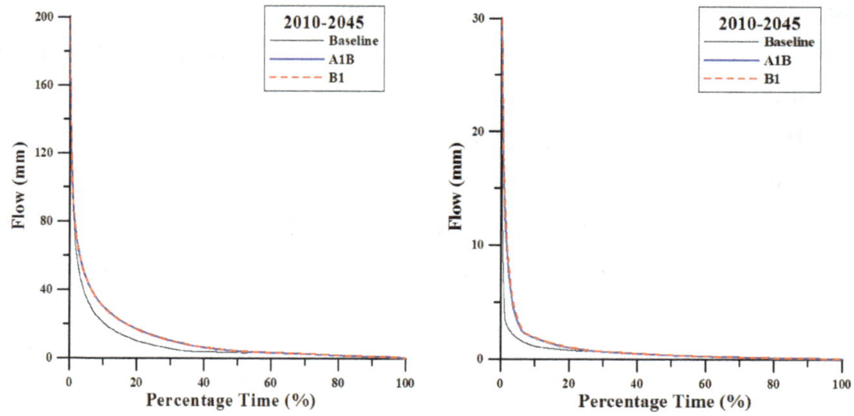

(**a**) (**b**)

Figure 6. Flow duration curves of the (**a**) wet season and (**b**) the dry season during the baseline period and the future period (2010–2045).

The tendency of drought magnitude is similar to that of drought duration. The magnitudes of projected droughts are between 102 mm and 196 mm for different GCMs and different future periods. The ensemble averages of projected drought magnitudes are about 140 mm for different scenarios and periods. The future drought magnitude decrease slightly, as compared to the magnitude of baseline drought (148 mm).

To sum up the above analysis, the duration and magnitude of scenario droughts may increase or decrease, as they depend on different GCMs. However, from the ensemble average of GCMs, the results reveal that the duration and magnitude of scenario droughts tend to decrease. Overall, streamflow droughts under future scenarios become more frequent, but their duration and magnitude become more diverse than the baseline droughts.

Figures 7 and 8, respectively, are the time distributions of the start and the end of streamflow droughts, under the A1B scenario for the period of 2010–2045. In the figures, the solid line indicates the baseline drought, and the dash line stands for the ensemble mean of scenario droughts. Figure 7 indicates that baseline droughts more often start during the period from the middle of October (the 2nd 10-day period in October) to early November (the 1st 10day period in November). However, scenario droughts are projected to come earlier and have a high likelihood to start in the middle of October (the 2nd 10-day period in October). Figure 8 reveals that scenario droughts are more likely to end during March that is earlier than the end of baseline droughts more likely in the early May. The time distributions of the start and the end of streamflow droughts pertaining to the B1 scenario and to the period of 2081–2100 are similar. Totally speaking, droughts may start and end earlier in the future.

Table 3. Impact of climate change on streamflow drought characteristics.

Drought Characteristics	**Baseline**	**Future Period**	**Scenario**	**Ensemble of GCMs**	**GCM-1**	**GCM-2**	**GCM-3**	**GCM-4**	**GCM-5**	**GCM-6**
Frequency (time/year)	0.77	2010–2045	A1B	0.94	0.87	1.02	0.99	0.98	0.89	0.90
			B1	0.96	0.89	0.99	0.95	0.99	0.96	0.95
		2081–2100	A1B	0.97	0.88	1.04	0.99	0.98	0.93	1.00
			B1	0.97	0.91	1.00	0.97	1.00	0.97	0.96
Duration (day)	173.65	2010–2045	A1B	161.2	131.8	193.3	158.8	181.9	149.7	151.6
			B1	161.6	132.9	197.2	160.7	181.3	147.3	150.4
		2081–2100	A1B	160.4	128.7	193.4	151.9	180.0	145.2	163.0
			B1	163.3	132.4	205.9	158.6	184.1	147.4	151.6
Magnitude (mm)	148.44	2010–2045	A1B	139.7	105.3	178.5	137.2	164.6	125.7	126.7
			B1	139.9	105.7	183.0	138.7	162.8	122.3	126.8
		2081–2100	A1B	139.3	102.1	181.6	130.3	161.9	121.1	138.5
			B1	142.9	105.5	195.5	137.7	166.0	125.0	127.9

To check the streamflow during the dry period (January to April) under different scenarios during the periods of 2010–2045 and 2081–2100, respectively, Figure 9 shows the hydrographs (*i.e.*, temporal distributions of streamflow). Concerning both scenarios (A1B and B1) and both periods (2010–2045 and 2081–2100), the streamflow under future scenarios in Figure 9 is the ensemble average of all six GCMs. The streamflow under future scenarios tends to decrease in January and February (at about the beginning of the first season of paddy rice growth in Taiwan), but to increase in March and April (at about the end of the first season of paddy rice growth). The 90% confidence interval of streamflow under future scenarios is wider than baseline streamflow after February, indicating the future projected streamflows are more scattered.

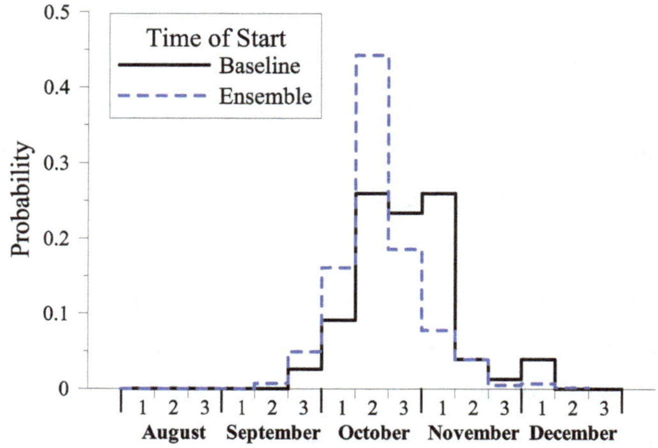

Figure 7. Distribution of drought start time.

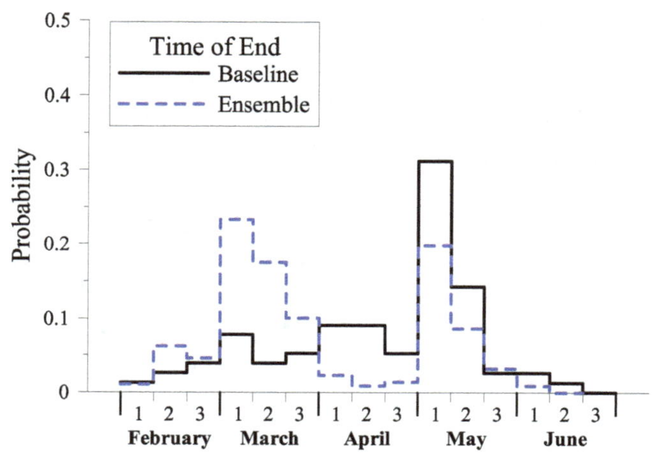

Figure 8. Distribution of drought end time.

4.4 Drought Frequency Analysis

As an important random phenomenon in hydrology, the frequency analysis is necessary in the aim to know about the drought's regime. The frequency analysis was performed in the present study for drought duration and magnitude respectively. The estimation steps of the drought frequency analysis are as follows. (1) The annual maximum values of drought duration and magnitude were selected during the baseline and two future periods, respectively; (2) Different probability distributions (including Extreme value type-I, Generalized extreme value, Pearson type-III, Log-Pearson type-III, Three parameters log-normal, Gamma, and Exponential distributions) were used to fit the annual maximum series of drought duration and magnitude, respectively, for each period. This work applied the Kolmogorov-Smirnov test to examine the goodness of fit of the distributions and the root mean square error was used as an index to compare the fitting performances of different distributions; (3) Based on the optimal probability distribution, the drought duration and magnitude with different return periods for a period (baseline or future) can be estimated.

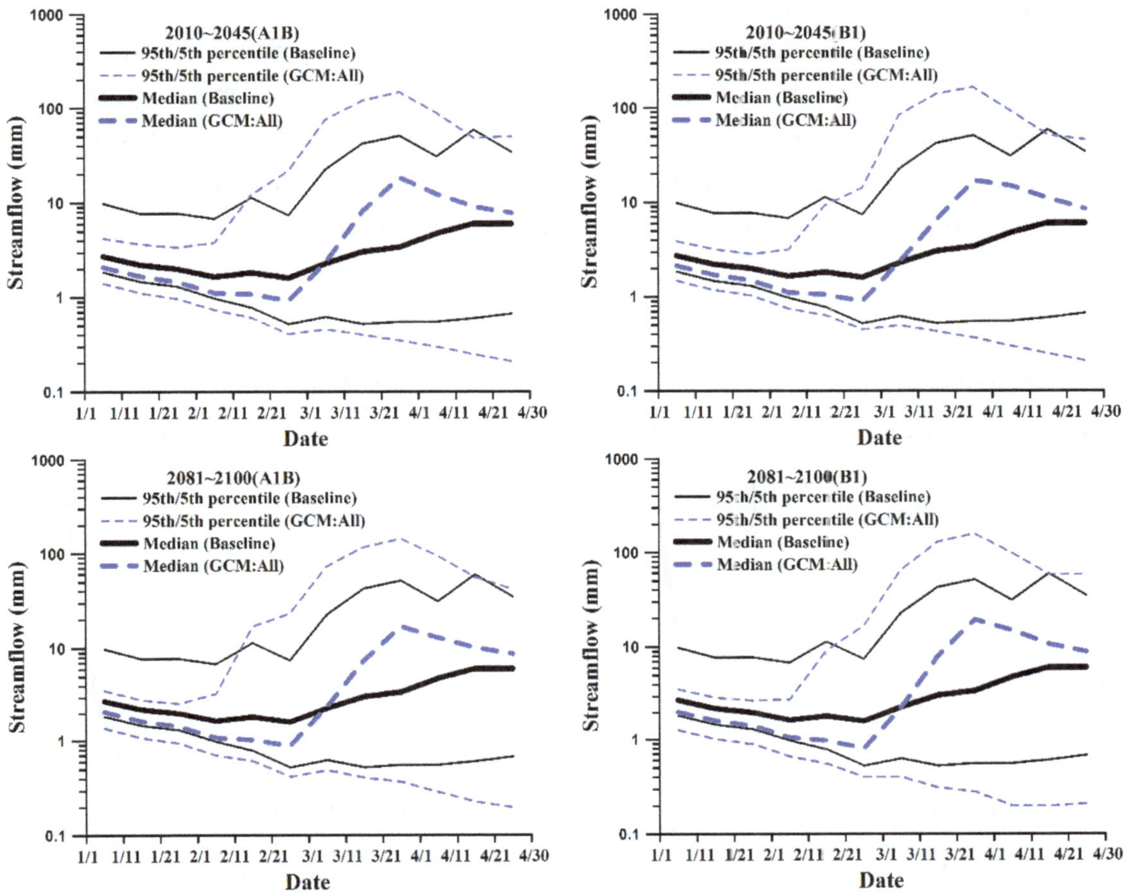

Figure 9. Hydrographs during the dry season. (Thick-solid line: the median during baseline period; Upper and lower thin-solid lines: the 95th and 5th percentiles during baseline period; the same notations are used for scenario data by thick-dash and thin dash lines).

Different probability distributions were tested in the study, and results showed that the Gamma distribution best fits the distributions of drought duration and magnitude. Therefore, the Gamma distribution was used to implement the frequency analysis for drought duration and magnitude respectively. Figure 10 shows the probability distribution of drought duration for each GCM pertaining to two future scenarios and two periods. In the figure, the abscissa indicates the drought duration (day) and the ordinate denotes the return period (year). Results reveal that four GCMs (*i.e.*, GCM-1, GCM-3, GCM-5 and GCM-6) have the same tendency that the drought duration regarding future scenarios will be less than the baseline drought duration for a given return period. However, the other two GCMs (*i.e.*, GCM-2 and GCM-4) have the opposite tendency to the former four GCMs. Therefore, most of GCMs' results support that there exists a decreasing tendency in drought duration for a given return period; and for a given drought duration, the return period tends to increase which means the occurrence probability of this drought duration tends to decrease.

Figure 11 displays the probability distribution of drought magnitude. In the figure, the abscissa represents the drought magnitude (mm) and the ordinate denotes the return period (year). Results show the drought magnitude of four GCMs (*i.e.*, GCM-1, GCM-3, GCM-5 and GCM-6) in future scenarios have the decreased tendency for a given return period. However, the other two GCMs (*i.e.*, GCM-2 and GCM-4) have the opposite tendency to the former four GCMs. Therefore, most of GCMs' results support that there exists a decreasing tendency in drought magnitude for a given return period; and for a given drought magnitude, the return period tends to increase which means the occurrence probability of this drought magnitude tends to decrease. From the aforementioned descriptions, most of GCMs' results support that both the drought duration and magnitude tend to decrease for a given return period in the future, and the occurrence probability of severer drought tends to decrease.

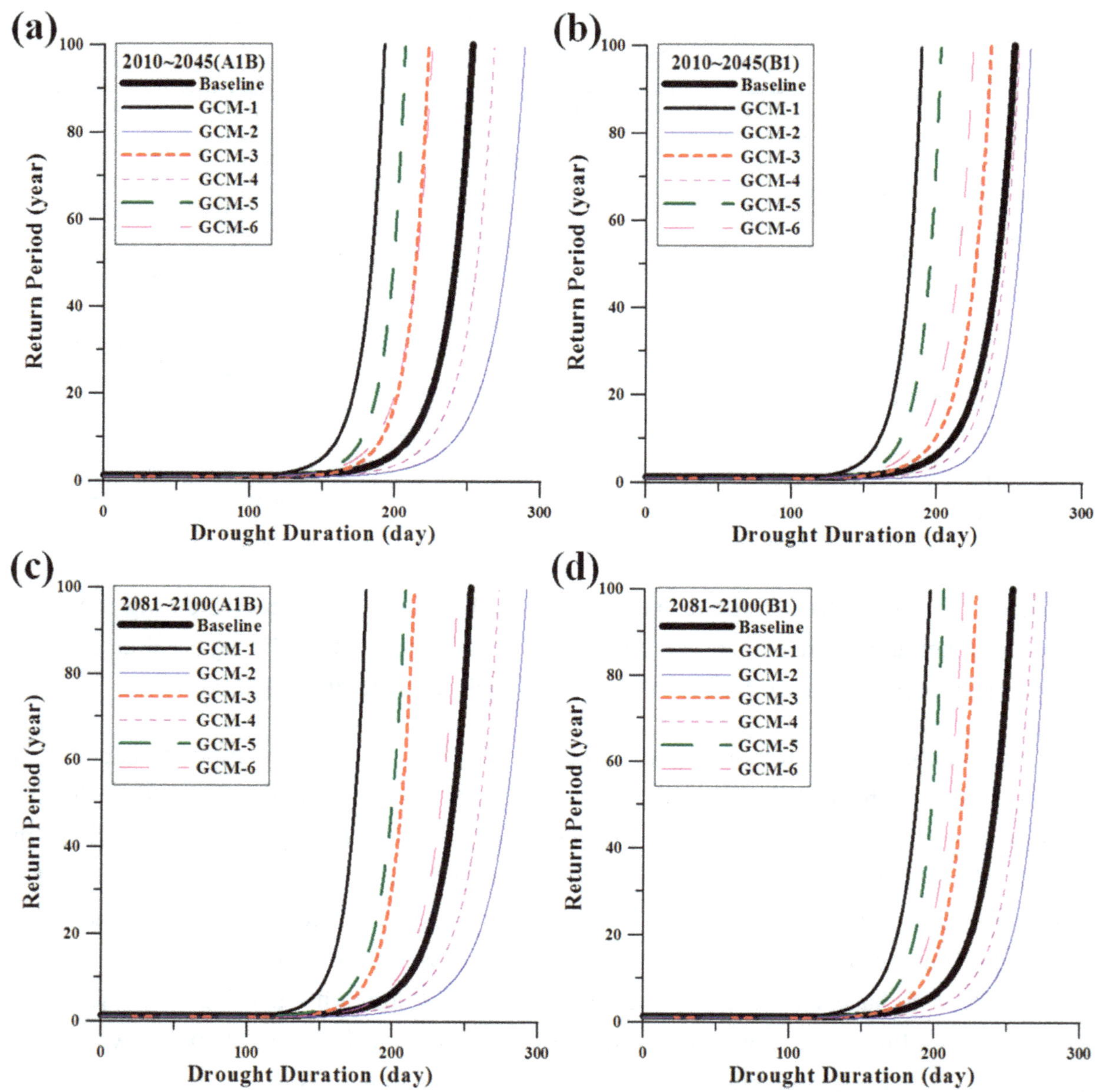

Figure 10. The probability distributions of drought duration for different GCMs, (**a**) during 2010–2045 under A1B scenario, (**b**) during 2010–2045 under B1 scenario, (**c**) during 2081–2100 under A1B scenario, and (**d**) during 2081–2100 under B1 scenario.

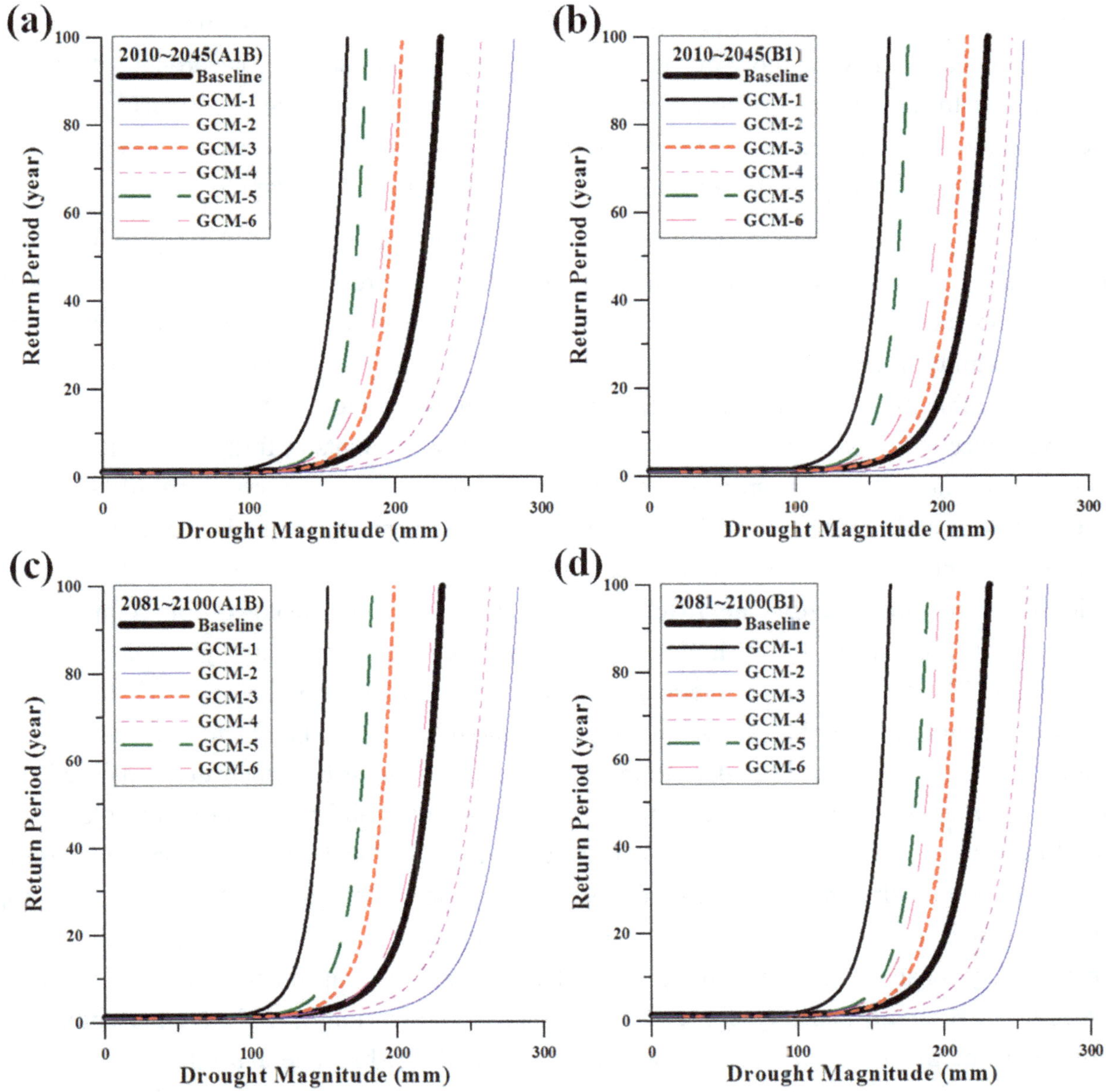

Figure 11. The probability distributions of drought magnitude for different GCMs, (**a**) during 2010–2045 under A1B scenario, (**b**) during 2010–2045 under B1 scenario, (**c**) during 2081–2100 under A1B scenario, and (**d**) during 2081–2100 under B1 scenario.

5. Conclusions and Future Work

This work applied downscaling methods and a hydrological model to derive streamflow under future scenarios and to assess the effect of climate change on streamflow droughts in southern Taiwan. Based on the definition of streamflow drought and the results of ensemble average of the six GCMs used in the present study, it is revealed that scenario droughts may become more frequent, and the duration and magnitude of scenario droughts may become more varied than those of baseline droughts. Analyzing time distributions of streamflow drought under future scenarios found that streamflow drought events will start and end earlier in the future. The streamflow under future scenarios tends to decrease in January and February (at the beginning of paddy rice growth season) but to increase in March and April (at the end of paddy rice growth season). The 90% confidence interval of streamflow under future scenarios is wider than the baseline streamflow after February, because more diverse streamflows are projected. From the frequency analysis, most of GCMs' results support the streamflow drought duration and magnitude tend to decrease for a given return period in the future, and the occurrence probability of severer drought tends to decrease. Due to the changed pattern of future streamflow drought, the authorities are suggested to plan to moderately adjust irrigation period to adapt to climate change.

From the analytical results of the present study, some future work as the following should be made for reinforcing the study and its conclusions. (1) The combined result of more frequent streamflow drought events and lower duration and magnitude, which is consistent with a change in the temporal variability, is a preliminary conclusion based on the definition of streamflow drought and the results of ensemble average of the six GCMs used in the present study. Future work may adopt more GCMs to analyze the impact of climate change on streamflow droughts to further test this conclusion; (2) The study aimed to investigate the impact of climate change on streamflow droughts. The development and comparisons of downscaling methods were not the main work. Therefore, the downscaling methods used in Taiwan previously were adopted in the present study. Future work may consider the other downscaling approaches (e.g., bias correction methods) for comparing their performances; (3) In Taiwan, the mean flow of a river is about to Q_{20} (=6.9 mm). Q_{50} is not a high value during a year. The threshold level method used in the present study is an approach to "define" a streamflow drought for drought studies. The set of threshold level, Q_{50} and Q_{90} for the baseline period, was used in the study only for the definition of streamflow drought by which the impact of climate change on this kind of streamflow drought was assessed. The other different sets of threshold level may define different streamflow droughts and can be considered in the future; (4) The study presently considered only three drought indices, *i.e.*, frequency, duration and magnitude, for studying the impact of climate change on streamflow droughts. Since the study area is in the catchment of the reservoir, future work may consider the analysis of operational drought and evaluate other indices that may provide more useful for reservoir management (e.g., the sequent-peak algorithm); (5) In the present study, the parameters of the hydrological model were fixed during both the baseline and future periods. This is because there is no observed data for model calibration during the future period that makes the parameter modification become a big challenge. The fixed parameters for the baseline and future periods made an assumption of the same mechanism of hydrological process in the study area during those periods. Since the study area is small and the catchment characteristics (e.g., the land use, cover, and soil type) in the study area will not obviously change, this assumption can be acceptable. A better way to modify model parameters in the future is suggested to use a fully-physical hydrological model and establish a link between the physical model parameters with climate change; (6) The latest climate change projections are available in the Coupled Model Intercomparison Project Phase 5 (CMIP5). Hence, a further investigation based on these new projections is suggested to include all potential future scenarios for assessing climate change impacts on streamflow drought in the study area.

Acknowledgments

The authors sincerely thank the Ministry of Science and Technology of Republic of China (Taiwan) for financially supporting this research under Contract No. NSC 97-2625-M-006-014, and the Global Change Research Center, National Taiwan University for offering projected temperature data.

Author Contributions

Pao-Shan Yu conceptualized and supervised the research. Tao-Chang Yang and Chen-Min Kuo conceived the research, wrote the manuscript, revised, and finalized the paper. Hung-Wei Tseng and Shien-Tsung Chen conceived the research and processed all the data.

References

1. Tallaksen, L.M; van Lanen, H.A.J. *Hydrological Drought: Processes and Estimation Methods for Streamflow and Groundwater*; Elsevier B.V.: Amsterdam, The Netherlands, 2004.
2. Yu, P.S.; Yang, T.C.; Kuo, C.C. Evaluating long-term trends in annual and seasonal precipitation in Taiwan. *Water Resour. Manag.* **2006**, *20*, 1007–1023.
3. Chen, S.T.; Kuo, C.C.; Yu, P.S. Historical trends and variability of meteorological droughts in Taiwan. *Hydrol. Sci. J.* **2009**, *54*, 430–441.
4. Yevjevich, V. *An Objective Approach to Definition and Investigations of Continental Hydrologic Droughts*; Hydrology Paper No. 23; Colorado State University: Fort Collins, CO, USA, 1967.
5. Burke, E.J.; Perry, R.H.J.; Brown, S.J. An extreme value analysis of UK drought and projections of change in the future. *J. Hydrol.* **2010**, *388*, 131–143.
6. Fowler, H.J.; Blenkinsop, S.; Tebaldi, C. Review linking climate change modeling to impacts studies: Recent advances in downscaling techniques for hydrological modeling. *Int. J. Climatol.* **2007**, *27*, 1547–1578.
7. Wilby, R.L.; Wigley, T.M.L. Downscaling general circulation model output: A review of methods and limitations. *Progress Phys. Geogr.* **1997**, *21*, 530–548.
8. Chen, S.T.; Yu, P.S.; Tang, Y.H. Statistical downscaling of daily precipitation using support vector machines and multivariate analysis. *J. Hydrol.* **2010**, *385*, 13–22.
9. Chu, J.L.; Kang, H.; Tam, C.Y.; Park, C.K.; Chen, C.T. Seasonal forecast for local precipitation over northern Taiwan using statistical downscaling. *J. Geophys. Res.* **2008**, *113*, D12118.
10. Lin, S.H.; Liu, C.M.; Huang, W.C.; Lin, S.S.; Yen, T.H.; Wang, H.R.; Kuo, J.T.; Lee, Y.C. Developing a yearly warning index to assess the climatic impact on the water resources of Taiwan, a complex–terrain island. *J. Hydrol.* **2010**, *390*, 13–22.
11. Yu, P.S.; Yang, T.C. Fuzzy multi-objective function for rainfall-runoff model calibration. *J. Hydrol.* **2000**, *238*, 1–14.
12. Yu, P.S.; Yang, T.C.; Wu, C.K. Impact of climate change on water resources in southern Taiwan. *J. Hydrol.* **2002**, *260*, 161–175.
13. Bergström, S. *Development and Application of a Conceptual Runoff Model for Scandinavian Catchments*; SMHI Norrköping: Norrköping, Sweden, 1976.
14. Bergström, S. *The HBV Model—Its Structure and Applications*; SMHI Norrköping: Norrköping, Sweden, 1992.
15. Bathia, P.K.; Bergström, S.; Persson, M. *Application of the Distributed HBV-6 Model to the Upper Narmada Basin in India*; SMHI Norrköping: Norrköping, Sweden, 1984.
16. Häggström, M.; Lindström, G.; Cobos, C.; Martinez, J.R.; Merlos, L.; Alonzo, R.D.; Castillo, G.;

Sirias, C.; Miranda, D.; Granados, J.I.; *et al. Application of the HBV Model for Flood Forecasting in Six Central American Rivers*; SMHI Norrköping: Norrköping, Sweden, 1990.
17. Bergström, S.; Carlsson, B.; Gardelin, M.; Lindström, G.; Pettersson, A.; Rummukainen, M. Climate change impacts on runoff in Sweden—Assessments by global climate models, dynamical downscaling and hydrological modeling. *Clim. Res.* **2001**, *16*, 101–112.
18. Menzel, L.; Bürger, G. Climate change scenarios and runoff response in the Mulde catchment (Southern Elbe, Germany). *J. Hydrol.* **2002**, *267*, 53–64.
19. Dibike, Y.B.; Coulibaly, P. Hydrologic impact of climate change in the Saguenay watershed: Comparison of downscaling methods and hydrologic models. *J. Hydrol.* **2005**, *307*, 145–163.
20. Yu, P.S.; Wang, Y.C. Impact of climate change on hydrological processes over a basin scale in northern Taiwan. *Hydrol. Process.* **2009**, *23*, 3556–3568.
21. IPCC. *Climate Change 2007—The Physical science Basis*; Fourth Assessment Report of the Intergovernmental Panel on Climate Change; Cambridge University Press: Cambridge, UK, 2007.
22. Caron, L.P.; Jones, C.G. Analysing present, past and future tropical cyclone activity as inferred from an ensemble of coupled global climate models. *Tellus* **2008**, *60A*, 80–96.
23. Min, S.K.; Park, E.H.; Kwon, W.T. Future projections of East Asian climate change from multi—AOGCM ensembles of IPCC SRES scenario simulations. *J. Meteorol. Soc. Jpn.* **2004**, *82*, 1187–1211.
24. Kitoh, A.; Uchiyama, T. Changes in onset and withdrawal of the East Asian summer rainy season by multi-model global warming experiments. *J. Meteorol. Soc. Jpn.* **2006**, *84*, 247–258.
25. Kim, M.K.; Kang, I.S.; Park, C.K.; Kim, K.M. Superensemble prediction of regional precipitation over Korea. *Int. J. Climatol.* **2004**, *24*, 777–790.
26. Feddersen, H.; Andersen, U. A method for statistical downscaling of seasonal ensemble predictions. *Tellus* **2005**, *57A*, 398–408.
27. Richardson, C.W. Stochastic simulation of daily precipitation, temperature, and solar radiation. *Water Resour. Res.* **1981**, *17*, 182–190.
28. Selker, J.S.; Haith, D.A. Development and testing of single-parameter precipitation distributions. *Water Resour. Res.* **1990**, *26*, 2733–2740.
29. Tung, C.P.; Haith, D.A. Global-warming effects on New York streamflows. *J. Water Resour. Plan. Manag.* **1995**, *121*, 216–225.
30. Coe R.; Stern, R.D. Fitting models to rainfall data. *J. Appl. Meteorol.* **1982**, *21*, 1024–1031.
31. Woolhiser, D.A.; Roldan, J. Stochastic daily precipitation models: 2. A comparison of distribution of amounts. *Water Resour. Res.* **1982**, *18*, 1461–1468.
32. Schubert, S. A weather generator based on the European "Grosswetterlagen". *Clim. Res.* **1994**, *4*, 191–202.
33. Corte-Real, J.; Xu, H.; Qian, B. A weather generator for obtaining daily precipitation scenarios based on circulation patterns. *Clim. Res.* **1999**, *13*, 61–75.
34. Woolhiser, D.A.; Pegram, G.G.S. Maximum likelihood estimation of Fourier coefficients to describe seasonal variation of parameters in stochastic daily precipitation models. *J. Appl. Meteorol.* **1979**, *18*, 34–42.
35. Woolhiser, D.A.; Roldan, J. Seasonal and regional variability of parameters for stochastic daily precipitation models. *Water Resour. Res.* **1986**, *22*, 965–978.

36. Hamon, W.R. Estimating potential evapotranspiration. *J. Hydraul. Div. Proc. Am. Soc. Civil Eng.* **1961**, *871*, 107–120.
37. Tate, E.L.; Gustard, A. Drought definition: A hydrological perspective. In *Drought and Drought Mitigation in Europe*; Vogt, J.V., Somma, F., Eds.; Kluwer Academic Publishers: Derdrecht, The Netherlands, 2000; pp. 23–48.
38. Kjeldsen, T.R.; Lundorf, A.; Rosbjerg, D. Use of a two-component exponential distribution in partial duration modeling of hydrological droughts in Zimbabwean rivers. *Hydrol. Sci. J.* **2000**, *45*, 285–298.
39. Shiau, J.T.; Shen, H.W. Recurrence analysis of hydrological droughts of differing severity. *J. Water Resour. Plan. Manag.* **2001**, *127*, 30–40.
40. Hisdal, H.; Stahl, K.; Tallaksen, L.M.; Demuth, S. Have streamflow droughts in Europe become more severe and frequent. *Int. J. Climatol.* **2001**, *21*, 317–333.
41. Shiau, J.T. Return period of bivariate distributed extreme hydrological events. *Stoch. Environ. Res. Risk Assess.* **2003**, *17*, 42–57.
42. Fleig, A.K.; Tallaksen, L.M.; Hisdal, H.; Demuth, S. A global evaluation of streamflow drought characteristics. *Hydrol. Earth Syst. Sci.* **2006**, *10*, 535–552.
43. Shukla, S.; Wood, A.W. Use of a standardized runoff index for characterizing hydrologic drought. *Geophys. Res. Lett.* **2008**, *35*, L02405.
44. Tallaksen, L.M. Streamflow drought frequency analysis. In *Drought and Drought Mitigation in Europe*; Vogt, J.V., Somma, F. Eds.; Kluwer Academic Publishers: Derdrecht, The Netherlands, 2000; pp. 103–117.
45. Chu, J.L.; Yu, P.S. A study of the impact of climate change on local precipitation using statistical downscaling. *J. Geophys. Res.* **2010**, *115*, D10105.
46. Gerrity, J.P. A note on Gandin and Murphy's equitable skill score. *Mon. Weather Rev.* **1992**, *120*, 2709–2712.
47. World Meteorological Organization. *Standardized Verification System (SVS) for Long-Range Forecasts (LRF)*; World Meteorological Organization: Geneva, Switzerland, 2002.
48. Duan, Q.; Sorooshian, S.; Gupta, V.K. Optimal use of the SCE-UA global optimisation method for calibrating watershed models. *J. Hydrol.* **1994**, *158*, 265–284.

2

Scenario-Based Impacts of Land Use and Climate Change on Land and Water Degradation from the Meso to Regional Scale

Aymar Y. Bossa [1,2,*], Bernd Diekkrüger [2] and Euloge K. Agbossou [3]

[1] West African Science Service Center on Climate Change and Adapted Land Use—WASCAL, Ouagadougou 06 P.O. Box 9507, Burkina Faso

[2] Department of Geography, University of Bonn, Meckenheimer Allee 166, Bonn 53115, Germany; E-Mail: b.diekkrueger@uni-bonn.de

[3] Faculty of Agricultural Sciences, University of Abomey-Calavi, Cotonou 01 BP. 526, Benin; E-Mail: agbossou.euloge@yahoo.fr

* Author to whom correspondence should be addressed; E-Mail: bossa.a@wascal.org;

External Editor: Julia Piantadosi

Abstract: Scale-dependent parameter models were developed and nested to the Soil and Water Assessment Tool-SWAT to simulate climate and land use change impacts on water-sediment-nutrient yields in Benin at a regional scale (49,256 km²). Weighted contributions of relevant landscape attributes characterizing the spatial pattern of ongoing hydrological processes were used to constrain the model parameters to acceptable physical meanings. Climate change projections (describing a rainfall reduction of up to 25%) simulated throughout the Regional Model-REMO, very sensitive to a prescribed degradation of land cover, were considered. Land use change scenarios in which the population growth was translated into a specific demand for settlements and croplands (cropland increase of up to 40%) according to the development of the national framework, were also considered. The results were consistent with simulations performed at the meso-scale (586 km^2) where local management operations were incorporated. Surface runoff, groundwater flow, sediment and organic N and P yields were affected by land use change (as major effects) of −8% to +50%, while water yield and evapotranspiration were dominantly affected by climate change of −31% to +2%. This tendency was more marked at the regional scale as response to higher scale-dependent rates of natural vegetations with higher conversions to croplands.

Keywords: SWAT; scaling; regionalization; nutrient loss; climate change; land use change

1. Introduction

Unsustainable land use is driving land degradation, which in the form of soil erosion, nutrient depletion, water scarcity, salinity and disruption of biological cycles is a fundamental and persistent problem, diminishing productivity, biodiversity, other ecosystem services, and contributing to climate change [1]. A global survey suggests that 40% of agricultural land is already degraded to the point that yields are greatly reduced, and a further 9% is degraded to the point that it cannot be reclaimed for productive use by farm level measures [2]. According to the Global Assessment of Human-induced Soil Degradation (GLASOD) [1,3,4] estimates, degradation of cropland appears to be most prevalent in Africa, affecting already in the 1990s 65% of cropland areas, compared with 51% in Latin America and 38% in Asia. Many studies have been conducted in parts of Africa to understand the processes as well as determinant and promoting causes, related to the specific climatological, meteorological and soil conditions [5–16], and have been capitalized into this current work.

Vulnerability to change, whether climate-induced or related to anthropogenic-induced changes in land use/land cover, is a major threat, consisting at the same time of a water dimension (reduced water availability), an agrarian dimension (falling yields) and an environmental dimension (weakening of the soil and increasing erosion) [7]. Several investigations on the global change processes and its impact on the hydrological cycle [9,17–26] have shown that global climate change has a significant influence on the regional water and soil resources. Neumann *et al.* [23] investigated climate trends of temperature, precipitation and river discharge in the Volta Basin (West Africa) and have concluded on weak trends towards a decrease in rainfall with no clear trend on discharge as the anthropogenic influences (e.g., building of dams, intensified irrigation) were not quantified. Legesse *et al.* [21] evaluated the hydrological response of a catchment to climate and land use changes in Tropical Africa and found that a 10% decrease in rainfall produced a 30% reduction on the simulated hydrologic response of the catchment, while a 1.5 °C increase in air temperature would result in a decrease in the simulated discharge of about 15%. Moreover, they indicated that a conversion of the present day dominantly cultivated/grazing land by woodland would decrease the discharge at the outlet by about 8%. Chaplot [20] examined the effects of increasing CO_2 concentrations and rainfall changes associated with changes in average daily rainfall intensity, and surface air temperature on loads of water, NO_3-N and sediments from watersheds exhibiting different environmental conditions. He found over a 100-year simulated period: (1) flow and sediment discharges affected by precipitation changes while temperature and changes in atmospheric CO_2 concentration had a smaller effect; (2) CO_2 concentration was the main controlling factor of NO_3-N loads; and (3) global changes in the humid watershed had a greater effect on the water and soil resources. Ward *et al.* [25] evaluated the impact of land use and climate change on future suspended sediment yields and found an increase in all simulations due to conversion of forest to agricultural land. Conjoint sensitivity analyses have shown that although land use change acts as the primary control on long-term changes in sediment yield, the sensitivity of sediment yield to changes in climate increases as the percentage of deforested land increases. Mahe *et al.* [24] modeled the impact of land use change on soil water holding capacity and river flow in West Africa and found that the total reduction in water holding capacity is estimated to range from 33% to 62% between 1965 and 1995. This was explained by the decline in the extent of natural vegetation from 43% to 13% of the total basin area, whilst the cultivated areas increased from 53% to 76% and the area of bare soil nearly tripled from 4% to 11%. Li *et al.* [22] modeled the hydrological impact of land use change (West Africa) and pointed out that total deforestation (clearcutting) increases the simulated runoff ratio from 0.15 to 0.44, and the annual streamflow by 35%–65%, depending on location in the basin, although forests occupy only a small portion (<5%) of the total basin area. They mentioned that there is no significant impact on the water yield and river discharge when the deforestation (thinning) percentage is below 50% or the overgrazing percentage below 70% for savanna and 80% for grassland areas; however, the water yield is increased dramatically when land cover change exceeds these thresholds. Faramarzi *et al.* [19] modeled the impacts of climate change on freshwater availability in Africa and the results indicated that the mean total quantity of water resources is likely to increase, but for individual catchments and countries, variations are substantial. Cornelissen *et al.* [18] assessed the suitability of different hydrological model types for simulating scenarios of future discharge behavior in West Africa in the context of climate and land use change. They found that all models simulate an increase in surface runoff due to land use change. The application of climate change scenarios resulted in considerable variation between the models and points not only to uncertainties in climate change scenarios. The conclusions drawn out from the above-presented studies are concordant with the different methodologies rolled out. They demonstrated significant sensitivities of soil and water resources to increase CO_2, temperature, rainfall and land use depending on the study locations, the catchment characteristics, the modeling approaches, the land use change drivers, the structure of the atmosphere-ocean global climate models as well as the regional climate models behind the climate projections used. None of them has addressed the chemical dimension of soil degradation, notably the soil organic N and P loads and delivered together with sediment at catchment outlet. Organic N and P loads highly depend on landscape heterogeneity and spatial patterns of hydrological processes which are well known to smooth out with increasing catchment size resulting in more uncertain model parameters (without physical meaning) and more uncertain impact calculations at large scale, but almost none of the available impact studies have addressed this scaling problem, what is also offered in the present study.

It is widely accepted that the complexity of hydrological processes as well as land and water degradation depend on the environmental heterogeneity such as soil pattern, topography, geology, vegetation and anthropogenic impacts. Thus, the process-dependent hierarchization of landscape elements from local to regional scales [27] is recently reflected in the development of several computationally efficient conceptual and distributed physical-based models (e.g., SWAT), attempting to quantify the hydrological variability occurring at a range of scales. Therefore, catchments may be subdivided into a number of smaller units such as sub-catchments, hillslopes, hydrological response units, contour-based elements, and square grid elements [6,28–31], but opened discussions on the modeling uncertainty issue, including scaling-effects in model internal aggregation [32,33] for large scale applications. This often affects the magnitudes of model parameters which may finally have no consistent physical meanings, carrying too poor information [33]. A scale dependent parameterization approach may significantly reduce these sources of uncertainty and the problem of lack or non-accurate measurement data (e.g., stream water-sediment-nutrient measurements) at large catchment scale. This is crucial and of high interest for impact assessment of climate and land use change at large-scale in a data-poor environment like Benin.

Previous integrated modeling works in West Africa and in the upper Ouémé catchment of roughly 15,000 km^2 in size in Benin [9,16], have largely contributed to improving knowledge of recent land degradation processes from local to regional scales. The results have shown that amongst others silt and clay particle loads totaling 0.5 ton ha^{-1} a^{-1} with an associated organic nitrogen load of 0.8 kg·ha^{-1} a^{-1} [16]. These results clearly indicate that the study area is impacted by land and water degradation processes, primary seen as human-induced or natural processes that negatively affect ecosystem as for resources storing and recycling. Scenario-based land use and climate change [34–37] may impact the degradation process to a level of +50% of the observations [15,16]. The current work attempts to expand all above-mentioned dimensions of erosion-related degradation under different land use and climate scenarios to large catchment scales (up to 49,256 km^2), while minimizing biases due to scaling-up processes such as model internal aggregation processes in the SWAT model. These thorough modeling exercises are still challenging the world scientific community and have never been parts of previous studies in Benin, as transferability of results through spatial scales stays an important underlying question.

This study aims to investigate how global change impacts on water/land degradation at different spatial scales in Benin. It specifically means to investigate the degradation trends at different catchment scales: (1) a meso-scale catchment (586 km^2) investigation incorporating local-scale farming practices [16]; and (2) a large-scale (49,256 km^2) investigation based on a regionalized model parameterization [38] based on scale-dependent model parameters for simulating water-sediment-nutrient fluxes. This latter approach makes use of physical catchment properties depending on the spatial scale as explanatory variables for model parameters using regression techniques.

2. Materials and Methods

2.1. Study Area

Located at about 90% in the Republic of Benin between 6°48' and 10°12' N of latitude and as part of the stable margin of the West African Craton, the Ouémé catchment (49,256 km^2, cf. Figure 1) is mainly characterized by a Precambrian basement, consists predominantly of complex migmatites granulites and gneisses, including less abundant mica shists, quarzites and amphibolites [9]. Syn-and post-tectonic intrusions of mainly granites, diorites, gabbros and volcanic rocks are present [39]. With a topographic relief generally low (highest elevation point of 617 meter) the land surface is slightly ondulating (granitic-gneissic plateau), strongly fractured (granitic peneplain) with typical seasonally waterlogged linear depressions (inland valleys) [9].

At a regional scale, fersialitic soils (ferruginous tropical sols) are predominant, characterized by clay translocation and iron segregation (ferruginous tropical sols with iron segregation), which lead to a clear horizon differentiation [40]. A local scale description has shown a typical catena with lixisols/acrisols on the upper and middle slopes, following by plinthosols on the downslopes, gleysols in the inland valleys and fluvisols on the fluvial plain [41].

Situated in a wet (Guinean coast) and a dry (Northern Soudanian zone) tropical climate, the Ouémé catchment records annual mean temperatures of 26 °C to 30 °C, annual mean rainfalls of 1280 mm (from 1950 to 1969) and 1150 mm (from 1970 to 2004) at a climatic station close to 9° N latitude [9]. As shown in the Figure 1, the Soudanian zone has a unimodal rainfall season that peaks in August whereas the Guinean zone exhibits a bimodal rainfall season that peaks in June and October.

Figure 1. Location and climate condition of the study area, after Speth *et al.* [9]. The investigated catchments are Donga-Pont (586 km^2), Vossa (1935 km^2), Térou-Igbomakoro (2344 km^2), Zou-Atchérigbé (6978 km^2), Kaboua (9459 km^2), Bétérou (10,072 km^2), Savè (23,488 km^2), Ouémé-Bonou (49,256 km^2).

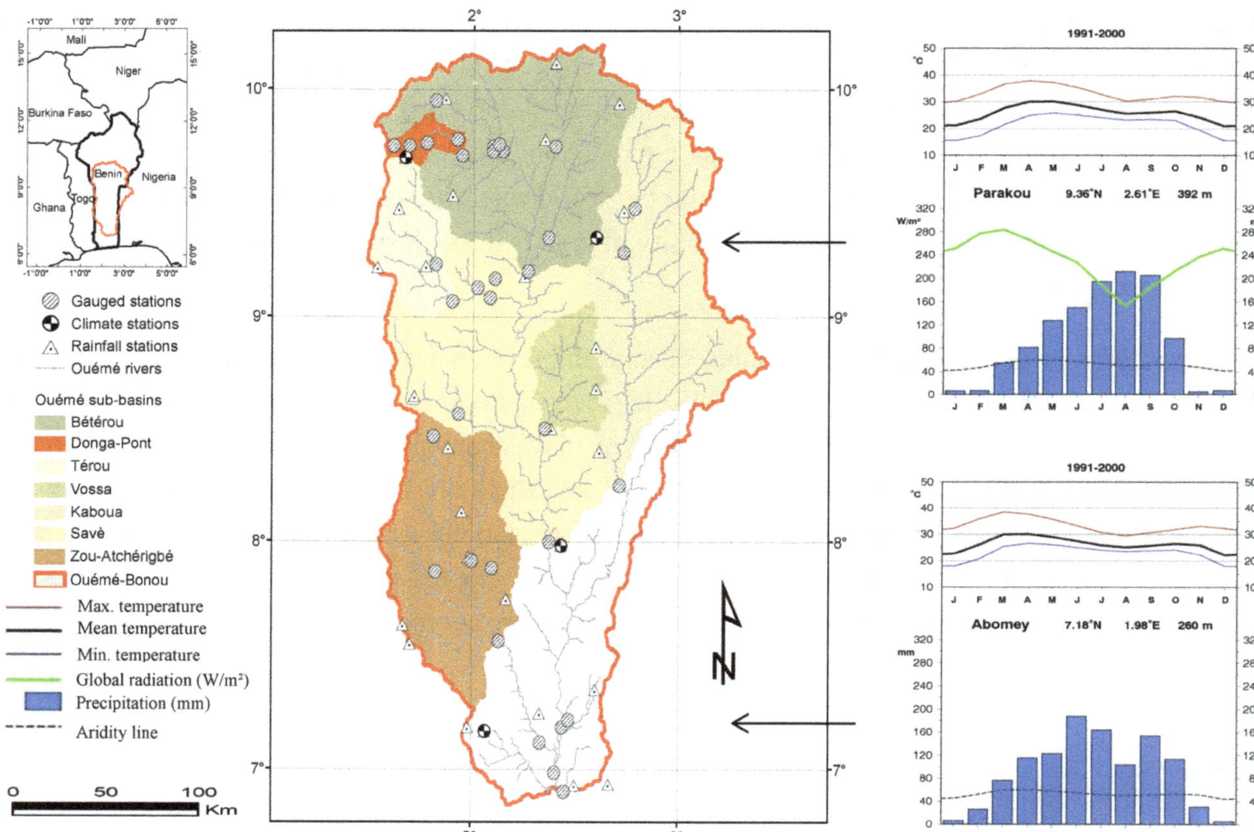

The catchment landscape is characterized by forest islands, gallery forest, savannah, woodlands, agricultural lands and pastures. Agriculture and other human activities have led to large-scale deforestation and fragmentations leaving only small relicts of the natural vegetation types within a matrix of degraded secondary habitats [9].

With a length of about 510 km and with two most important tributaries, Zou (150 km) and Okpara (200 km), the Ouémé river drains into Lake Nokoué (150 km^2) and flows through the coastal lagoon system into the sea. Rainfall-runoff variability is high in the catchment, leading to runoff coefficients varying from 0.10 to 0.26 (of the total annual rainfall), with the lowest values for the savannahs and forest landscapes [9].

2.2. Modeling Approach

The SWAT (Soil and Water Assessment Tool) model is an eco-hydrological model developed by the United States Department of Agricultural-Research-Service (USDA-ARS) [31]. It is a continuous-time model that operates at a daily time-step. It allows the assessment of various subsurface flows and storages and related sediment and nutrient loads, taking into account the feedback between plant growth, water, and nutrient cycle, and helps to understand land management practice effects on water, sediment, and nutrient dynamics. It is a catchment scale model which can be applied from small (km²) to regional (100,000 km²) scale. SWAT subdivides the catchment into sub-catchments based on a Digital Elevation Model (DEM). Each sub-catchment consists of a number of Hydrological Response Units (HRUs) which are homogeneous concerning soil, relief, and vegetation. The HRUs are not georeferenced and not linked to each other within the sub-catchment.

In SWAT surface runoff is simulated using a modified version of the SCS CN method [42]. Lateral flow is simulated using the kinematic method of Sloan and Moore [43]. Percolation occurs when the soil field capacity is exceeded, recharging two aquifer systems: an unconfined aquifer generating base flow to the catchment streams, and a confined (deep) aquifer generating base flow to streams outside the catchment. The mass of nitrate lost from the soil horizons is determined using the nitrate concentration in the mobile water multiplied by the water volume flowing in each pathway. Rainfall-runoff erosion is estimated using the Modified Universal Soil Loss Equation (MUSLE) [44]. Organic N attached to sediments is estimated based on the loading function of McElroy *et al.* [45] and modified by Williams and Hann [46] to consider each runoff events. The model computes evaporation from soils according to Ritchie [47]. Actual soil water evaporation is estimated using exponential functions of soil depth and water content. Plant transpiration is computed as a linear function of potential evapotranspiration and leaf area index.

The overall modeling approach is summarized in Figure 2, showing the nature and source of the different data layers, their scales and types of parameters and investigations [48,49], in the structure as required for applying the SWAT model. A 90 m resolution Digital Elevation Model from the Shuttle Radar Topography Mission-SRTM was used. A SOil and TERrain (SOTER) digital database established at the scale 1:200.000 for the whole Ouémé catchment, in corporation with INRAB (Institut National de la Recherche Agricole du Bénin) is considered in this study (cf. Bossa *et al.* [50] for more details and an overview of the map and soil properties). This database includes different soil properties that were determinant for the model setup and parameterization: saturated hydraulic conductivity, organic CNP, bulk density, texture, erodibility factor, available water content, hydrology group, *etc.* The land use/cover map considered in this study has been established at 250 m resolution from 3 scenes satellite images LANDSAT ETM+ of 2003 [37] with an overall accuracy of 87%. More than 650 observation points were checked during the ground checks and 17 land use/cover classes were defined. Agricultural calendars depending on rainy season onsets, rainfall rhythms, and crop growth cycles/management over the period 2004–2009 (activity reports from the Regional Center of Agricultural Promotion-CeRPA and Ministry of Agriculture, Livestock and Fisheries-MAEP) have been used for the baseline agricultural practices introduced in SWAT (cf. Tables 1 and 2 for an example). Climate data (rainfall, temperature, solar radiation, wind speed and air humidity) were collected from 35 stations managed by the German Research Project IMPETUS, IRD (Institut de Recherche pour le Développement, France), and DMN (Direction de la Météorologie Nationale).

Figure 2. Schematization of the modeling approach. Soil and land use data are from IMPETUS [48] and INRAB (Institut National de la Recherche Agricole du Bénin [49]), Climate data are from IMPETUS, IRD (Institut de Recherche pour le Développement, France), and DMN (Direction de la Météorologie Nationale), Geology data is from OBEMINES (Office Béninoise des MINES). CountryStat: Benin National Statistics (Food and Agriculture data network). CeRPA: Regional Center of Agricultural Promotion. MAEP: Ministry of Agriculture, Livestock and Fisheries. SSC means suspended sediment concentration.

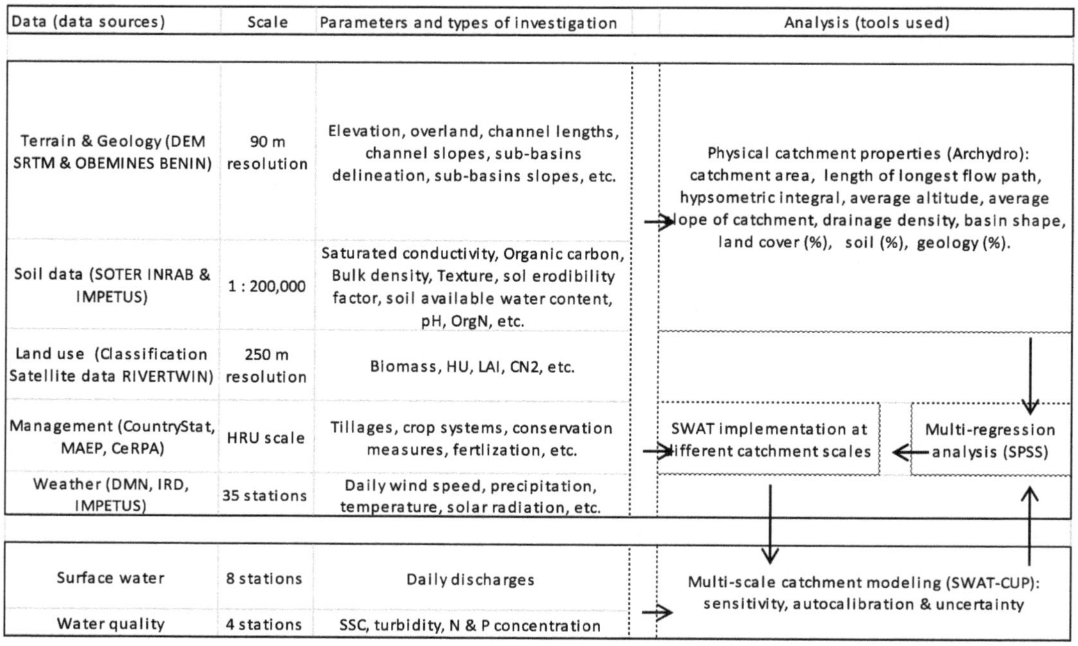

Table 1. Management operations considered for croplands in the Donga-Pont catchment. T1, T2 and T3: Tillage operation; F1 and F2: Fertilization.

Croplands	T1	F1	T2	T3	F2
Elemental N-P (kg ha^{-1})		10.5–6			
Elemental N (kg ha^{-1})					10.5
Tillage depth (cm)	25		10	10	
Mixing efficiency	0.5		0.25	0.25	

Table 2. Management operations considered for pastures in the Donga-Pont catchment. NH3N is the dissolved nitrogen, easily convertible into nitrate.

Pastures	Grazing
Grazing days	300
Biomass eaten: beef/dairy-sheep-goat (kg ha^{-1} d^{-1})	76-24-28
Biomass trampled: beef/dairy-sheep-goat (kg ha^{-1} d^{-1})	15-5-6
Manure beef (1%N-0.4%P-3%ORGN-0.7%ORGP-95%NH3N) (kg ha^{-1} d^{-1})	38
Manure sheep (1%N-0.4%P-3%ORGN-0.7%ORGP-95%NH3N) (kg ha^{-1} d^{-1})	12
Manure goat (1%N-0.4%P-3%ORGN-0.7%ORGP-95%NH3N) (kg ha^{-1} d^{-1})	64

Besides discharge data continuously available for more or less 10 years (1998–2008) at 8 gauging stations, water samples (9 liters per day) were collected in 2004, 2005, 2008, 2009 and 2010 at 4 gauging stations (Donga-Pont, Bétérou, Térou and Zou-Atchérigbé, cf. Figure 1) and filtered in order to calculate daily suspended sediment concentration. Multi-parameter probes YSI 600 OMS (including one turbidity-broom sensor YSI 6136) were installed at the same stations to register turbidity at a high temporal resolution (used to calculate continuous time series of suspended sediment concentrations) to consider the hysteresis effects on the relationship between sediment and discharge. After filtration the obtained sediments were analyzed in the laboratory for organic Nitrogen and non-soluble/organic Phosphorus content. Weekly water samples were collected (2008–2010) for analyzing Nitrate and soluble Phosphorus.

The general input data were used to compute selected physical catchment attributes and beyond six individual Ouémé sub-catchments (Donga-Pont, Térou, Bétérou, Zou-Atchérigbé, Vossa and Kaboua, cf. Figure 1) were considered for SWAT calibration. Multi-scale auto-calibration and uncertainty analysis were performed applying the SUFI-2 procedure (Sequential Uncertainty Fitting version 2, SWAT-CUP interface [51]) so discharge, sediment, nitrate and organic N and P were simultaneously calibrated. SPSS software was used to statistically analyze two different matrixes of calibrated parameter sets and computed catchment attributes. A correlation analysis was performed to identify physical catchment attributes meaningful for each calibrated model parameter. Multiple regression analyses were later on performed to establish the regionalization rules which are in fact assumed to highly capture the catchment heterogeneity as well as the spatial pattern of the hydrological processes. Table 3 shows the calibration and validation periods as well as quality measures of the simulations for the different sub-catchments investigated.

Table 3. Model goodness of fit to measurements for the different sub-catchments involved in the multiple regression analysis, for model calibration. Information concerning validation are provided in brackets.

		Donga	Vossa	Térou	Atchérigbé	Kaboua	Bétérou
Discharge	Period	2006–2008 (1998–2005)	1998–2000 (1995)	2002–2005 (1998–2001, 2006)	2007–2008 (2001–2006, 2009)	2004–2006 (1995–1998)	2006–2009 (1998–2005)
	R^2	0.72 (0.58)	0.75 (0.63)	0.75 (0.61)	0.89 (0.71)	0.73 (0.55)	0.75 (0.64)
	NS	0.72 (0.51)	0.75 (0.34)	0.74 (0.51)	0.82 (0.62)	0.67 (0.40)	0.60 (0.59)
Sediment	Period	2008 (2005)	– –	2004–2005 (2006)	2008 (2009)	– –	2008–2009 (2004–2005)
	R^2	0.69 (0.58)	–	0.44 (0.33)	0.66 (0.67)	–	0.43 (0.27)
	NS	0.67 (0.55)	–	0.41 (0.32)	0.64 (0.67)	–	0.30 (0.14)
Nitrate	Period	2008 (2008–2009)	– –	– –	2008 (2009)	– –	2008–2009 (2008)
	R^2	0.99 (0.95)	–	–	0.86 (0.62)	–	0.73 (0.52)
	NS	0.99 (0.78)	–	–	0.81 (0.54)	–	(0.46)

Regarding discharge simulations, poor model efficiency were obtained for validation in Vossa and Kaboua sub-catchments (0.34 and 0.40 respectively) due mainly to peak overestimation caused partly by land use map derived from 2003 Landsat images, which considered more agricultural areas than the reality of the validation period (1995–1998). Critical model performances were obtained for sediment simulation in the Bétérou sub-catchments, where the model efficiency decreased even to 0.14. This may be mainly caused by strong hysteresis effects observed at this station, which was not equipped of turbidity probe as used at the Donga-Pont and Atchérigbé gauging stations to minimize this effect. Nitrate load was in general well represented in the model with coefficients of determination ranging from 0.62 to 0.99 and model efficiencies ranging from 0.54 to 0.99. Higher performances were observed for smaller sub-catchments. Calibrated model parameters are presented in Table 4 for all investigated sub-catchments.

Table 4. Calibrated model parameter matrix involved in the multiple regression analysis. The letters v, r and a mean values, relative change and absolute change, respectively. CN2: SCS Curve Number, ALPHA_BF: base flow recession constant; SOL_K: soil hydraulic conductivity, RCHRG_DP: aquifer percolation coefficient, GWQMN: minimum water level for base flow generation, REVAPMN: threshold water level in a shallow aquifer for capillary rise, ESCO: Soil evaporation compensation factor, GW_DELAY: groundwater delay, Ch_K2: Effective channel hydraulic conductivity, USLE_P: Practice factor, USLE_K: Soil erodibility factor, SPEXP: Exponent for calculating max sediment retrained, SURLAG: Surface runoff lag coefficient, NPERCO: Nitrate percolation coefficient.

Parameter	Description	Donga	Vossa	Térou	Atchérigbé	Kaboua	Bétérou
ESCO (v)	Soil evaporation compensation factor (–)	0.38	0.28	0.49	0.35	0.28	0.43
SOL_Z (r)	Soil depth (mm)	0.27	0.01	0.37	0.16	0.06	0.03
CN2 (r)	Curve Number (–)	6.65	3.86	5.55	6.24	3.97	2.51
GWQMN (v)	Threshold depth for ground water flow to occur (mm)	38.75	7.50	47.50	28.50	30.50	43.50
REVAPMN (v)	Threshold water level in shallow aquifer for revap (mm)	15.25	6.50	45.50	18.50	26.50	26.50
Ch_K2 (v)	Effective channel hydraulic conductivity (mm/hr)	3.95	1.00	12.77	10.65	1.00	12.52
Sol_K (r)	Saturated hydraulic conductivity (mm/hr)	−0.78	−0.65	−0.73	−0.35	−0.82	−0.66
GW_DELAY (v)	Ground water delay (day)	15.08	17.12	23.25	10.87	16.04	24.80
USLE_P (v)	Practice factor (–)	0.13	0.10	0.07	0.15	0.18	0.00
USLE_K (r)	Soil erodibility factor (0.013 t m^2h/(m^3 t cm))	0.03	0.16	0.08	0.14	0.25	−0.57
SPEXP (v)	Exponent for calculating max sediment retrained (–)	1.35	1.07	1.21	1.20	1.28	1.38
SURLAG (v)	Surface runoff lag coefficient (–)	0.19	0.35	0.24	0.25	0.17	0.10
ALPHA_BF (v)	Base flow recession factor (day)	0.06	0.17	0.07	0.12	0.11	0.15
NPERCO (v)	Nitrate percolation coefficient (–)	0.49	0.88	0.74	0.71	0.32	0.67
RCHRG_DP (v)	Fraction of deep aquifer percolation (–)	0.25	0.20	0.17	0.22	0.15	0.29

In the following paragraphs of this section, details of the calibration and validation issues are presented as follows for the Atchérigbé sub-catchment (6978 km^2) to provide a complete overview on the measurements involved into the multi-scale modeling step. The validation of the regionalization rules is presented for the Savè sub-catchment (23,488 km^2) in this same section and we should highlight that orders of magnitude of impacts of climate and land use change scenarios are presented and compared for the two targeted spatial scales (Donga-Pont: 586 km^2 and Ouémé-Bonou: 49,256 km^2) in the result section.

Simulated *versus* observed daily water discharge and sediment yield are shown in Figures 3 and 4 for the Atchérigbé sub-catchment (cf. Figure 1). Recession periods were generally well represented.

Less accurate predictions of single peaks are also shown in some years, partly due to the measurement errors during exceptional flooding years (2003 and 2007) in which over bank full discharge was observed at the gauging station. Differences are also usually caused by the SWAT structure, since it is a continuous time model with a daily time step and sub-scale processes such as single-event flood routing cannot be efficiently predicted. In addition, the daily measured precipitation for 24 h starts at 6:00 am and may not well match to the daily average discharge values, which were measured for 24 h from midnight on [52]. As it can be seen from the figures in the year 2008, discharge measurement gaps of even more than 10 days can happen due mainly to technical problems.

Figure 3. Simulated *vs.* observed daily discharge for the Atchérigbé sub-catchment (6978 km^2). Calibration period was 2007 to 2008 ($R^2 = 0.89$ and ME = 0.83), validation period was 2001–2006, and 2009 ($R^2 = 0.71$ and ME = 0.62).

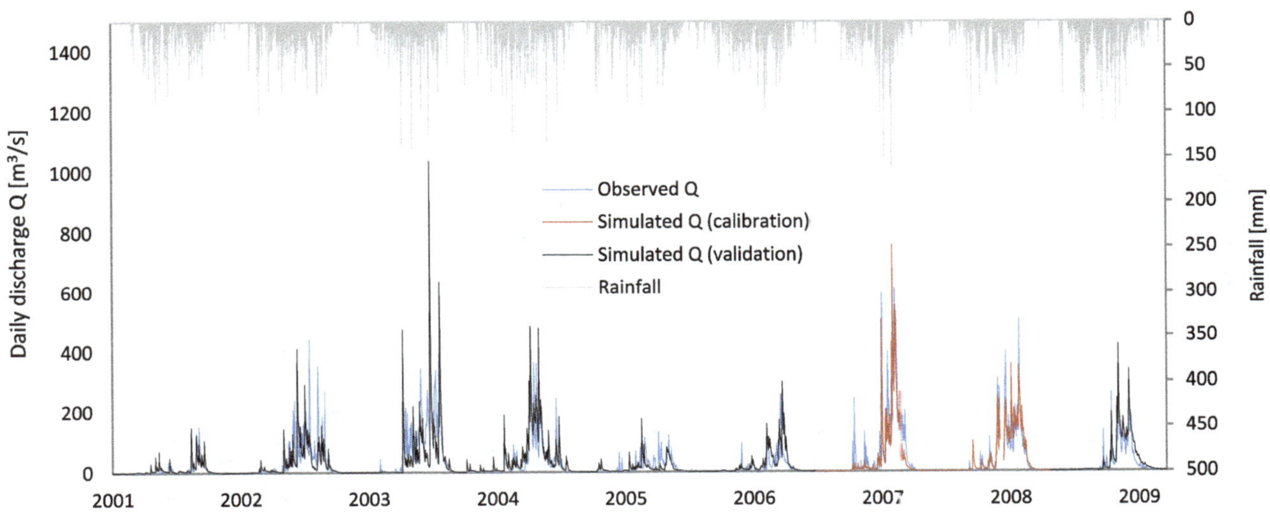

Figure 4. Simulated *vs.* Observed daily sediment yield for Atchérigbé sub-catchment (6978 km^2). Calibration period was 2008 ($R^2 = 0.66$ and ME = 0.64), validation period was 2009 ($R^2 = 0.67$ and ME = 0.67). SSC = suspended sediment.

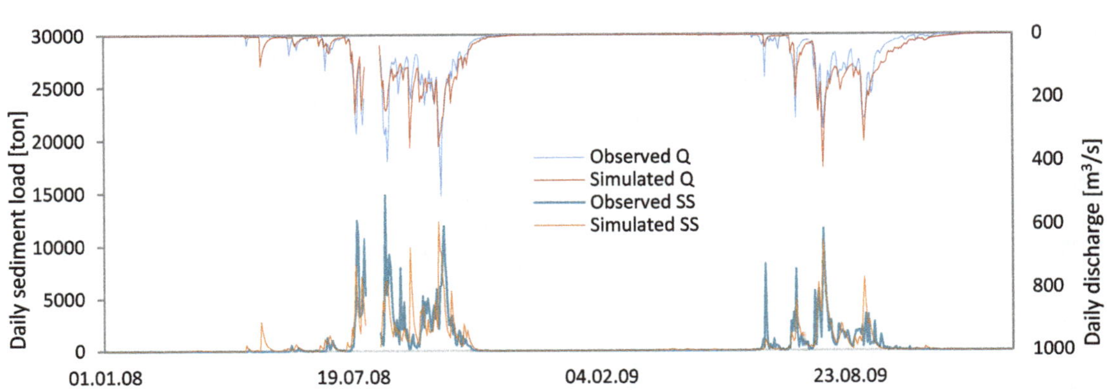

Simulated *versus* observed daily stream water nitrate load are shown in Figure 5 for the same Atchérigbé sub-catchment. Similarly to the sediment yield, nitrate peaks accompanied discharge peaks mainly caused by combined effects of increase nitrate loading and increase in water volume. Due to the sampling time scale (one time a week) several peaks were missed, but did not affect the model calibration.

According to FAO [53], water degradation by sediment has a chemical dimension—the silt and clay fraction, primary carrier of adsorbed chemicals, like nitrogen and phosphorus, which are transported by sediment into the aquatic system. Figure 6 shows weekly simulated *versus* observed organic N and P delivery at the Atchérigbé gauging station. Organic N and P were not calibrated. Since it was assumed that a good adjustment of soil nutrient pools, nitrate and sediment loads would be reflected in their simulations, only a validation was performed. Model goodness-of-fit were acceptable: 0.58 (R^2) and 0.78 (NS) for organic Nitrogen and 0.89 (R^2) and 0.96 (NS) for organic Phosphorus.

Figure 5. Simulated *vs.* observed daily nitrate load for Atchérigbé sub-catchment (6978 km²). Calibration period was 2008 (R^2 = 0.86 and ME = 0.81), validation period was 2009 (R^2 = 0.62 and ME = 0.54).

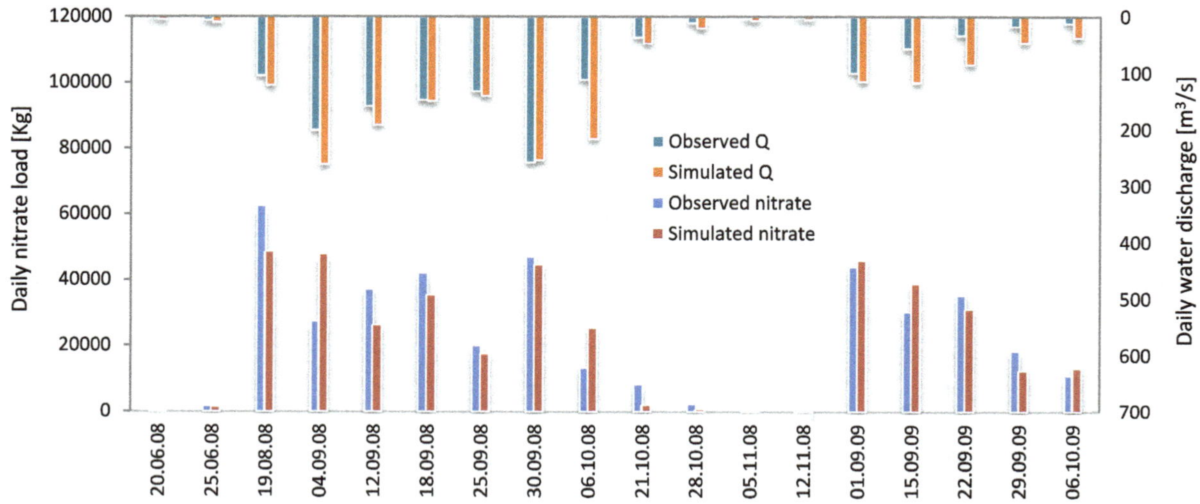

Figure 6. Simulated *vs.* observed weekly organic N and P load for the Atchérigbé sub-catchment (6978 km²). Only validation was performed from 2008 to 2009 with R^2 = 0.58 and ME = 0.78 for organic Nitrogen and R^2 = 0.89 and ME = 0.96 for organic Phosphorus.

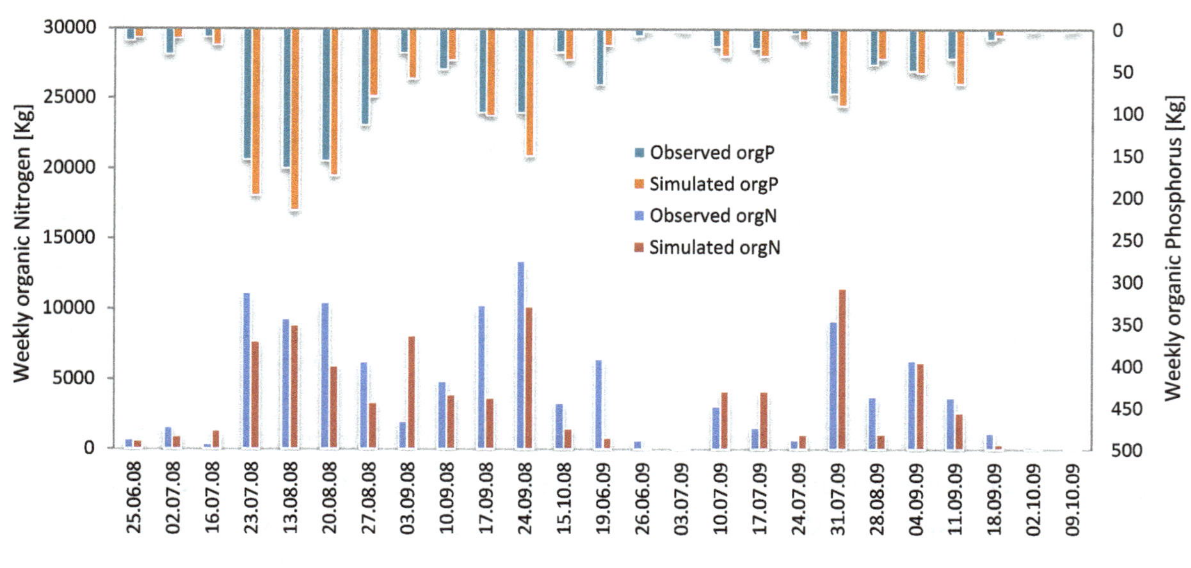

Table 5 shows the computed regionalization rules and derived parameter sets for the Savè (23,488 km², cf. Figure 1) sub-catchment and Ouémé-Bonou (49,256 km², cf. Figure 1) catchment. Physical catchment attributes depending on spatial scale were used as explanatory variables of SWAT model parameters. With respect to discharge, validation was performed for the Savè sub-catchment with a goodness-of-fit around 0.7 for model efficiency and R^2 (Figure 7).

Table 5. Best regression-based parameter model and resulting values for three independent catchments (Savè: 23,488 km², Ouémé-Bonou: 49,256 km²). CN2: SCS Curve Number, ALPHA_BF: base flow recession constant; SOL_K: soil hydraulic conductivity, RCHRG_DP: aquifer percolation coefficient, GWQMN: minimum water level for base flow generation, REVAPMN: threshold water level in a shallow aquifer for capillary rise, ESCO: Soil evaporation compensation factor, GW_DELAY: groundwater delay, Ch_K2: Effective channel hydraulic conductivity (mm/h), USLE_P: Practice factor (–), USLE_K: Soil erodibility factor [0.013 t m²h/(m³ t cm)], SPEXP: Exponent for calculating max sediment retrained (–), SURLAG: Surface runoff lag coefficient (–), NPERCO: Nitrate percolation coefficient (–).

Parameters	Equations	R^2	Savè	Ouémé-Bonou
ESCO	= 0.935 – 0.217 (Average slope of catchment) + 0.00327 (% Alterites)	0.92	0.34	0.37
SOL_Z	= 0.758 – 0.01 (% Migmatites)	0.81	0.02	0.08
SOL_K	= 26.991 – 0.278 (% Percentage of level)	0.92	–0.58	–0.76
CN2	= 10.0 – 0.0824 Migmatites (%)	0.49	3.94	4.4
GWQMN	= 185 – 49.2 (Average slope of catchment) – 0.255 (% Migmatites)	0.85	26.89	37.28
REVAPMN	= 16.5 + 0.769 (% Alterites)	0.6	20.37	18.56
Ch_K2	= 56.1 – 16.0 (Average slope of the catchment) – 0.461 (% Granites)	0.98	8.12	11.53
ALPHA_BF	= – 0.0794 + 0.00300 (% Migmatites)	0.87	0.14	0.12
GW_DELAY	= 19.0 – 0.248 (% Crop land) + 0.165 (% Savannah)	0.98	22.17	16.82
USLE_P	= 0.129 – 0.0143 (% Lateritic consolidated soil layer)	0.51	0.06	0.07
USLE_K	= 0.162 – 0.0848 (% Lateritic consolidated soil lay)	0.85	–0.24	–0.18
NPERCO	= 1.72 – 3.80 (% Hypsometric integral) + 0.00779 (% Circularity Index) – 0.033 (% Elongation ratio)	0.85	0.08	0.47
RCHRG_DP	= – 0.758 + 0.462 (Drainage density (km/ km²))	0.55	0.24	0.99
SPEXP	= 1.47 – 0.00454 Lixisol (%) + 0.00011 Migmatites (%)	0.7	1.22	1.47
SURLAG	= 0.109 + 0.003 Lixisol (%) – 0.016 Lateritic consolidated soil layers (%)	0.93	0.19	0.22

2.3. Climate and Land Use Change Scenarios

The climate scenarios used in this study were computed by Paeth *et al.* [36] for a part of Africa from −15° S to 45° N latitude using the regional climate model REMO driven by the IPCC (Intergovernmental Panel on Climate Change) SRES (Special Report on Emission Scenarios) scenarios A1B and B1. The IPCC SRES scenario A1B characterizes a globalized world of rapid economic growth and comparatively low population growth. The SRES scenario B1 also characterizes a future globalized world with a low population growth. REMO is a regional climate model that is nested in the global circulation model ECHAM5/MPI-OM Paeth *et al.* [54]. REMO was forced on a grid of 50 km resolution throughout the first half of the 21st century over West Africa.

Figure 7. Observed *vs.* simulated total discharge (validation) using the regression-based parameters for the Savè catchment (23,488 km^2), with 0.71 for R^2 and 0.67 for model efficiency (ME). Savè was chosen for the validation because measurements at Ouémé-Bonou are not reliable.

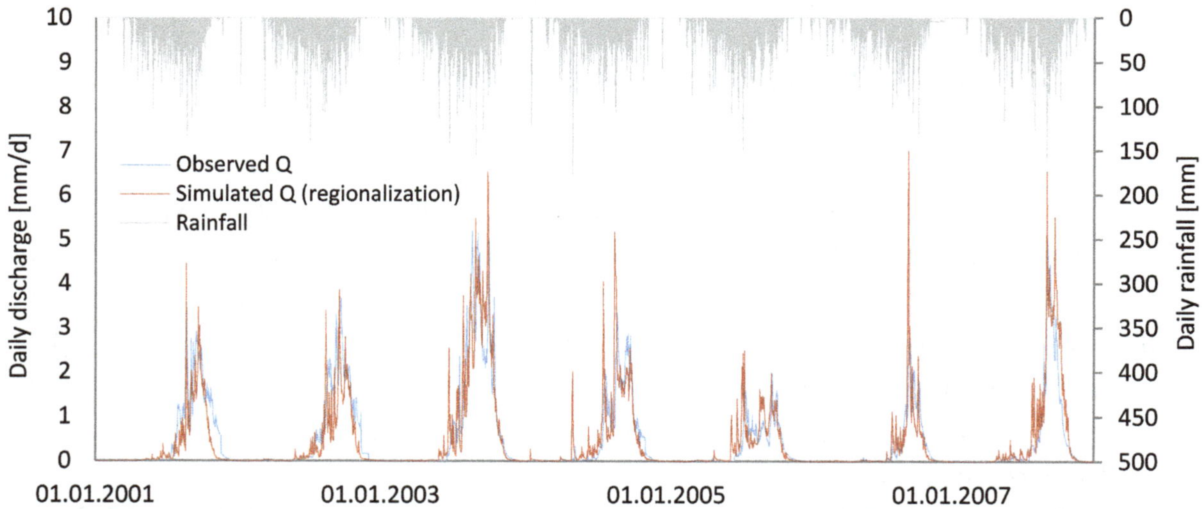

Initial runs of REMO over West Africa have shown systematically underestimated rainfall amounts and variability with a shift in the pattern towards more weak events and fewer extremes. This was addressed by applying the Model Output Statistics—MOS to correct monthly bias using other near-surface parameters such as temperature, sea level pressure and wind. Since the regional-mean (precipitation) strongly differed from the observed spatial patterns of daily rainfall events, a conversion of the MOS-corrected regional-mean from REMO to local rainfall event patterns has been done. Virtual station data, matching the rainfall stations in Benin, were useful to adjust the results to the statistical characteristics of observed daily precipitation at the rainfall stations by probability matching.

Figure 8a shows mean monthly REMO rainfall amounts over 1960–2000 compared with measurements over 1998–2005 for the upper Ouémé catchment (14,500 km^2 including Donga-Pont (586 km^2), Térou-Igbomakoro (2344 km^2) and Bétérou (10,072 km^2)), while Figure 8b presents mean monthly water discharges simulated with SWAT using REMO outputs over 1960–2000 compared with measurements over 1998–2005. These figures suggested that REMO and SWAT represent correctly the observations.

Climate change projections as simulated throughout REMO are very sensitive to a prescribed degradation of land cover. This sensitivity in addition to an increasing greenhouse gases concentrations have resulted in distinctly warmer and drier climates (with frequent droughts) for the investigated period 2000–2050 over West Africa, reductions in annual rainfall amounts of about 20%–25% of the 20th century annual amounts.

For the Ouémé-Bonou catchment REMO projects a decrease of annual rainfall between 9% and 12% for the scenario B1 and for the period 2010–2030. It increases of up to 4% for the scenario A1B over the period 2010–2014, before decreasing of up to 14% between 2015 and 2029. Maximum and minimum temperatures are expected to increase of up to 2.5 °C over the next 40 years (Figure 9).

Figure 8. (**a**) Mean monthly rainfall from REMO output (period 1960–2000) compared with measurements (period 1998–2005) for the upper Ouémé catchment [15]; (**b**) Simulated mean monthly water discharge with SWAT using REMO output (period 1960–2000) compared measurements (period 1998–2005) for the upper Ouémé catchment [15].

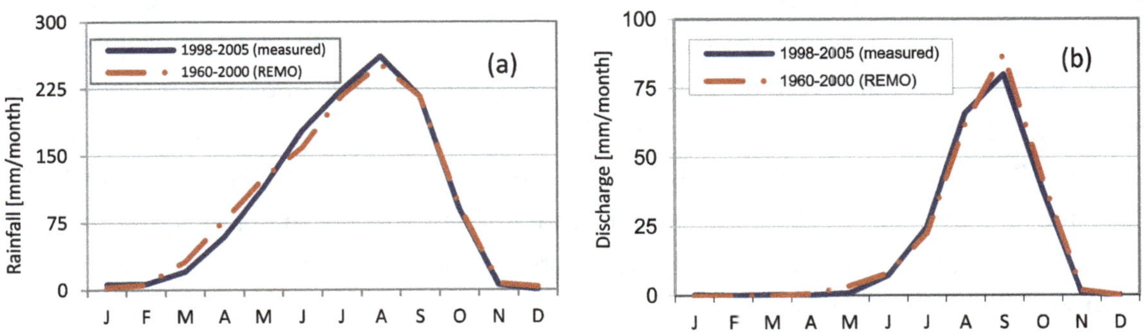

Figure 9. Projected changes in annual precipitation and near-surface temperatures until 2050 over tropical and northern Africa due to increasing greenhouse gas concentrations and man-made land cover changes [55]. The scenario A1B describes a globalized world of rapid economic growth and comparatively low population growth. The scenario B1 also characterizes a future globalized world with a low population growth.

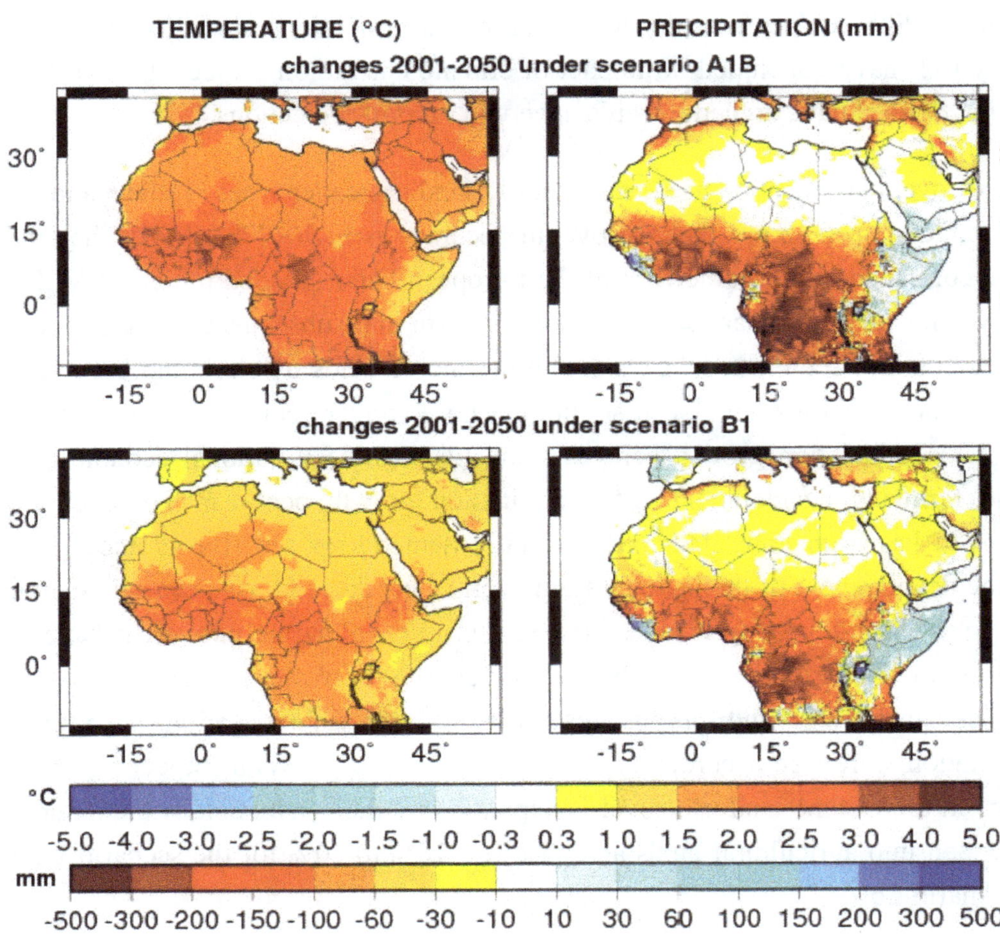

Many recent research studies attempted to simulate West African future rainfall and climate parameters throughout the 21st century using atmosphere-ocean global climate models and relying on greenhouse gas emissions scenarios as outlined in the Intergovernmental Panel on Climate Change archives for Assessment Reports (AR3 & 4). The recent Coordinated Regional Climate Downscaling Experiment-CORDEX initiative from the World Climate Research Program promotes running multiple RCM simulations at 50 km resolution for multiple regions including West Africa, highly expected to bring clarifications and improve the projections [56]. The CORDEX initiative includes the study of uncertainty due to structural errors of different GCMs and/or RCMs. Beyond CORDEX and apart from Paeth *et al.* [36,54], who nested REMO in the global circulation model ECHAM5/MPI-OM as described above, Patricola and Cook [57] also attempt to overcome the limitations of global models by nesting a higher resolution regional model, the Weather Research and Forecasting (WRF) model on a grid of 90 km resolution, over West Africa and for the second half of the 21st century. They found a very mixed rainfall change signal characterized by June–July drought, followed by copious rainfall towards the end of the summer [58]. Although focused on different time periods, both studies [36,54,57] favor desiccation, albeit with caveats regarding intraseasonal and spatial variability.

As an alternative to the above-described studies using relatively high-resolution regional climate models, many other studies [59–61] used atmosphere-ocean global climate models to run climate change experiments and have rather concluded a wetter climate for the first half of the 21st century in reference to the 20th century contrary to REMO-based projections presented in Paeth *et al.* [36]. Cook and Vizy [62] have concluded no impact of climate change on projected West Africa rainfall, while Biasutti *et al.* [63] have argued towards uncertain rainfall projections.

The land use/cover classification used, considers 17 land use/cover types (Table 6 and Figure 10). A subsequent accuracy check shows that the overall accuracy is high (87%) [64]. The land use/cover scenarios were computed in the framework of the European Union funded project RIVERTWIN [37]. The major driver for land use change is population growth and subsequent conversion of the natural savannah vegetation into settlements, roads, and a mosaic of fields by slash and burn clearance [65]. Two socio-economic scenarios have been set up: (1) La, stronger economic development, controlled urbanization, 3.2% population growth per year; and (2) Lb, weak national economy, uncontrolled settlement and farmland development, 3.5% population growth per year. For each scenario, the population growth has been translated into a specific demand for settlements and agricultural area according to the development of the national framework. This demand has been satisfied according to the proximity to roads and existing villages, new settlements and agricultural areas have been created leading to the land use distribution. The General Directorate for Water has selected several potential sites for future construction of multi-purpose reservoirs for large scale irrigation. Therefore, large areas of natural vegetation were also accordingly converted to croplands. With respect to the scenarios La and Lb, change in the Ouémé land use/cover is expressed by the conversion of the natural vegetation including savannah into agricultural lands and pastures: 10% to 20% for the scenario La and 20% to 40% for the scenario Lb.

Table 6. Land use/cover categories, their area and percentage of total area for the Ouémé-Bonou catchment (49,256 km²). Values displayed in brackets are related to the Donga-Pont catchment (586 km²) [64]. SWAT model was adapted to consider almost all land use classes mentioned in the table.

Land Use Categories	Land Use Code	Area (km²)	Percentage of Total Area
Galery forest	GF	1759 (8.6)	3.98 (1.47)
Humid and dry dense forest	FD	1220 (0.4)	2.76 (0.06)
Swamp formations	FM	17 (0)	0.04 (0)
Riverine formations	FR	107 (0)	0.24 (0)
Woodland and woodland savannah	FCSB	6716 (8.6)	15.2 (1.47)
Flooding savannah	SM	222 (0)	0.5 (0)
Tree and shrub savannah	SA	17231 (285.8)	38.99 (48.74)
Saxicolous savannah	SS	313 (0)	0.71 (0)
Grassland	PH	14 (0)	0.03 (0)
Mosaic of cropland and bush fallow	CJ	13713 (280.1)	31.03 (47.77)
Mosaic of cultivation with Parkia and Cashew trees	CJNA	32 (0)	0.07 (0)
Mosaic of cultivation with palm trees	CJP	1189 (0)	2.69 (0)
Industrial plantations	PI	127 (0)	0.29 (0)
Village plantations	PV	1209 (0.5)	2.74 (0.08)
Barren lands/area without vegetation	BAR	5 (0)	0.01 (0)
Urban and built-up	AG	277 (2.4)	0.63 (0.4)
Water bodies	PE	47 (0)	0.11 (0)

Figure 10. Land use/cover of the Ouémé-Bonou (49,256 km²) and the Donga-Pont catchments (586 km²). The legend is fully explained in Table 3. (**a**) reference map (2003); (**b**) La 2015–2019; (**c**) La 2025–2029; (**d**) Lb 2015–2019; (**e**) Lb 2025–2029 [37].

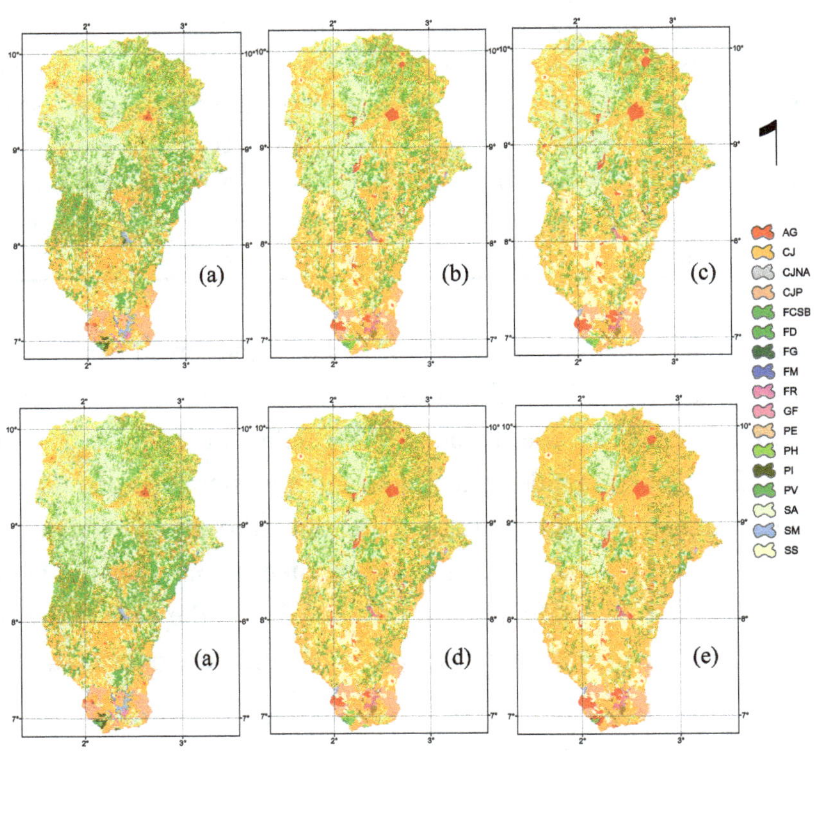

3. Results and Discussion

3.1. Impacts of Climate Change

Climate change impacts on key water balance components and sediment-organic N yield over time windows of 10 to 20 years for the Donga-Pont sub-catchment (Table 7 and Figure 11a) were computed to be consistent with other climate change impact studies. The Table and Figure show a decrease in rainfall all over the 10-year and 20-year time windows, resulting in a reduction of up to 22% of the simulated water-nutrient components for all scenarios except the scenario A1B over the time window 2010–2019. These results suggest significant impacts of year-to-year climate variability. To be consistent with the used land use scenarios 5-year time windows (Tables 8 and 9, Figure 11b,c) will be considered in the discussion as provided in the following paragraphs.

As shown in Tables 8 and 9, the simulated absolute surface runoff ranges from 100 to 140 mm per year for the Donga-Pont catchment (586 km²), while varying from 60 to 80 mm per year for the Ouémé-Bonou catchment (49,256 km²), due to the significantly different catchment sizes with a likely slight effect of the higher rate of agricultural lands in the Donga-Pont catchment compared to the Ouémé-Bonou catchment (cf. Table 6). In fact, due to unsustainable agriculture practices over the region, higher surface runoff may be associated to larger rate of agricultural lands.

Table 7. Simulated SWAT components under climate scenarios (with unchanged land use map derived from 2003 Landsat image). Deviation (in %) from the reference scenario (2000–2009) are shown in brackets.

Donga-Pont (586 km²)	Reference (2000–2009)	A1B (2010–2019)		A1B (2020–2029)		A1B (2010–2029)		B1 (2010–2019)		B1 (2020–2029)		B1 (2010–2029)	
Rainfall (mm a^{-1})	1233.4	1196.6	(−3)	1107.2	(−10)	1151.9	(−7)	1145.5	(−7)	1126.2	(−9)	1135.8	(−8)
Water yield (mm a^{-1})	254.7	282.8	(11)	206.6	(−19)	244.7	(−4)	238.2	(−6)	219.4	(−14)	228.8	(−10)
Groundwater flow (mm a^{-1})	117.6	122.7	(4)	91.4	(−22)	107.0	(−9)	105.3	(−10)	99.5	(−15)	102.4	(−13)
Surface runoff (mm a^{-1})	137.2	158.8	(16)	113.8	(−17)	136.3	(−1)	131.5	(−4)	118.5	(−14)	125.0	(−9)
Act. Evapotranspiration (mm a^{-1})	923.7	840.9	(−9)	849.4	(−8)	845.1	(−9)	848.6	(−8)	846.7	(−8)	847.6	(−8)
Sediment yield (ton ha^{-1} a^{-1})	0.4	0.5	(17)	0.4	(−5)	0.4	(6)	0.4	(−2)	0.4	(−8)	0.4	(−5)
Organic N load (kg ha^{-1} a^{-1})	0.7	0.8	(11)	0.6	(−10)	0.7	(0)	0.7	(−5)	0.6	(−9)	0.7	(−7)

Sediment yield was more important at the Donga-Pont catchment scale, ranging from 0.3 to 0.4 ton·ha^{-1}·a^{-1} against 0.3 ton·ha^{-1}·a^{-1} for the Ouémé-Bonou catchment. This can be explained by the higher rate of agricultural lands in the Donga-Pont catchment compared to the Ouémé-Bonou catchment (cf. Table 6). Thus, due to inadequate tillage and unsustainable practices, sediments are very susceptible to loading. Conversely, the simulated organic nitrogen yields for the Ouémé-Bonou catchment (roughly 1.2 ton·ha^{-1}·a^{-1}) were twice the computed amount for the Donga-Pont catchment (0.6 to 0.7 ton·ha^{-1}·a^{-1}). This can be easily understood since due to the higher rate of agricultural lands in the Donga-Pont catchment, the topsoils are more degraded and very poor humus and organic matters are available for loading.

Figure 11. Simulated trends under climate scenarios with land use from the year 2003. (**a**) Donga-Pont: trend over 20 years; (**b**) Donga-Pont: trend over a 5-year time window; (**c**) Ouémé-Bonou: trend over a 5-year time window.

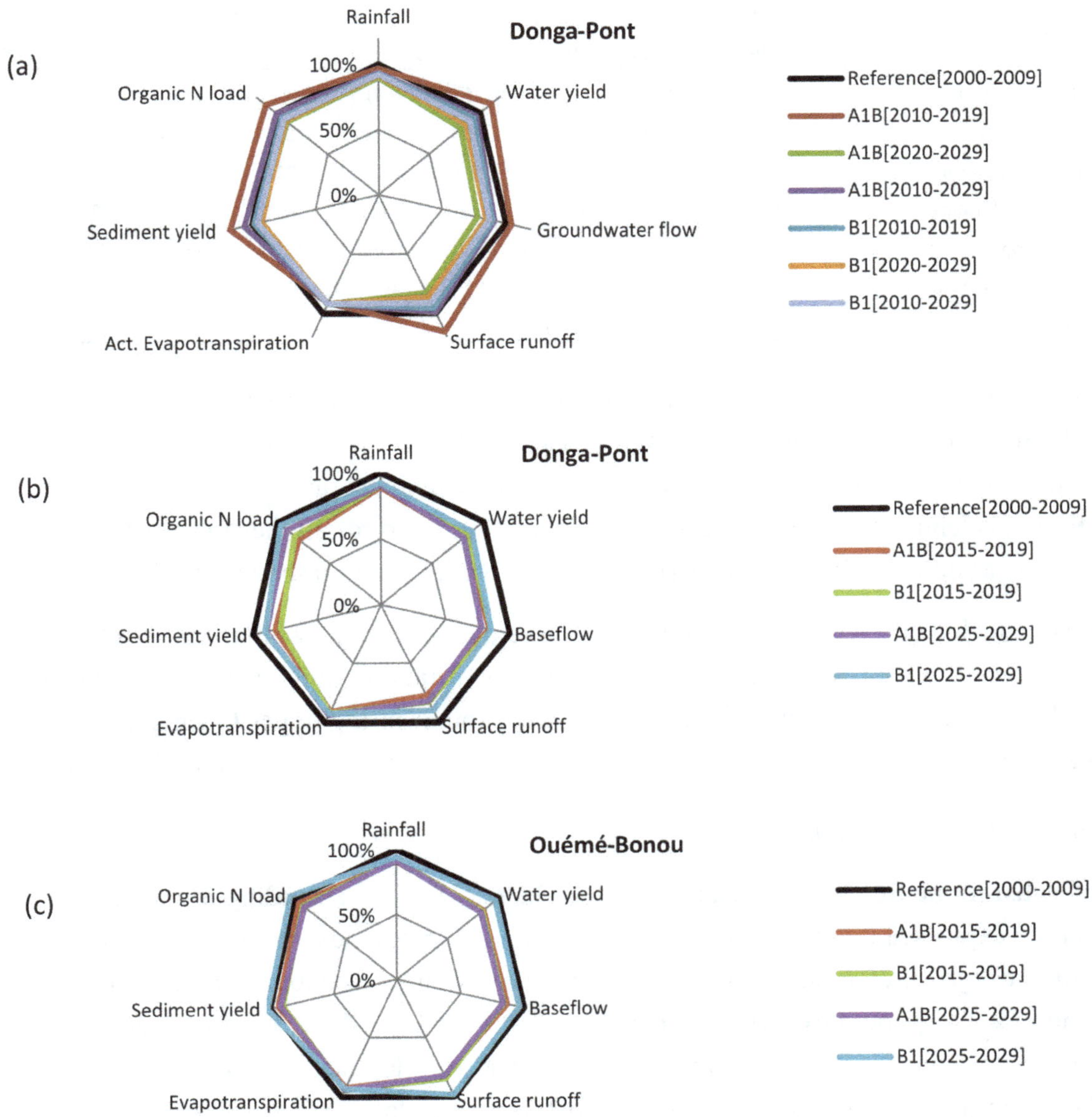

Table 8. Simulated SWAT components under climate scenarios (with unchanged land use map derived from 2003 Landsat image). Deviation (in %) from the reference scenario (2000–2009) are shown in brackets.

Donga-Pont (586 km²)	Reference (2000–2009)	A1B (2015–2019)		B1 (2015–2019)		A1B (2025–2029)		B1 (2025–2029)	
Rainfall (mm a^{-1})	1233.4	1096.7	(−11)	1113.9	(−10)	1116.0	(−10)	1138.0	(−8)
Water yield (mm a^{-1})	254.7	206.9	(−19)	215.5	(−15)	205.6	(−19)	225.3	(−12)
Groundwater flow (mm a^{-1})	117.6	98.9	(−16)	100.3	(−15)	91.6	(−22)	100.5	(−15)
Surface runoff (mm a^{-1})	137.2	106.7	(−22)	113.7	(−17)	112.6	(−18)	123.4	(−10)
Act. Evapotranspiration (mm a^{-1})	923.7	839.1	(−9)	849.2	(−8)	863.6	(−7)	854.1	(−8)
Sediment yield (ton ha^{-1} a^{-1})	0.4	0.3	(−25)	0.3	(−25)	0.4	(0)	0.4	(0)
Organic N load (kg ha^{-1} a^{-1})	0.7	0.6	(−14)	0.6	(−14)	0.6	(−14)	0.7	(0)

Table 9. Simulated SWAT components under climate scenarios (with unchanged land use map derived from 2003 Landsat image). Deviation (in %) from the reference scenario (2000–2009) are shown in brackets.

Ouémé-Bonou (49,256 km²)	Reference	A1B		B1		A1B		B1	
	(2000–2009)	(2015–2019)		(2015–2019)		(2025–2029)		(2025–2029)	
Rainfall (mm a^{-1})	1138.9	1035.2	(−9)	1045.2	(−8)	1041.5	(−9)	1074.9	(−6)
Water yield (mm a^{-1})	224.6	191.4	(−15)	189.0	(−16)	186.2	(−17)	218.0	(−3)
Groundwater flow (mm a^{-1})	147.6	127.5	(−14)	123.7	(−16)	122.8	(−17)	142.2	(−4)
Surface runoff (mm a^{-1})	77.0	63.9	(−17)	65.3	(−15)	63.4	(−18)	75.8	(−2)
Act. Evapotranspiration (mm a^{-1})	794.8	734.3	(−8)	747.5	(−6)	745.8	(−6)	740.2	(−7)
Sediment yield (ton ha^{-1} a^{-1})	0.3	0.3	(0)	0.3	(0)	0.3	(0)	0.3	(0)
Organic N load (kg ha^{-1} a^{-1})	1.2	1.1	(−8)	1.1	(−8)	1.1	(−8)	1.2	(0)

In both investigated scales, annual sediment yield and actual evapotranspiration have significantly decreased (of up to 20%) over the simulated years (2000 to 2029). Groundwater flow decreased significantly from 15% to 22% for the Donga-Pont catchment and from 4% to 17% for the Ouémé-Bonou catchment. Figure 11b,c shows a decreasing trend for all the simulated components (of up to 20%), regardless the different scenarios, but more pronounced for the Donga-Pont catchment.

One may conclude that differences in water balance components as well as nutrient load rate while moving between different catchment scales are functions of rate of agricultural lands, which are more sensitive than natural vegetations. Similar results were found in the upper Ouémé catchment (about 15,000 km²) using the SWAT model and the same IPCC SRES scenarios A1B and B1 [9,15].

3.2. Impacts of Land Use Change

As indicated in the Section 2.3, changes in the Ouémé land use according to the scenarios La and Lb are mainly expressed by the conversion of savannah into croplands and pastures in a range of 10% to 20% of the agricultural lands for the scenario La (stronger economic development, controlled urbanization, 3.2% population growth per year) and 20% to 40% for the scenario Lb (weak national economy, uncontrolled settlement and farmland development, 3.5% population growth per year). Accordingly, an increasing surface runoff was simulated (from 2000 to 2029) for both scenarios La and Lb, but in a pronounced way for the scenario Lb (Table 10 and Figure 12). These increases range from 9% to 27% for the Donga-Pont catchment and between 22% and 57% for the Ouémé-Bonou catchment.

In spite of this simulated increase for the surface runoff, water yield has change in a very low rate (roughly ±5%), forcing a decrease of the groundwater flow (between −3% and −33% for the Donga-Pont catchment and between −7% and −18% for the Ouémé-Bonou). This reveals the proof of decrease of the infiltration rate over the study area, and a severe threat to its groundwater systems. Sediment yield has increased from 25% to 75% for the Donga-Pont catchment and from 33% to 66% for the Ouémé-Bonou catchment. Organic nitrogen load has increased from 14% to 43% for the Donga-Pont catchment and from 17% to 58% for the Ouémé-Bonou due to conversion of natural vegetation to new cropland, which are more important at the Ouémé-Bonou scale.

Actual evapotranspiration has decreased slightly from 0.5% to 1.1% for the Donga-Pont catchment and 0.8% to 2% for the Ouémé-Bonou catchment. For this trend the land use scenario Lb differs from the other by showing the highest underlying simulated rates, as already discussed above for the other components.

Table 10. Simulated SWAT components under land use scenarios (with unchanged climate condition of the period 2000 to 2009). Deviation (in %) from the reference scenario (2000–2009) are shown in brackets.

Donga-Pont (586 km²)	**Reference (2000–2009)**	**La (2015–2019)**		**Lb (2015–2019)**		**La (2025–2029)**		**Lb (2025–2029)**	
Rainfall (mm a^{-1})	1233.4	1233.4	(0)	1233.4	(0)	1233.4	(0)	1233.4	(0)
Water yield (mm a^{-1})	254.7	228.7	(−10)	236.7	(−7)	267.0	(5)	263.4	(3)
Groundwater flow (mm a^{-1})	117.6	78.8	(−33)	79.0	(−33)	113.6	(−3)	88.8	(−24)
Surface runoff (mm a^{-1})	137.2	149.1	(9)	157.0	(14)	152.4	(11)	174.3	(27)
Act. Evapotranspiration (mm a^{-1})	923.7	913.7	(−1)	918.0	(−1)	915.3	(−1)	919.1	(0)
Sediment yield (ton ha^{-1} a^{-1})	0.4	0.5	(25)	0.6	(50)	0.5	(25)	0.7	(75)
Organic N load (kg ha^{-1} a^{-1})	0.7	0.8	(14)	0.9	(29)	0.8	(14)	1.0	(43)
Ouémé-Bonou (49,256 km²)									
Rainfall (mm a^{-1})	1138.9	1138.9	(0)	1138.9	(0)	1138.9	(0)	1138.9	(0)
Water yield (mm a^{-1})	224.6	231.5	(3)	238.8	(6)	233.6	(4)	243.0	(8)
Groundwater flow (mm a^{-1})	147.6	137.9	(−7)	128.8	(−13)	133.9	(−9)	121.7	(−18)
Surface runoff (mm a^{-1})	77.0	93.7	(22)	110.0	(43)	99.7	(29)	121.3	(58)
Act. Evapotranspiration (mm a^{-1})	794.8	788.9	(−1)	782.1	(−2)	787.4	(−1)	778.8	(−2)
Sediment yield (ton ha^{-1} a^{-1})	0.3	0.4	(33)	0.4	(33)	0.4	(33)	0.5	(67)
Organic N load (kg ha^{-1} a^{-1})	1.2	1.4	(17)	1.7	(42)	1.5	(25)	1.9	(58)

Figure 12. Simulated trends under land use scenarios (with unchanged climate: 2000–2009). (**a**) Donga-Pont; (**b**) Ouémé-Bonou.

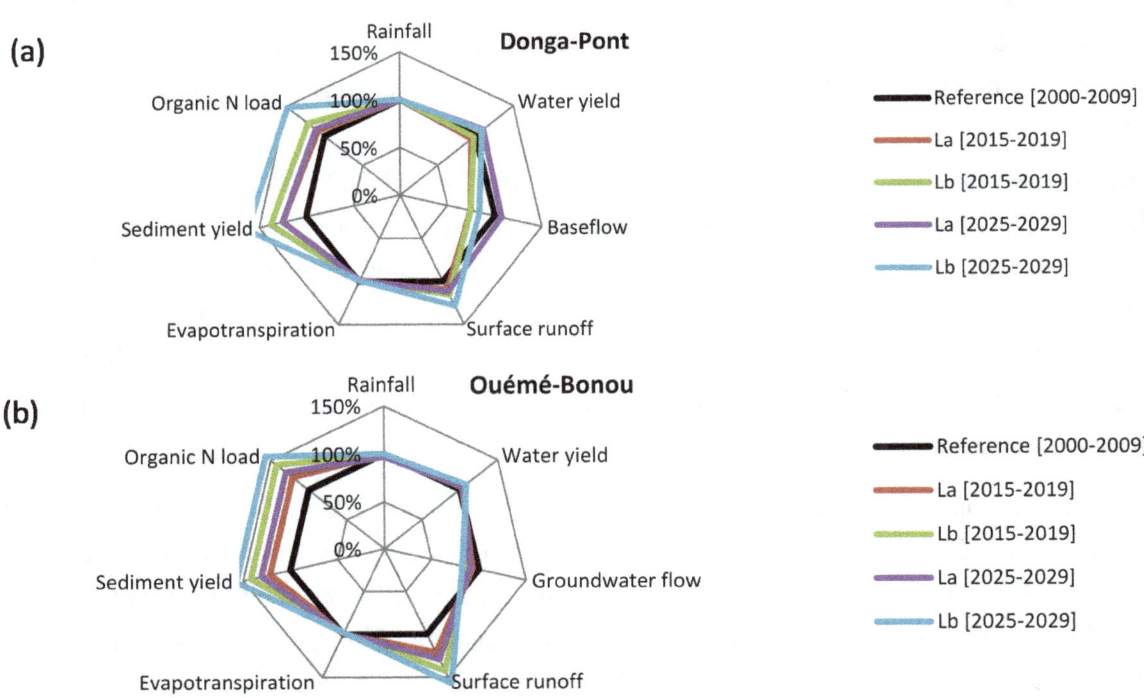

In summary (combining the Sections 3.1 and 3.2), surface runoff was found more sensitive to land use change (+22% to +75% of changes) than climate change (−5% to −20% of changes), evapotranspiration was more sensitive to climate change (−8 to −12% of changes) than land use change (−0.5% to −2% of changes) and groundwater flow was less sensitive to climate change (−4% to −22%) than land use change (−9% to −57%). Sediment yield was more sensitive to land use (+25% to +75%) than climate change (−25% to 0%). Organic nitrogen load was more sensitive to land use change (+14% to +58%) than climate change (−8% to −14%).

At the large catchment scale (Ouémé-Bonou), sediment yield has decreased (−25% for the climate scenarios and −57% for the land use scenarios) and organic nitrogen load has increased (100% for the climate scenarios and 175% for the land use scenarios) in reference to the meso scale catchment (Donga-Pont).

Li *et al.* [22] modeled the hydrological impact of land use change in the Niger and the Lake Chad basins (West Africa) and found that a deforestation percentage below 50% has no significant impact on the stream flow, but a total deforestation increases the annual stream runoff by 35%–65%. This finding is very contrasting the results obtained here, where a progressive conversion of natural vegetation to cropland to a maximum of 40% over 30 years has increased the surface runoff by 22%–57% and decreased the groundwater flow by 7%–18%.

3.3. Impacts of Combined Climate and Land Use Scenarios

In reference to the results discussed in the Sections 3.1 and 3.2, an overview of combined land use and climate change scenario effects on the degradation trend presented in Table 11 and Figure 13 identified climate change as the major driver of the changes in water yield and actual evapotranspiration. In the meantime, land use changes raises as the major driver affecting surface runoff, groundwater flow, sediment yield and organic nitrogen load. Stronger effects of climate change were computed for the time period 2015–2019 in the Donga-Pont catchment scale, which have been significantly but not completely compensated by the land use change effects over the time period 2025–2029.

At the Donga-Pont scale (Table 12 and Figure 14, time period 2025–2029), annual surface runoff may change from −8% to +17%. The decrease (−8%) (scenarios La + A1B) was mainly driven by climate change, while land use change effects have mainly resulted in an increase (+17%) (scenarios La + B1, Lb + A1B and Lb + B1). Sediment yield increased for all the scenarios from 12.5% to 50%, and organic nitrogen load also has increased from 1.4% to 46% (driven by land use changes). Under climate change effects, water yield has changed from −9% to +1.6%, groundwater flow from −29% to 1% and evapotranspiration from −9% to −7.5%.

At the Ouémé-Bonou scale (Table 12 and Figure 14, time period 2025–2029), annual surface runoff may change from +5.7% to +42% for all the scenarios, driven by land use change. Sediment yield increased for all the scenarios from 15.6% to 41%, and organic nitrogen load also has increased from 15% to 47% (driven by land use changes). Under climate change effects, water yield has changed from −14% to +0.6%, groundwater flow from −30.5% to −16% and evapotranspiration from −8.4% to −7%.

Table 11. Simulated SWAT components under land use and climate scenarios (2015 to 2019). Deviation (in %) from the reference scenario (2000–2009) are shown in brackets.

Donga-Pont (586 km²)	Reference (2000–2009)	La & A1B (2015–2019)		La & B1 (2015–2019)		Lb & A1B (2015–2019)		Lb & B1 (2015–2019)	
Rainfall (mm a^{-1})	1233.4	1096.7	(−11)	1113.9	(−10)	1096.7	(−11)	1113.9	(−10)
Water yield (mm a^{-1})	254.7	191.7	(−21)	201.1	(−17)	199.1	(−18)	206.8	(−15)
Groundwater flow (mm a^{-1})	117.6	74.1	(−30)	76.6	(−28)	74.5	(−30)	75.1	(−29)
Surface runoff (mm a^{-1})	137.2	116.4	(−14)	123.1	(−9)	123.5	(−9)	130.4	(−4)
Act. Evapotranspiration (mm a^{-1})	923.7	831.7	(−10)	841.5	(−9)	833.6	(−10)	843.7	(−9)
Sediment yield (ton ha^{-1} a^{-1})	0.4	0.5	(20)	0.4	(−8)	0.5	(15)	0.4	(7)
Organic N load (kg ha^{-1} a^{-1})	0.7	0.5	(−29)	0.6	(−10)	0.7	(0)	0.7	(3)
Ouémé-Bonou (49,256 km²)									
Rainfall (mm a^{-1})	1138.9	1035.2	(−9)	1045.2	(−8)	1035.2	(−9)	1045.2	(−8)
Water yield (mm a^{-1})	224.6	195.9	(−13)	194.1	(−14)	201.1	(−10)	200.1	(−11)
Groundwater flow (mm a^{-1})	147.6	120.2	(−19)	116.0	(−21)	113.9	(−23)	109.5	(−26)
Surface runoff (mm a^{-1})	77.0	75.7	(−2)	78.1	(1)	87.3	(13)	90.6	(18)
Act. Evapotranspiration (mm a^{-1})	794.8	730.8	(−8)	743.4	(−6)	726.2	(−9)	738.1	(−7)
Sediment yield (ton ha^{-1} a^{-1})	0.3	0.4	(9)	0.3	(6)	0.4	(25)	0.4	(22)
Organic N load (kg ha^{-1} a^{-1})	1.2	1.3	(13)	1.3	(9)	1.5	(29)	1.5	(26)

Figure 13. Simulated trends under land use and climate scenarios (2015 to 2019). **(a)** Donga-Pont; **(b)** Ouémé-Bonou.

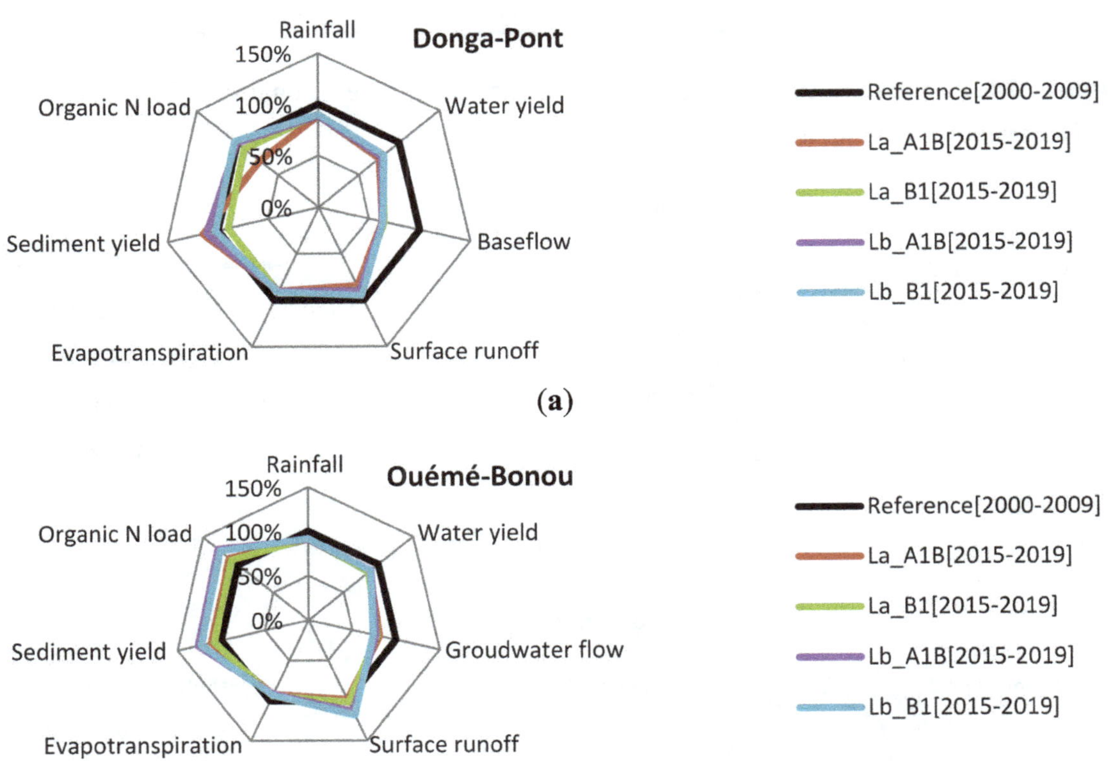

(b)

Regardless the modeling scale, the simulated impacts for the land use and climate change scenarios, over the simulated period 2000 to 2029 (30 years) may be summarized as follow: (1) surface runoff, groundwater flow, sediment and organic nitrogen load are found mainly sensitive to land use change with roughly −8% to 50% of changes; and (2) water yield and evapotranspiration are found more sensitive to climate change with roughly −31% to +2% of changes as consequence of rainfall reduction (cf. Section 2.3).

Table 12. Simulated SWAT components under land use and climate scenarios (2025 to 2029). Deviation (in %) from the reference scenario (2000–2009) are shown in brackets.

Donga-Pont (586 km²)	**Reference (2000–2009)**	**La & A1B (2025–2029)**		**La & B1 (2025–2029)**		**Lb & A1B (2025–2029)**		**Lb & B1 (2025–2029)**	
Rainfall (mm a^{-1})	1233.4	1116.0	(−10)	1138.0	(−8)	1116.0	(−10)	1138.0	(−8)
Water yield (mm a^{-1})	254.7	221.9	(−9)	244.9	(1)	221.8	(−9)	246.7	(2)
Groundwater flow (mm a^{-1})	117.6	95.5	(−10)	107.0	(1)	75.0	(−29)	87.4	(−18)
Surface runoff (mm a^{-1})	137.2	125.1	(−8)	136.5	(1)	145.9	(7)	158.4	(17)
Act. Evapotranspiration (mm a^{-1})	923.7	857.4	(−8)	847.3	(−9)	854.7	(−8)	845.2	(−9)
Sediment yield (ton ha^{-1} a^{-1})	0.4	0.5	(13)	0.5	(15)	0.6	(45)	0.6	(50)
Organic N load (kg ha^{-1} a^{-1})	0.7	0.7	(1)	0.8	(9)	1.0	(36)	1.0	(46)
Ouémé-Bonou (49,256 km²)									
Rainfall (mm a^{-1})	1138.9	1041.5	(−9)	1060.9	(−7)	1041.5	(−9)	1060.9	(−7)
Water yield (mm a^{-1})	224.6	193.3	(−14)	215.8	(−4)	201.6	(−10)	223.3	(−1)
Groundwater flow (mm a^{-1})	147.6	111.9	(−24)	124.1	(−16)	102.6	(−30)	113.8	(−23)
Surface runoff (mm a^{-1})	77.0	81.4	(6)	91.7	(19)	99.0	(29)	109.6	(42)
Act. Evapotranspiration (mm a^{-1})	794.8	740.4	(−7)	734.6	(−8)	733.0	(−8)	728.3	(−8)
Sediment yield (ton ha^{-1} a^{-1})	0.3	0.4	(16)	0.4	(22)	0.4	(34)	0.5	(41)
Organic N load (kg ha^{-1} a^{-1})	1.2	1.4	(15)	1.5	(25)	1.6	(37)	1.7	(47)

Figure 14. Simulated trends under land use and climate scenarios (2025 to 2029). (**a**) Donga-Pont; (**b**) Ouémé-Bonou.

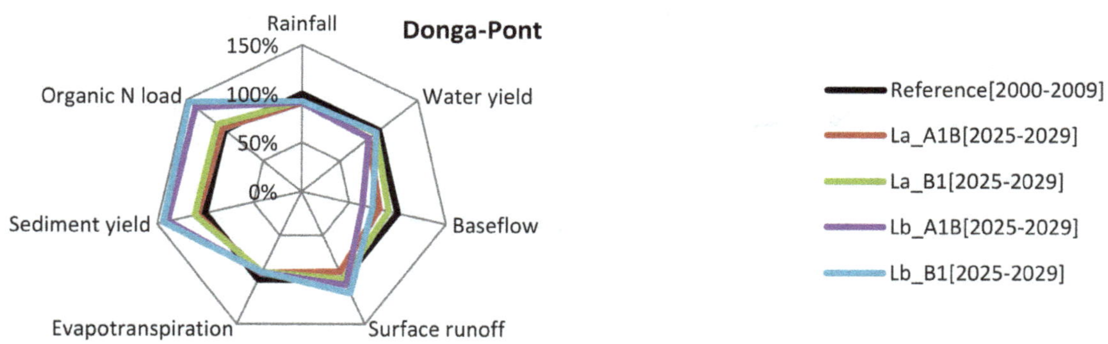

(**a**)

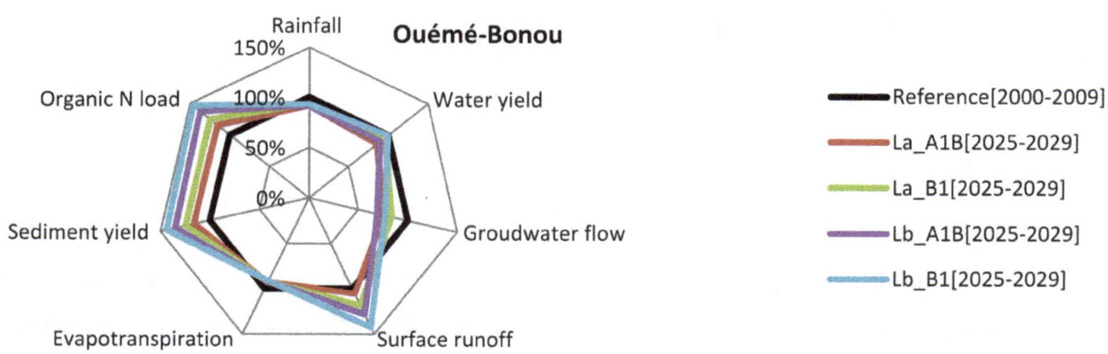

(**b**)

In addition, the results suggest that variables such as surface runoff, groundwater flow, sediment and transported nutrients, mainly sensitive to land use change are significantly affected by scale-dependent rate of agricultural lands, while variables such as water yield and evapotranspiration, mainly sensitive to climate change, have changed almost similarly at both scales. An application of the conceptual semi-distributed model UHP-HRU (Universal Hydrological Program-Hydrological Response Unit) in the same study area, using the physical-based model SIMULAT-H as benchmark [9], has led to a similar conclusion that water yield is more influenced by climate than land use, which more affects runoff components (surface runoff *vs.* interflow and base flow). Results of scenario analysis from the same work [9] revealed that the amount of renewable water decreases during the period 2001–2049 in both climate IPCC SRES scenarios A1B and B1, which were also used in the present study.

Sediment yield was more important at the Donga-Pont catchment scale, ranging from 0.4 to 0.6 ton·ha^{-1}·a^{-1} against 0.32 to 0.45 ton·ha^{-1}·a^{-1} at the Ouémé-Bonou catchment scale. This may be explained by decreases of retrained eroded particles at large scale, due to sequestration in the inland valleys, which are important at large scales [9]. According to Vanmaercke *et al.* [66] and de Vente *et al.* [67], high erosion rates are often observed at the local scale, but sediment yield, which is the net transport out of the catchment, decreases with increasing size of the catchment. This is consistent with the results obtained.

The simulated organic nitrogen yield for the Ouémé-Bonou catchment (roughly 1.2 to 1.7 ton·ha^{-1}·a^{-1}) was almost twice the computed amount for the Donga-Pont catchment (roughly 0.5 to 1 ton·ha^{-1}·a^{-1}). This may be the consequence of the higher rate of agricultural lands in the Donga-Pont catchment, resulting in poorer topsoils in terms of humus and organic matters rates available.

Figure 15. Impacts of land use and climate change on Organic Nitrogen patterns in the Ouémé-Bonou catchment (49,256 km^2).

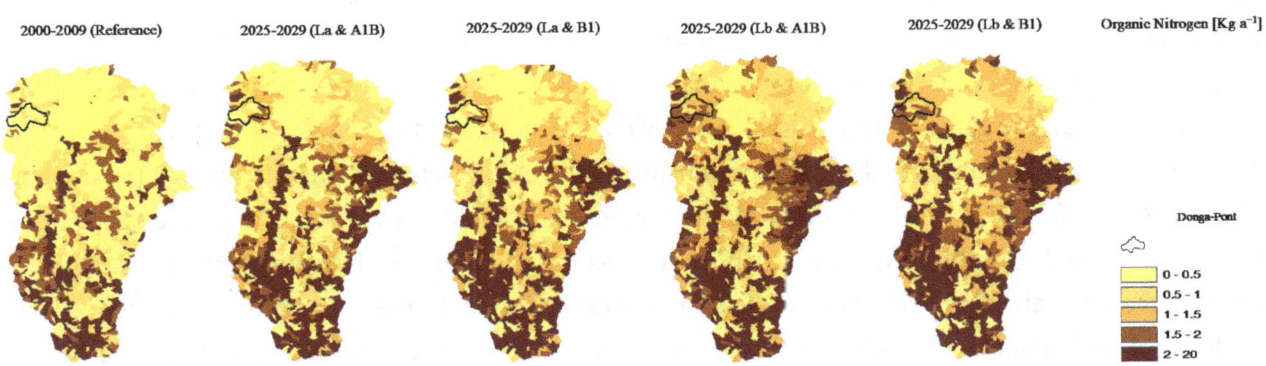

Figure 15 shows organic nitrogen load patterns for the combined scenarios (La & A1B), (La & B1), (Lb & A1B) and (Lb & B1) from 2000 to 2030 within the Ouémé-Bonou catchment. In accordance with the land use dynamic, the simulated pattern reflects higher and increasing degradation in the agricultural land. Surface runoff varies from 0 to 350 mm·a^{-1}, sediment yield varies from 0 to 10 ton·ha^{-1}·a^{-1}, and lost soil organic nitrogen varies from 0 to 20 kg·ha^{-1}·a^{-1}. Lost soil organic nitrogen shows higher dynamic and more threatened areas compared to sediment yield. This is completely consistent with the findings in section 3.3. As land use change was identified as the major driver of the ongoing land degradation, the scenario Lb (pessimistic, weak national economy, uncontrolled settlement and farmland development, 3.5% population growth per year) must be avoided in order to exclude the effects simulated by the combinations (Lb & A1B) and (Lb & B1) for which larger threatened areas are shown in the figures.

In another study, SWAT was applied in the upper Ouémé catchment to evaluate the effects of climate change and land use change on soil erosion [9,15]. In their combined land use and climate scenarios, the soil erosion rates increases with a large variability within the study area, which shows the high impact of land use change. The study has concluded that in areas with a high potential of cropland expansion, future sediment yield will be driven by land use change and may therefore strongly increase. Conversely, in the areas with a low potential for cropland expansion and strong reductions in rainfall, future sediment yield may decrease.

These findings are consistent with the results presented in this study, which furthermore discussed aspects of soil organic nutrients.

4. Conclusions

Soil and water are essential natural resources, available in limited quantities, but are nowadays dangerously exposed to climate and land use change [33]. Balancing the future degradation requires a sufficient understanding of processes behind at different scales [33], a task that may not be possible without an overriding of difficulties related to data availability in a data-poor environment such as Benin.

In this work, a regionalization methodology is used to overcome two difficulties in model setup in the Ouémé catchment: parameter scale-effects and associated uncertainty issues for large scale model application [32] and the lack and non-accurateness of boundary condition data (e.g., stream water-sediment-nutrient measurements). The SWAT model was applied using scale dependent and regression-based parameter models to simulate climate and land use change impacts on water yield, sediment and nutrient loads in Benin at the meso and the regional scale (49,256 km^2).

The results revealed significant and increasing impacts over years. Surface runoff, groundwater flow, sediment and organic nitrogen load were affected by land use change (as dominant effects) of −8% to +50%, while water yield and evapotranspiration were affected by climate change (as dominant effects) of −31% to +2%. These rates may be reached gradually over years and according to the scenario data used. It was found that variables such as sediment and soil nutrients, mainly sensitive to land use change were likely functions of scale-dependent cropland rates. Furthermore, higher sediment yields were associated with higher scale-dependent rate of croplands, while higher organic nitrogen loads were associated with higher scale-dependent rate of natural vegetations. Partly the effect of climate change (decrease in surface runoff) and land use change (increase in surface runoff) balance out, resulting in a complex reaction of the system to Global Change. These results are consistent with findings of Hiepe [15] and Speth *et al.* [9]. Nevertheless, it is clear that land use change acts on shorter time scales than climate change causing a higher impact of land use change in the next years.

This work is a significant contribution for supporting sustainable management strategies to drive stronger economic development that considers controlled settlements, controlled farmland extension, less pressures on natural vegetations, and sustainable farming system managements. The study reveals the relevancy and the efficiency of the modeling strategy used for bridging the data gap, since the results are consistent with previous findings within the study area. Future works should focus on the interaction between uncertainties associated with the scenario data and the model structural errors. This will help to clearly quantify the uncertainties in the simulated impacts which are not investigated in this work.

Acknowledgments

This work has been funded by the German Federal Ministry of Education and Research (BMBF) as part of the West African Science Service Center on Climate Change and Adapted Land Use (WASCAL) research program. Many thanks to our partners in Benin and all colleagues of the GLOWA-IMPETUS project-Global Wandel des Wasserkreislaufs-Integratives Management Projekt für einen effizienten und tragfähigen Umgang mit Süßwasser in Westafrika (Global Cange and the Hydrological Cycle—An integrated approach to the efficient management of scarce water resources in West Africa) who provided data and assistance. The first author thanks the German Academic Exchange Service for funding.

Authors Contribution

Aymar Y. Bossa, Euloge K. Agbossou and Bernd Diekkrüger designed the study, developed the methodology, and wrote the manuscript. Aymar Y. Bossa performed the field work, collected the data, and conducted the computer analysis while Euloge K. Agbossou and Bernd Diekkrüger supervised this part of the work.

References

1. United Nations Environment Programme (UNEP). *Global Environment Outlook 4*; Progress Press Ltd: Valleta, Malta, 2007.
2. Bossio, D.; Geheb, K.; Critchley, W. Managing water by managing land: Addressing land degradation to improve water productivity and rural livelihoods. *Agric. Water Manag.* **2010**, *97*, 536–542.
3. Oldeman, L.R. *Guidelines for General Assessment of the Status of Human-induced Soil Degradation*; Working Paper and Reprint 88/4; International Soil Reference and Information Centre: Wageningen, the Netherlands, 1991.
4. Scherr, S.J.; Yadav, S. *Land degradation in the Developing World: Implications for Food, Agriculture, and the Environment to 2020*; Food, Agriculture, and the Evironment Discussion Paper 14; International Food Policy Research Institute (IFPRI): Washington, DC, USA, 1996.
5. Lal, R. Soil Erosion in the Tropics. Principles & Management; McGraw-Hill: New York, NY, USA, 1990.
6. Giertz, S.; Diekkrüger, B.; Steup, G. Physically-based modeling of hydrological processes in a tropical headwater catchment in Benin (West Africa)—Process representation and multi-criteria validation. *Hydrol. Earth Syst. Sci.* **2006**, *10*, 829–847.

7. Laouina, A.; Coelho, C.; Ritsema, C.; Chaker, M.; Nafaa, R.; Fenjiro, I.; Antari, M.; Ferreira, A.; van Dijck, S. Dynamique de l'eau et gestion des terres dans le contexte du changement global, analyse agro-hydrologique dans le bassin du Bouregreg (Maroc). *Sécheresse* **2004**, *15*, 66–77. (In French)
8. Roose, E.; de Noni, G. Recherches sur l'érosion hydrique en Afrique: Revue et perspectives. *Sci. Chang. Planétaires Sécheresse* **2004**, *15*, 121–129. (In French)
9. Speth, P., Christoph, M., Diekkrüger, B., Eds. *Impacts of Global Change on the Hydrological Cycle in West and Northwest Africa*; Springer Publisher: Heidelberg, Germany, 2010.
10. Chen, L.; Zhu, A.X.; Qin, C. Identification of critical source areas of soil erosion on moderate fine spatial scale in Loess Plateau in China. *African J. Agric. Res.* **2012**, *7*, 2962–2970.
11. Ouyang, W.; Skidmore, A.K.; Hao, F.; Wang, T. Soil erosion dynamics response to landscape pattern. *Sci. Total Environ.* **2010**, *408*, 1358–1366.
12. United Nations World Water Assessment Programme. *The Impact of Global Change on Erosion and Sediment Transport by Rivers: Current Processes and Future Challenges*; the United Nations World Water Development Report 3; United Nation Educational: Paris, France, 2009.
13. Yang, D.; Kanae, S.; Oki, T.; Koike, T.; Musiake, K. Global potential soil erosion with reference to land use and climate changes. *Hydrol. Process.* **2003**, *17*, 2913–2928.
14. Lawal, O.; Gaiser, T.; Nuga, B. Estimation of potential soil losses on a regional scale: A case of Abomey-Bohicon Region, Benin Republic. *Agric. J.* **2007**, *2*, 1–8.
15. Hiepe, C. Soil Degradation by Water Erosion in a Sub-Humid West-African Catchment, a Modelling Approach Considering Land Use and Climate Change in Benin. Ph.D. Thesis, University of Bonn, Bonn, Germany, 19 December 2008.
16. Bossa, A.Y.; Diekkrüger, B.; Giertz, S.; Steup, G.; Sintondji, L.O.; Agbossou, E.K.; Hiepe, C. Modeling the effects of crop patterns and management practices on N and P loads to surface water and groundwater in a semi-humid catchment (West Africa). *Agric. Water Manag.* **2012**, *115*, 20–37.
17. Keyzer, M.A.; Sonneveld, B.G.J.S.; Pande, S. The Impact of Climate Change on Crop Production and Health in West Africa: An Assessment for the Oueme River Basin in Benin; Centre for World Food Studies: Amsterdam, the Netherlands, 2007.
18. Cornelissen, T.; Diekkrüger, B.; Giertz, S. A comparison of hydrological models for assessing the impact of land use and climate change on discharge in a tropical catchment. *J. Hydrol.* **2013**, *498*, 221–236.
19. Faramarzi, M.; Abbaspour, K.C.; Ashraf Vaghefi, S.; Farzaneh, M.R.; Zehnder, A.J.B.; Srinivasan, R.; Yang, H. Modeling impacts of climate change on freshwater availability in Africa. *J. Hydrol.* **2013**, *480*, 85–101.
20. Chaplot, V. Water and soil resources response to rising levels of atmospheric CO_2 concentration and to changes in precipitation and air temperature. *J. Hydrol.* **2007**, *337*, 159–171.
21. Legesse, D.; Vallet-Coulomb, C.; Gasse, F. Hydrological response of a catchment to climate and land use changes in Tropical Africa: Case study South Central Ethiopia. *J. Hydrol.* 2003, 275, 67–85.
22. Li, K.Y.; Coe, M.T.; Ramankutty, N.; de Jong, R. Modeling the hydrological impact of land-use change in West Africa. *J. Hydrol.* **2007**, *337*, 258–268.

23. Neumann, R.; Jung, G.; Laux, P.; Kunstmann, H. Climate trends of temperature, precipitation and river discharge in the Volta Basin of West Africa. *Int. J. River Basin Manag.* **2007**, *5*, 37–41.
24. Mahe, G.; Paturel, J.; Servat, E.; Conway, D.; Dezetter, A. The impact of land use change on soil water holding capacity and river flow modelling in the Nakambe River, Burkina-Faso. *J. Hydrol.* **2005**, *300*, 33–43.
25. Ward, P.J.; van Balen, R.T.; Verstraeten, G.; Renssen, H.; Vandenberghe, J. The impact of land use and climate change on late Holocene and future suspended sediment yield of the Meuse catchment. *Geomorphology* **2009**, *103*, 389–400.
26. Oguntunde, P.G.; Abiodun, B.J. The impact of climate change on the Niger River Basin hydroclimatology, West Africa. *Clim. Dyn.* **2013**, *40*, 81–94.
27. Blöschl, G.; Sivapalan, M. Scale issues in hydrological modeling: A review. *Hydrol. Proc.* **1995**, *9*, 251–290.
28. Abbott, M.B.; Bathurst, J.C.; Cunge, J.A.; O'Connell, P.E.; Rasmussen, J. An introduction to the European Hydrological System—Systeme Hydrologique Europeen, "SHE", 2: Structure of a physically-based, distributed modeling system. *J. Hydrol.* **1986**, *87*, 61–77.
29. Sivapalan, M.; Viney, N.R. Large scale catchment modeling to predict the effects of land use and climate. *Water J. Aust. Wat. Wastewat. Assoc.* **1994**, *21*, 33–37.
30. Sivapalan, M.; Viney, N.R. Application of a nested catchment model for predicting the effects of changes in forest cover. In *Forest Hydrology*, Proceedings of International Symposium, Tokyo, Japan, October 1994; pp. 315–322.
31. Arnold, J.G.; Srinivasan, R.; Muttiah, R.S.; Williams, J.R. Large area hydrologic modeling and assessment part I: Model development. *J. Am. Water Resour. Assoc.* **1998**, *34*, 73–89.
32. FitzHugh, T.W.; Mackay, D.S. Impacts of input parameter spatial aggregation on an agricultural nonpoint source pollution model. *J. Hydrol.* **2000**, *236*, 35–53.
33. De Vente, J.; Poesen, J.; Verstraeten, G.; Govers, G.; Vanmaercke, M.; van Rompaey, A.; Arabkhedri, M.; Boix-Fayos, C. Predicting soil erosion and sediment yield at regional scales: Where do we stand? *Earth Sci. Rev.* **2013**, *127*, 16–29.
34. Thamm, H.P.; Judex, M.; Menz, G. Modeling of Land-Use and Land-Cover Change (LUCC) in Western Africa using remote sensing. *Photogramm. Fernerkund. Geoinf.* **2005**, *3*, 191–199.
35. Judex, M. Modellierung der Landnutzungsdynamik in Zentralbenin mit dem XULU-Framework. Ph.D. Thesis, University of Bonn, Bonn, Germany, 11 April 2008.
36. Paeth, H.; Born, K.; Podzun, R.; Jacob, D. Regional dynamic downscaling over West Africa: Model evaluation and comparison of wet and dry years. *Meteorol. Zeit.* **2005**, *14*, 349–367.
37. Regional Modelfor Integrated WaterManagement inTwinned River Basins (RIVERTWIN). *Adapted and Integrated Model for the Ouémé River Basin, Final Report*; Institute for Landscape Planning and Ecology: Stuttgart, Germany, 2007.
38. Bossa, A.Y.; Diekkrüger, B. Estimating scale effects of catchment properties on modeling soil and water degradation. In *Managing Resources of a Limited Planet: Pathways and Visions under Uncertainty*, Proceedings of the International Environmental Modelling and Software Society (iEMSs), Sixth Biennial Meeting, Leipzig, Germany, 1–5 July 2012; Seppelt, R., Voinov, A.A., Lange, S., Bankamp, D., Eds.; International Environmental Modelling and Software Society: Leipzig, Germany, 2012; pp. 2974–2981.

39. Wright, E.P.; Burgess, W. The hydrogeology of crystalline basement in Africa. *Geol. Soc. Spec.* **1992**, *66*, 1–27.
40. Faure, P.; Volkoff, B. Some factors affecting regional differentiation of the soils in the Republic of Benin (West Africa). *Catena* **1998**, *32*, 281–306.
41. Junge, B. Die Böden des oberen Ouémé-Einzugsgebietes in Benin/Westafrika—Pedologie, Klassifizierung, Nutzung und Degradierung. Ph.D. Thesis, Institut für Bodenkunde, Bonn, Germany, 23 July 2004.
42. Soil Conservation Service Engineering Division. *Urban Hydrology for Small Watersheds*; Technical Release 55; U.S. Department of Agriculture: Washington, DC, USA, 1986.
43. Sloan, P.G.; Moore, I.D. Modelling subsurface stormflow on steeply sloping forested watersheds. *Water Resour. Res.* **1984**, *20*, 1815–1822.
44. Williams, J.R. Sediment-yield prediction with universal equation using runoff energy factor. In *Present and Prospective Technology for Predicting Sediment Yield and Sources*. Proceedings of the sediment yield workshop, USDA Sedimentation Lab, Oxford, MS, USA, 28–30 November 1972; U.S. Department of Agriculture: Washington, DC, USA, 1975; pp. 244–252.
45. McElroy, A.D.; Chiu, S.Y.; Nebgen, J.W.; Aleti, A.; Bennett, F.W. *Loading Functions for Assessment of Water Pollution from Nonpoint Sources*; EPA document EPA 600/2-76-151; Midwest Research Institute: Kansas City, KS, USA, 1976.
46. Williams, J.R.; Hann, R.W. *Optimal Operation of Large Agricultural Watersheds with Water Quality Constraints*; Tech. Rep. No. 96; Texas Water Resources Institute: Temple, TX, USA, 1978.
47. Ritchie, J.T. A model for predicting evaporation from a row crop with incomplete cover. *Water Resour. Res.* **1972**, *8*, 1204–1213.
48. Christoph, M.; Fink, A.; Diekkrüger, B.; Giertz, S.; Reichert, B.; Speth, P. IMPETUS: Implementing HELP in the Upper Ouémé Basin. *Water SA* **2008**, *34*, 481–490.
49. Igué, A.M. Soil Information System for the Oueme Basin; Institut National de la Recherche Agricole du Bénin (INRAB): Cotonou, Benin, 2005.
50. Bossa, A.Y.; Diekkrüger, B.; Igué, A.M.; Gaiser, T. Analyzing the effects of different soil databases on modeling of hydrological processes and sediment yield in Benin (West Africa). *Geoderma* **2012**, *173–174*, 61–74.
51. Abbaspour, K.C. *SWAT Calibration and Uncertainty Programs—A User Manual*; Department of Systems Analysis, Integrated Assessment and Modeling (SIAM), Eawag, Swiss Federal Institute of Aquatic Science and Technology: Duebendorf, Switzerland, 2008.
52. Lam, Q.D.; Schmalz, B.; Fohrer, N. Modelling point and diffuse source pollution of nitrate in a rural lowland catchment using the SWAT model. *Agric. Water Manag.* **2010**, *97*, 317–325.
53. Food and Agriculture Organization (FAO). *Control of Water Pollution from Agriculture—FAO Irrigation and Drainage Paper 55*; Food and Agriculture Organization of the United Nations Rome: Rome, Italy, 1996.
54. Paeth, H.; Born, K.; Girmes, R.; Podzun, R.; Jacob, D. Regional climate change in tropical Africa under greenhouse forcing and land-use changes. *J. Clim.* **2009**, *22*, 114–132.
55. Patricola, C.M.; Cook, K.H. Northern African climate at the end of the twenty-first century: An integrated application of regional and global climate models. *Clim. Dyn.* **2010**, *37*, 1165–1188.

56. Druyan, L.M. Studies of 21st-century precipitation trends over West Africa. *Int. J. Climatol.* **2011**, *31*, 1415–1424.
57. Held, I.M.; Delworth, T.L.; Lu, J.; Findell, K.L.; Knutson, T.R. Simulation of Sahel drought in the twentieth- and twenty-first centuries. *Proc. Natl. Acad. Sci. USA* **2005**, *102*, 17891–17896.
58. Kamga, A.F.; Jenkins, G.S.; Gaye, A.T.; Garba, A.; Sarr, A.; Adedoyin, A. Evaluating the National Center for Atmospheric Research climate system model over West Africa: Present-day and the 21st century A1 scenario. *J. Geophys. Res.* **2005**, *110*, doi:10.1029/2004JD004689.
59. Hoerling, M.; Hurrell, J.; Eischeid, J.; Phillips, A. Detection and attribution of 20th century Northern and Southern African rainfall change. *J. Clim.* **2006**, *19*, 3989–4008.
60. Cook, K.H.; Vizy, E.K. Coupled model simulations of the West African monsoon system: Twentieth- and twenty-first-century simulations. *J. Clim.* **2006**, *19*, 3681–3703.
61. Biasutti, M.; Held, I.M.; Sobel, A.H.; Giannini, A. SST forcings and Sahel rainfall variability in simulations of the twentieth and twenty- first centuries. *J. Clim.* **2008**, *21*, 3471–3486.
62. Maraun, D.; Wetterhall, F.; Ireson, A.M.; Chandler, R.E.; Kendon, E.J.; Widmann, M.; Brienen, S.; Rust, H.W.; Sauter, T.; Themeßl, M.; *et al.* Precipitation downscaling under climate change: Recent developments to bridge the gap between dynamical models and the end user. *Rev. Geophys.* **2009**, *48*, doi:10.1029/2009RG000314.
63. Paeth, H. Key factors in African climate change evaluated by a regional climate model. *Erdkunde* **2004**, *58*, 290–315.
64. Igue, A.M.; Houndagba, C.J.; Gaiser, T.; Stahr, K. Land use/cover map and its accuracy in the Oueme Basin of Benin (West Africa). *Int. J. Agri. Sci.* **2012**, *2*, 174–184.
65. Götzinger, J. Distributed Conceptual Hydrological Modeling-Simulation of Climate, Land Use Change Impact and Uncertainty Analysis. Ph.D. Thesis, University of Stuttgart, Stuttgart, Germany, 19 July 2007.
66. Vanmaercke, M.; Poesen, J.; Verstraeten, G.; de Vente, J.; Ocakoglu, F. Sediment yield in Europe: Spatial patterns and scale dependency. *Geomorphology* **2011**, *130*, 142–161.
67. De Vente, J.; Poesen, J.; Verstraeten, G. The application of semi-quantitative methods and reservoir sedimentation rates for the prediction of basin sediment yield in Spain. *J. Hydrol.* **2005**, *305*, 63–86.

3

Attribution of Decadal-Scale Lake-Level Trends in the Michigan-Huron System

Janel Hanrahan [1,*], Paul Roebber [2] and Sergey Kravtsov [2]

[1] Atmospheric Sciences Department, Lyndon State College, PO Box 919, Lyndonville, VT 05851, USA

[2] Department of Mathematical Sciences, Atmospheric Sciences Group, University of Wisconsin-Milwaukee, P.O. Box 413, Milwaukee, WI 53201, USA; E-Mails: roebber@uwm.edu (P.R.); kravtsov@uwm.edu (S.K.)

* Author to whom correspondence should be addressed; E-Mail: janel.hanrahan@lyndonstate.edu;

Abstract: This study disentangles causes of the Michigan-Huron system lake-level variability. Regional precipitation is identified as the primary driver of lake levels with sub-monthly time lag, implying that the lake-level time series can be used as a proxy for regional precipitation throughout most of the 1865–present instrumental record. Aside from secular variations associated with the Atlantic Multidecadal Oscillation, the lake-level time series is dominated by two near-decadal cycles with periods of 8 and 12 years. A combination of correlation analysis and compositing suggests that the 8-y cycle stems from changes in daily wintertime precipitation amounts associated with individual storms, possibly due to large-scale atmospheric flow anomalies that affect moisture availability. In contrast, the 12-y cycle is caused by changes in the number of instances, or frequency, of summertime convective precipitation due to a preferred upper-air trough pattern situated over the Great Lakes. In recent decades, the lake-level budget exhibited an abnormal—relative to the remainder of the instrumental record—evaporation-driven trend, likely connected to regional signatures of anthropogenic climate change. The latter effect must be accounted for, along with the effects of precipitation, when assessing possible scenarios of future lake-level variability.

Keywords: Great Lakes; Lake Michigan-Huron; water levels; precipitation; near-decadal cycles; climate variability

---

1. Introduction

The Laurentian Great Lakes have over 10,000 miles of shoreline which is subject to submersion or drought depending on the water level of the lakes at any given time. Extremes in these levels affect shoreline erosion, cargo ship capacities, hydroelectric power supplies, and recreation for the basins' inhabitants. While commonly considered two separate bodies of water, the second and third largest of the Great Lakes, Lake Michigan and Lake Huron, are connected by the deep Straits of Mackinac which ensures that the water levels of the separate reservoirs remain at equal elevations. Hence, they behave hydraulically like a single lake, and together, Lake Michigan-Huron has over 117,000 km^2 of surface water making it the world's largest freshwater lake [1]. Given this vast surface area, it can reasonably be assumed that climate fluctuations should be well represented in the Michigan-Huron lake-level time series, and indeed several studies have indicated that precipitation is most likely the primary interannual lake-level driver [2–6]. Hence, in addition to addressing immediate socioeconomic impacts that the water levels have on surrounding communities, an understanding of the 1865–present lake-level time series also contributes to a better understanding of historic climate modes and regional precipitation behavior whose impacts extend to other area lakes and groundwater [7], river flows, and agriculture. While this extensive time series likely contains a great deal of information pertaining to past climate behavior, clear illustrations linking interannual lake-level fluctuations to regional precipitation variability have been lacking.

To better isolate climatic connections implicit in the lake-level time series, Hanrahan *et al.* [8] subtracted the outflow-related damping effects from the full lake-level record and determined that the resulting time series exhibits variability over a range of time scales. The longest time scale—secular variations are anti-correlated with North Atlantic sea-surface temperatures (SSTs). The climate mode associated with this SST signal has been dubbed the Atlantic Multidecadal Oscillation (AMO) and is associated with dynamics of the oceanic thermohaline circulation [9–11]. Previous studies have linked AMO phases to precipitation anomalies over large portions of the U.S. [12–15], and this signal is likely transmitted to the lake levels through precipitation changes [8]. In our previous work using Singular Spectrum Analysis (SSA: see Ghil *et al.* [16,17]), we also identified two near-decadal cycles in the Michigan-Huron water levels, namely the 8-y and 12-y cycles [6]. The latter signal is in agreement with Watras *et al.* [7] who identified a ~13-y cycle in lakes and aquifers across the upper Great Lakes' region. The full lake-level time series can thus be decomposed as the sum of SSA reconstructed components associated with each decadal signal, and the residual variability is dominated by the AMO signal.

In this study, we aim to further our understanding of lake-level drivers responsible for decadal-scale lake-level changes. Using regional precipitation datasets, Section 2 establishes a solid connection between precipitation anomalies and lake-level variability, and addresses the issue of a possible time lag between precipitation forcing and lake-level response. In Sections 3 and 4, we explore seasonality of near-decadal lake-level cycles and examine the large-scale atmospheric patterns associated with the corresponding lake-level changes. Section 5 summarizes our results on the connection between regional precipitation and historic lake-level changes, and addresses a recent evaporative trend which is becoming increasingly apparent in the previously precipitation-driven lake-level time series.

2. Michigan-Huron Lake-Level Drivers

The interannual Michigan-Huron lake levels fluctuate in response to the sum of five primary drivers (see Table 1 for a summary of variables used throughout the text): over-lake precipitation P_y, runoff from tributary rivers and streams R_y, evaporative losses $-E_y$, inflow through the St. Marys River from Lake Superior I_y, and outflow to the Mississippi River through the Chicago Diversion and the lower Great Lakes through the St. Clair River $-O_y$; the subscript y denotes the year under consideration. The estimated values for all these quantities were obtained online from the Great Lakes Environmental Research Laboratory (GLERL [18])).

Table 1. Descriptions of variables used throughout the text.

Variable	Description
P_y	Annual total over-lake precipitation [1]
R_y	Annual total runoff [1]
E_y	Annual total evaporative losses [1]
I_y	Annual total inflow (from Lake Superior for Lake Michigan-Huron) [1]
O_y	Annual total outflow (through the St. Marys River and the Chicago Diversion for Lake Michigan-Huron) [1]
$L^G_{m,y}, L^U_{m,y}$	Observed beginning-of-month lake-level as reported by GLERL (G) and monthly-average level by USACE (U)
$dL_{m,y}$	Observed monthly lake-level change as estimated by $L_{m+1,y} - L_{m,y}$
dL^s_y, dL^w_y	Observed seasonal (3-month) lake-level change as estimated by $dL_{m-1,y} + dL_{m,y} + dL_{m+1,y}$, where $m = 7$ for summer (s) and $m = 12$ for winter (w)
dL^p_y	Computed annual lake-level change associated with the precipitation-driven components [1] P_y, R_y, and I_y; see Equation (1)
P^r_d, P^r_y	Daily (d) and annual (y) total regional precipitation depth [2]
dL^{pr}_y	Computed annual lake-level change associated with P^r_y; see Equation (2)
L^{rc}_y	Reconstructed lake-level components associated with the 8-y and 12-y cycles [3]
dL^{rc}_y	Annual change of each reconstructed component as estimated by $L^{rc}_{y+1} - L^{rc}_{y-1}$
p^T	Average total precipitation [2] defined as $\frac{\text{accumulated precipitation depth}}{\text{total number of days}}$
p^f	Average precipitation frequency [2], defined as $\frac{\text{number of days with precipitation}}{\text{total number of days}}$
p^a	Average precipitation amount [2], defined as $\frac{\text{accumulated precipitation depth}}{\text{number of precipitation days}}$
$\hat{p}^f, \hat{p}^a$	Precipitation frequency index $\hat{p}^f$, and amount $\hat{p}^a$, with multidecadal variability removed and averaged over 3 years

Notes: [1] Lake fluxes are provided in linear units (depth) over the lakes surface from GLERL; [2] Computed from the NOAA NCEP CPC gridded precipitation data; [3] Generated from statistically significant spectral peaks identified by Multi-taper method (MTM) analysis in Hanrahan *et al.* [6].

2.1. Drivers of Lake-Level Changes

Three of the lake-level drivers—precipitation, runoff, and inflow—can all be tied to changes in regional precipitation. The over-lake precipitation, along with the river and stream flows that make up the runoff component, all stem from precipitation that occurs within the Michigan-Huron catchment basin. The inflow for Michigan-Huron has been regulated since 1887 by way of navigation locks on the St. Marys River to keep Lake Superior within a specified range of historic water levels [19]. If precipitation is the primary lake-level driver of Lake Superior, long-term precipitation variability over the Superior basin will thus be correlated with the regulated Michigan-Huron inflow.

The fourth lake-level driver, evaporation, is a major contributor to seasonal lake-level variability, but it has historically played a minor role in the behavior of the year-to-year lake level fluctuations. Between 1948 and 2005, annual evaporation totals exhibited a standard deviation of about 0.08 m, while the sum of the precipitation-driven components (precipitation, runoff, and inflow) had a standard deviation of 0.24 m over the same time period. Thus, the expected year-to-year lake-level changes stemming from evaporation variability are considerably smaller than those which occur from precipitation changes. The final lake-level driver, outflow, is a function of the lake-level itself and does not directly respond to atmospheric processes. It instead tends to damp the climate-related lake-level fluctuations arising from the four remaining components [8].

It follows that the historic annual lake-level changes can be largely described by the precipitation-driven components alone. The annual precipitation-driven lake-level changes dL_y^p (Figure 1, black line) can thus be estimated as

$$dL_y^p = P_y + R_y + I_y - \bar{E} - \bar{O} \quad (1)$$

where P_y, R_y, and I_y values are all available for the period 1916–2005, while the historical averages of annual evaporative losses $\bar{E}$, and the outflow average $\bar{O}$ are computed from 1948–2005 values.

Figure 1. Annual lake-level changes estimated from observed levels (dL_y^U; blue x's) and as computed by Equation (1) (dL_y^p; black circles).

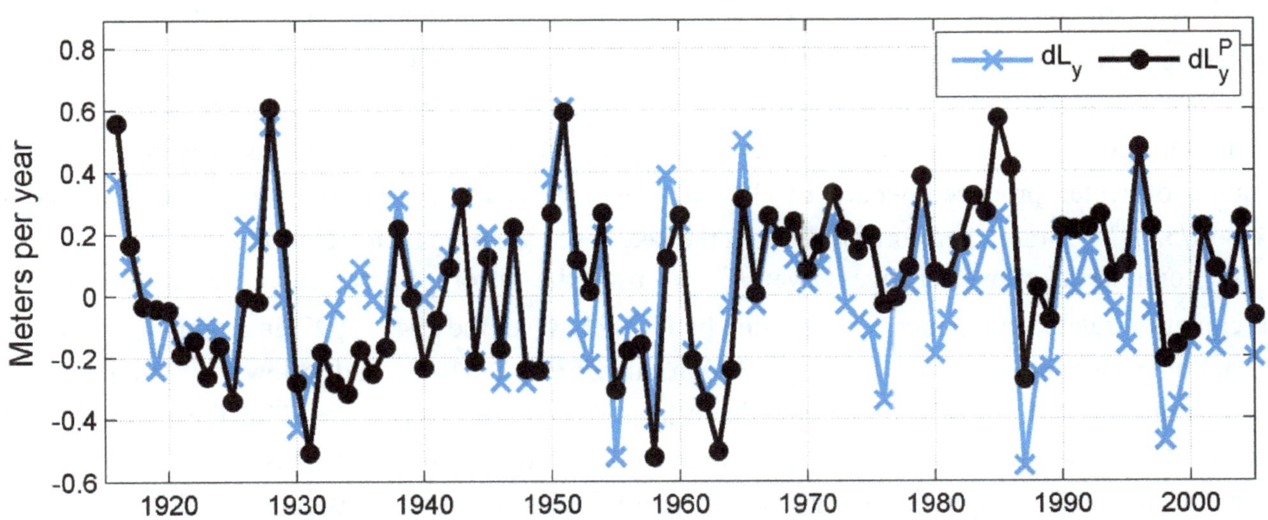

The year-to-year lake-level variability associated with the precipitation-driven components alone (Figure 1; black line), as computed from Equation (1), accounts for about 87% ($r = 0.93$) of the 1916–2005 lake-level behavior as estimated from the monthly-averaged levels obtained from the U.S. Army Corps of Engineers (USACE) dL_y^U (Figure 1; blue line). This indicates that precipitation has indeed had the greatest impact on interannual lake-level changes as it accounts for most of the historic lake-level variability. This precipitation–lake-level relationship is the primary focus of the present study.

2.2. Connecting Regional Precipitation to Lake-Level Fluctuations

While the precipitation-driven components (precipitation, runoff, and inflow) alone can be used to describe much of the historic lake-level behavior, the determination of these individual values is still quite complex. For example, as discussed above, the inflow rates ultimately stem from annual precipitation over the Lake Superior basin, but there are additional factors that affect the amount of water that actually runs into Lake Michigan-Huron. Furthermore, because of ground absorption, evaporation, and other factors, the amount of precipitation that falls onto the land which surrounds Lake Michigan-Huron, is not equal to the amount of water that runs into the lake as runoff. Both the inflow and runoff as reported by GLERL must therefore be directly measured as they flow into the lake, or estimated by nearby flows. Here, we simplify the precipitation–lake-level connection by estimating these fluxes from precipitation indices alone.

We obtained NOAA NCEP Climate Prediction Center (CPC) daily precipitation data from the IRI/LDEO Climate Data Library [20]. This gridded data precipitation data product was based on raw data from NCDC daily co-op stations, the CPC dataset, and hourly precipitation datasets. The precipitation data for overlake gridpoints (Figure 2) were estimated from surrounding land-based stations via Cressman Scheme gridding onto a 0.25 degree grid. For the present analysis, seven of the resulting gridded locations over the Great Lakes' basin were chosen as indicated by the red stars in Figure 2, for 1948–1997. Here, the over-Superior locations were selected to represent the Michigan-Huron inflow and the remaining locations represent the Michigan-Huron over-lake precipitation and runoff. Annual precipitation totals were computed from spatially averaged precipitation over the seven grid locations, resulting in a single 50-y time series of precipitation depths P_y^r. In order to estimate lake-level changes from these regional precipitation values, we must consider that the resulting Michigan-Huron lake-level variations are a multiple of P_y^r, because the surface area of Lakes Michigan and Huron makes up only a small fraction—about 1/5—of the total Michigan-Huron and Superior basin area A. For example, suppose that during a particular period there was an average rainfall of 3 mm in the region, and for simplicity, assume that all of it flowed directly into Lake Michigan-Huron. The total volume of water produced would be $3A$ mm^3, which would equate to a lake-level increase of $3A \times (1/5A)^{-1} = 15$mm, therefore amplifying the lake-level response by a factor of 5.

To compute lake-level changes given P_y^r, we used the historic instrumental record of individual precipitation-related components as reported by GLERL. Over the 1948–1997 time period, the average annual over-lake precipitation was $\bar{P} = 0.840$ m, runoff $\bar{R} = 0.617$ m, and inflow $\bar{I} = 0.604$ m, while the average annual regional precipitation depth from the gridded dataset was $\overline{P^r} = 0.697$ m. The total annual Michigan-Huron over-lake precipitation, runoff, and inflow values, thus equate to about 2.96 times the regional precipitation depth. If this ratio remains relatively constant through our period of interest, lake-level change dL_y^{pr} can be computed given only P_y^r, the historic average annual evaporative losses $\bar{E} = 0.64$ m, and outflow $\bar{O} = 1.48$ m (resulting in the subtraction of a constant value of 2.12 m). This results in the following simplified lake-level equation:

$$dL_y^{pr} = 2.96P_y^r - 2.12 \tag{2}$$

The lake-level changes estimated from Equation (2) are illustrated in Figure 3 (black line) in conjunction with the observed lake-level change dL_y^U (blue line). To account for possible delays related to snowpack storage, the annual rainfall totals illustrated here are defined with August as the beginning month.

Figure 2. Grid locations considered for regional precipitation estimate using NCEP CPC daily gridded precipitation data. Original image obtained from NOAA/GLERL.

Figure 3. Annual lake-level changes (July–August) estimated from observed levels (dL_y^U; blue x's) and as computed by Equation (2) (dL_y^{pr}; black circles).

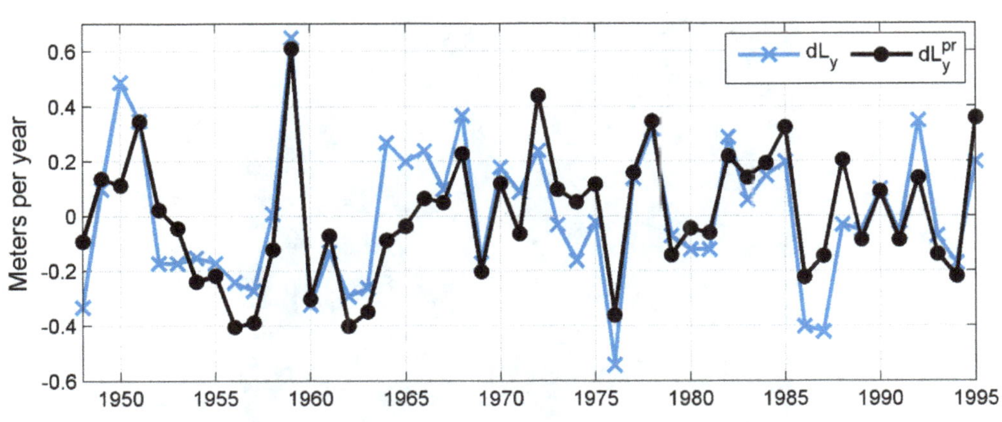

While numerous difficult-to-measure processes are occurring which affect the lake-level water supply (*i.e.*, over-lake and river evaporation, over-land evaporation, channel adjustments, vegetative consumption, groundwater intake, among several others), this simplified regional precipitation index dL_y^{pr} still accounts for the majority of lake-level variability ($r = 0.81$). This ability to connect the lake-level changes directly to a single atmospheric variable is ideal for furthering our understanding of underlying atmospheric behavior. In particular, one can use the lake-level time series, which begins at 1865, and Equation (2) to estimate regional precipitation prior to 1948, where the CPC gridded precipitation dataset is not available, thus providing a much extended period during which to study rainfall variability.

2.3. Lag Time of Precipitation Effects on Lake-Level Changes

To investigate the possibility of a sub-annual time lag between precipitation and lake-level changes, we compared monthly lake-level changes estimated from beginning-of-month lake level data provided by GLERL $dL^G_{m,y}$ to 30-day precipitation totals P^r_d. Correlation coefficients were computed between the 50-year-long monthly time series of $dL^G_{m,y}$ and daily time series of P^r_d. To account for potential end-of-year water-level lag times, the years were overlapped, so that correlations were computed for all months and days July 1948–December 1949 to July 1996–December 1997. The contour plot in Figure 4 illustrates the magnitude of these correlations where months along the horizontal axis refer to the first day of 30-day precipitation totals P^r_d, and those along the vertical axis indicate the month of lake-level change $dL^G_{m,y}$. For example, the lake-level changes in June (from left axis), correlate most strongly with the 30-day precipitation totals that begin on about June 1 of the same year (bottom axis), where $r \approx 0.8$; the points that correspond to this example are indicated by the white arrows. On average, statistically significant correlations ($p < 0.5$) occur where $r > 0.45$. Because the strongest correlations occur during concurrent times, we conclude that the lake levels primarily respond to regional precipitation with a sub-monthly time lag. Therefore, any season-specific variability in the water levels of Michigan-Huron should agree with the timing of regional precipitation drivers.

Figure 4. Correlations between regional precipitation P^r_d that begin on the dates along the horizontal axis, and monthly lake-level changes $dL^G_{m,y}$ that occur during the months along the vertical axis. The shading indicates the strength of correlation (r), as indicate by the bar on the right. The average r-values which significantly exceed zero are 0.45 ($p < 0.05$) and 0.55 ($p < 0.01$). See text for further explanation.

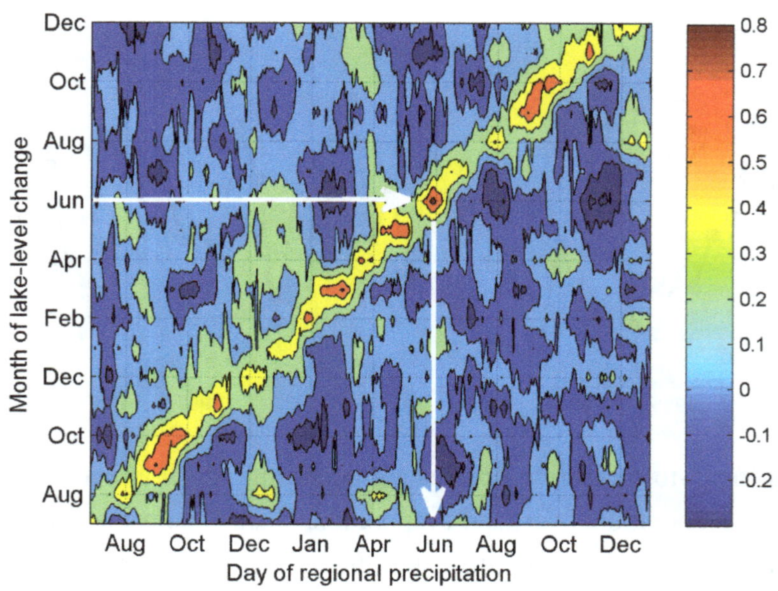

3. Timing of Lake-Level Changes and Precipitation Behavior

3.1. Seasonality of Lake-Level Periodicities

We now investigate potential seasonality in lake-level quasi-periodicities previously identified in Hanrahan *et al.* [6]. For this analysis, seasonal (3-month) lake-level changes $dL^G_{m,y}$ were compared to the derivatives dL^{rc8}_y and dL^{rc12}_y of the reconstructed lake-level components corresponding to the 8-y and 12-y signals identified by the SSA analysis of the 1865–1999 lake-level time series; see Hanrahan *et al.* [6], and Ghil *et al.* [16,17] for details pertaining to the computation of reconstructed components. Correlations between 24 consecutive months of lake-level changes $dL^G_{m,y}$ and the reconstructions dL^{rc}_y are illustrated in Figure 5 for the lake-level change periods of October 1900–1998 to November 1901–1999. For example, the first November correlations (left) are between the November 1900–1998 lake-level changes and the 1901–1999 reconstructions, dL^{rc8}_y (thick green) and dL^{rc12}_y (thin red), respectively. The second November correlations (right) are between the same reconstruction years, and the November 1901–1999 lake-level changes. Because artificially-high correlations may be generated among time series exhibiting red-noise variability, a Monte Carlo test was used to establish significance of the co-variance between the reconstructions and seasonal lake-level behavior by generating 1000 surrogate lake-level time series using the same length and lag-1 autocorrelation as the actual lake-levels. Correlations were computed between the surrogate time series and the reconstructions, which were sorted and then compared to the observed values. The observed values exceeding the 99th percentile of the synthetic values were determined to be significant at the 1% level (black stars in Figure 5).

Figure 5. Correlation between the 8-y (dL^{rc8}_y; thick green) and 12-y (dL^{rc12}_y; thin red) reconstructed lake-level changes, and the observed seasonal (3-month) lake-level changes ($dL^G_{m,y}$). Black stars indicate correlations significant at the 1% level.

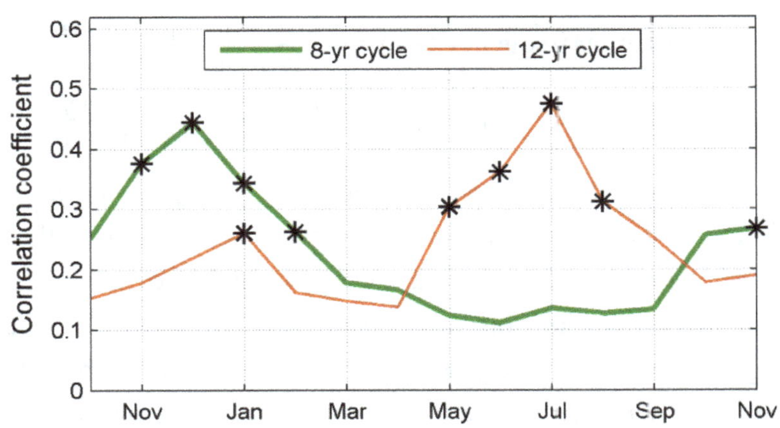

The 8-y reconstruction best correlates with lake-level changes during the preceding winter months of November to January. Conversely, the 12-y cycle is dominant during the summer months, as the peak correlations occur during June to August. The smoothed (3-y average) lake-level change time series that correspond to these seasons (blue) are illustrated in Figure 6a for the winter months dL^w_y (November–January) and Figure 6b for the summer months dL^s_y (June–August), along with the derivatives dL^{rc}_y of the 8-y and 12-y cycles (Figure 6a,b, green and red dashed lines, respectively). We have thus established that the two near-decadal lake-level periodicities occur during different times of the year; the 8-y cycle during winter months, and the 12-y cycle during summer months.

Figure 6. (**a**) Detrended wintertime (November–January) lake-level changes dL_y^w (blue) and the derivative of the 8-y lake-level reconstruction dL_y^{rc8} (green dashed); (**b**) Detrended summertime (June–August) lake-level changes dL_y^s (blue) and the derivative of the 12-y lake-level reconstruction dL_y^{rc12} (red dashed).

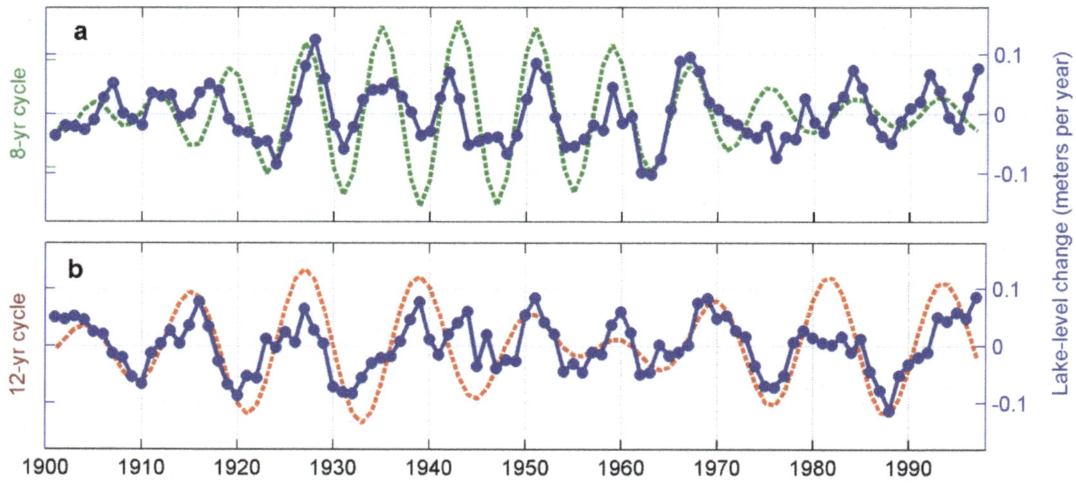

3.2. Seasonality of Precipitation Characteristics

Because regional precipitation drives lake-level changes with sub-monthly time lag (Section 2.3), it follows that the lake-level cycle seasonalities identified above (Section 3.1) should coincide temporally with those of precipitation. In this section, we examine potential seasonal precipitation periodicities by comparing the reconstructed lake-level cycles to two precipitation indices: frequency and amount. Both indices were computed from the daily gridded CPC precipitation data and are defined as follows:

$$\text{Precipitation frequency } (P^f) = \frac{\text{number of precipitation days}}{\text{total number of days}} \tag{3}$$

$$\text{Precipitation amount } (P^a) = \frac{\text{accumulated precipitation depth}}{\text{number of precipitation days}} \tag{4}$$

where a "precipitation day" is defined as having measurable precipitation (at least 1 mm) in one or more of the seven grid locations, and the sum of precipitation is computed as the average precipitation depth over all grid locations summed up over a specified number of days. Hence, P^f indicates the probability of precipitation occurring on any given day, irrespective of the intensity of the precipitation-producing system. Conversely, P^a is related to the intensity of precipitation during individual storm events, as it accounts for the average amount of precipitation that fell during a given precipitation day and location, not for the number of instances or spatial extent. These metrics are similar to those used by other authors where daily "frequency" and "amount" are sometimes referred to as "relative number of wet days" and "intensity," respectively [21].

To encompass full seasons, P^f and P^a were computed over 120-day moving segments (total number of days = 120) over 1948–1997. As identified in Hanrahan [22], both of these indices exhibit significant multidecadal variability possibly associated with the AMO. To filter out this low-frequency variability, we subtracted 20-y moving averages from both indices. Next, to minimize the highest-frequency variability for the purpose of concentrating on decadal time scales, we computed 3-y moving averages of the final time series. The resulting filtered indices $\hat{P}^f$ and $\hat{P}^a$ are illustrated in Figure 7a,b, respectively.

To evaluate the existence of near-decadal cycles in precipitation frequency and amount, correlation coefficients were computed between the two indices and the reconstructed derivatives. That is, 365 50-year-long time series of $\hat{P}^f$ and $\hat{P}^a$ (vertical axes in Figure 7a,b) which correspond to each day of the year (defined in months on the horizontal axes), were compared to each 50-year-long time series of the 8-y and 12-y reconstructed derivatives, dL_y^{rc8} and dL_y^{rc12}. Precipitation frequency $\hat{P}^f$ best correlates with the 12-y cycle during the summer months (Figure 7c, thin red; black stars indicate correlations significant at the 1% level), and the precipitation amount index $\hat{P}^a$ correlates best with the 8-y cycle during the winter months (Figure 7d, thick green). These results are consistent with the previously-discussed timing of periodicities, where the 12-y and 8-y cycles were linked to the summertime and wintertime lake-level changes, respectively (Section 3.1).

Figure 7. Identification of precipitation characteristics associated with the 8-y and 12-y lake-level cycles. Filtered 120-day moving averages of regional precipitation (**a**) frequency $\hat{P}^f$ and (**b**) amount $\hat{P}^a$; (**c**,**d**) Correlations between the precipitation indices (from **a** and **b**) and the 8-y cycle dL_y^{rc8} (thick green lines) and 12-y cycle dL_y^{rc12} (thin red lines). Black stars indicate significant correlations at 1% level.

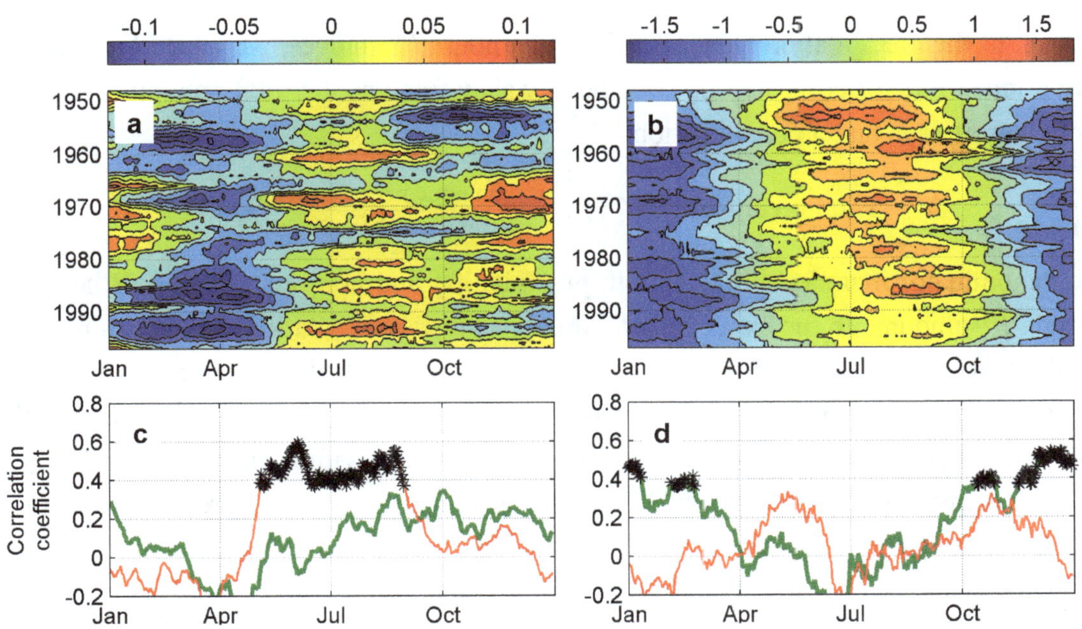

3.3. Verification of Seasonal Characteristics

To further establish significance of these findings, we compared monthly precipitation indices from Equations (3) and (4) during opposing phases of the 8-y and 12-y cycles. The 1948–1997 reconstruction derivatives dL_y^{rc} were divided into two categories: increasing and decreasing lake-level years, as illustrated in Figure 8 (blue circles and red squares, respectively). For the 12-y periodicity (Figure 8b, black line), summertime (June–August) monthly precipitation indices were categorized during 1948–1997, resulting in 25 years, or 75 individual months, per phase. For the 8-y periodicity

(Figure 8a, black line), wintertime (September–January) indices were categorized during 1949–1997 (the 1948 winter was not included because November–December data were not available for 1947), resulting in 24 years, or 72 months, per phase.

Figure 8. (**a**) Derivatives of 8-y lake-level reconstruction dL_y^{rc8} and (**b**) the 12-y reconstruction dL_y^{rc12} (black lines), and categorization of phases: increasing lake-level years (blue circles) and decreasing lake-level years (red squares).

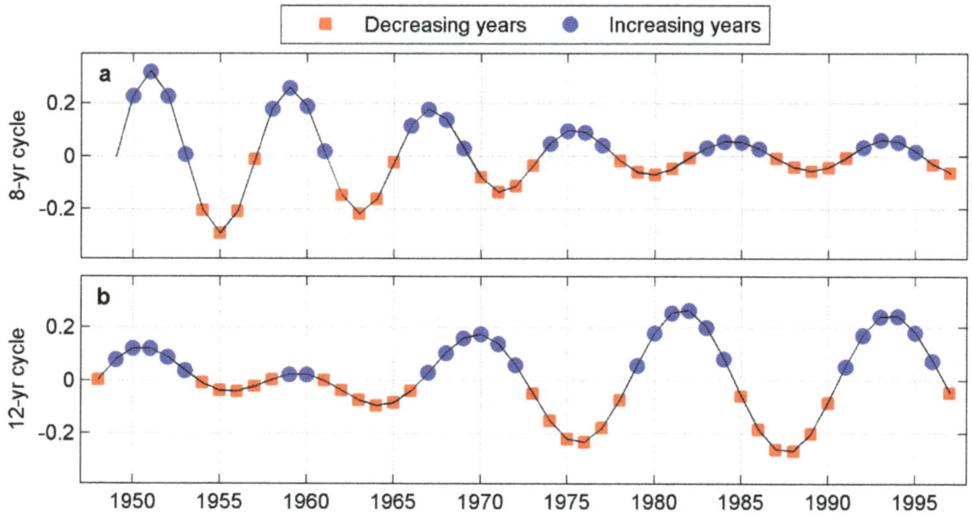

In addition to the precipitation frequency P^f and amount P^a indices, we have also included a total precipitation index P^T, for comparison. This index was defined as the total precipitation amount over the total number of days under consideration:

$$\text{Total precipitation } (P^T) = \frac{\text{accumulated precipitation depth}}{\text{total number of days}} \tag{5}$$

where the denominator includes all days, both precipitating and non-precipitating. When summed up over a month, P^T amounts to the expected total area-averaged precipitation, which can be translated to monthly lake-level changes with Equation (2). On the other hand, the precipitation frequency index P^f, which represents the probability of precipitation occurring during any given day through the month, results in the number of days with some amount of precipitation within the Superior/Michigan-Huron region. The precipitation amount index P^a represents the average daily precipitation amount only during precipitation days.

We computed each index value during years of increasing and decreasing lake levels, over winter (P_W^T, P_W^f, P_W^a) and summer (P_S^T, P_S^f, P_S^a) months. The means were tested for significant differences with a one-sided *t*-test, between the phases of the 8-y cycle (Table 2) and the 12-y cycle (Table 3). We find that P^T is significantly different between the phases of both periodicities—P_W^T during the 8-y cycle, and P_S^T during the 12-y cycle—indicating that more precipitation fell during the increasing phases of these periodicities than during the decreasing phases. This is in agreement with our results from Section 3.1 that compared seasonal lake-level changes to the lake-level reconstructions. Furthermore, the wintertime precipitation amount index P_W^a is significantly different between the phases of the 8-y cycle, and the summertime precipitation frequency index P_S^f is significantly different between phases of the 12-y cycle. This is in agreement with our findings from Section 3.2 that examined correlations between rainfall indices and lake-level reconstructions.

Table 2. Precipitation indices during winter months associated with the increasing and decreasing phases of the 8-y cycle*.

Index	Average	Increasing	Decreasing	Significance
Total precipitation P_w^T (daily average in mm)	1.58	1.67	1.49	$\mathbf{p = 0.03}$
Frequency P_w^f (daily probability)	0.72	0.73	0.72	$p = 0.44$
Amount P_w^a (daily average in mm)	2.18	2.29	2.06	$\mathbf{p = 0.01}$

Note: * p-values in bold indicate significant differences between the increasing and decreasing phases of the 8-y cycle as determined by a one-sided t-test ($p < 0.05$).

Table 3. Precipitation indices during summer months associated with the increasing and decreasing phases of the 12-y cycle*.

Index	Average	Increasing	Decreasing	Significance
Total precipitation P_S^T (daily average in mm)	2.38	2.50	2.26	$\mathbf{p = 0.02}$
Frequency P_S^f (daily probability)	0.72	0.75	0.70	$\mathbf{p < 0.01}$
Amount P_S^a (daily average in mm)	3.26	3.32	3.20	$p = 0.20$

Note: * p-values in bold indicate significant differences between the increasing and decreasing phases of the 12-y cycle as determined by a one-sided t-test ($p < 0.05$).

In summary, it is changes in seasonal precipitation totals P^T that are driving the near-decadal lake-level cycles, and these totals vary due to changes both in precipitation frequency P^f and precipitation amount P^a. Consistent with our findings discussed in the previous section, we conclude that precipitation fluctuations associated with the 8-y wintertime cycle are largely stemming from changes in precipitation amounts during the winter months P_W^a. Conversely, the summertime 12-y cycle is primarily being driven by changes in precipitation fluctuations through variations in summertime precipitation frequency P_S^f.

4. Climate Connections

To identify large-scale climate variability associated with the cyclic Michigan-Huron lake-level changes, we compared the winter and summer time series of lake-level tendencies dL_y^W and dL_y^S (Figure 6, blue lines) to the evolution of various atmospheric fields and SSTs. We hypothesize that the actual lake-level changes serve as a more accurate proxy for seasonal precipitation totals due to the spatial restrictions and coarse temporal resolution of the actual precipitation data.

We used monthly averaged 1949–1998 NOAA NCEP-NCAR reanalysis data for 850-mb air temperatures, sea-level pressures (SLPs), and 500-mb geopotential heights, and Kaplan *et al.* [23]

gridded SSTs analyzed over the same time period. The climatological monthly means were removed from all indices, and 20-y moving averages were subtracted from each location's time series to concentrate on decadal scales. The anomalous lake-level changes, dL_y^w and dL_y^s, were sorted into years from the largest negative to the largest positive lake-level change, and the first and last 12 years—approximately the upper and lower 25% of all available years—were selected. Next, we generated composite plots to connect the behavior of seasonal lake-level fluctuations to atmospheric and SST variability. The differences between the composites over the years with the greatest positive and negative lake-level changes during winter dL_y^w and summer dL_y^s, are illustrated in Figures 9 and 12, respectively, where dL_y^w was compared to November–January, and dL_y^s was compared to June–August anomalies of SSTs (°C), SLPs (mb), 850-mb air temperatures (°C), and 500-mb heights (m). The white lines encompass areas where the upper and lower means were determined to be significantly different at the 5% confidence level, as determined by a two-sided *t*-test.

4.1. The 8-y Wintertime Cycle

We concluded in Section 3 that the 8-y lake-level cycle is active during November–January, and found variations in wintertime precipitation amounts P_w^a to be the primary driver. Hence, atmospheric connections to Michigan-Huron lake-level changes during these months can help to identify climate modes associated with this periodicity.

4.1.1. Wintertime Anomalies

Composite North Atlantic and North Pacific SST anomalies (Figure 9a) associated with the wintertime lake-level changes dL_y^w have patterns similar to those of the 850-mb temperatures (Figure 9b) and upper-level heights (Figure 9d). The years of greatest wintertime lake-level increase correspond to a warm North Atlantic SST anomaly off the New England coast, which underlies a positive 850-mb temperature and 500-mb height anomaly, both extending back over part of the Great Lakes' region. Also during these years, below-average 850-mb temperatures and 500-mb heights are observed along the northwestern edge of the U.S. and western Canada. These climate patterns are characteristic of an increased precipitation-producing storm-track displacement, where the trough over the western U.S. indicates a southward displacement, and the ridge over northeastern U.S. indicates a northward displacement. Hence, it appears that wintertime cyclones that travel over the U.S. during increasing lake-level years tend to be of more southern origin, allowing more moisture from the Gulf of Mexico to reach the Great Lakes' region. Another way to interpret moisture transport is by examination of the SLP plot (Figure 9c). The anomalous cyclonic rotation of the low SLPs in the southwest (Colorado low), combined with the anomalous anticyclonic rotation of the anomalous high SLPs in the northeast, set up an anomalous northeastward flow (Panhandle hook) directly transporting moisture from the Gulf of Mexico toward the Great Lakes, resulting in greater precipitation amounts there.

We additionally examined potential differences in precipitable water (kg/m^2) associated with storm-track displacements (Figure 10; analogous to composite plots of Figure 9). A statistically significant positive anomaly in precipitable water includes most of the Great Lakes' region and is consistent with that which would be expected from a northward storm track displacement over the eastern U.S, further substantiating our findings that the wintertime lake-level changes are associated with moisture anomalies stemming from variations in storm track.

Figure 9. Composite differences between the 12 years with the greatest increase and 12 years with the greatest decrease of wintertime (NDJ) lake-level, for high-pass filtered seasonal anomalies of **(a)** SSTs; **(b)** 850-mb temperatures; **(c)** SLPs; and **(d)** 500-mb geopotential heights, for 1949–1998. Areas encompassed by white lines indicate significant differences at the 5% level.

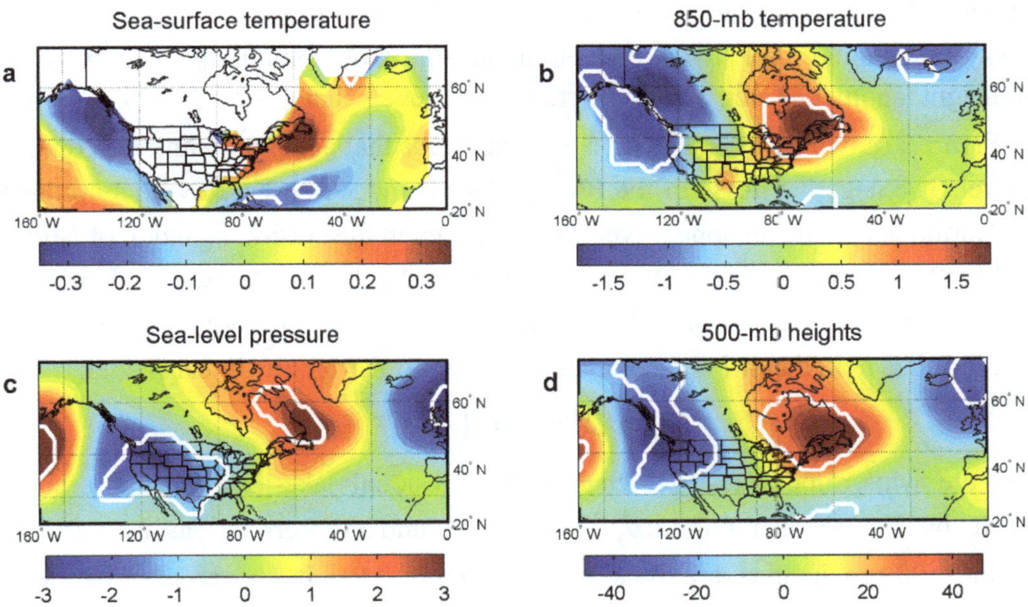

Figure 10. Composite differences between the 12 years with the greatest increase and 12 years with the greatest decrease of wintertime (NDJ) lake-level for high-pass filtered seasonal anomalies of precipitable water (kg/m^2). Areas encompassed by white lines indicate significant differences at the $p < 0.05$ level.

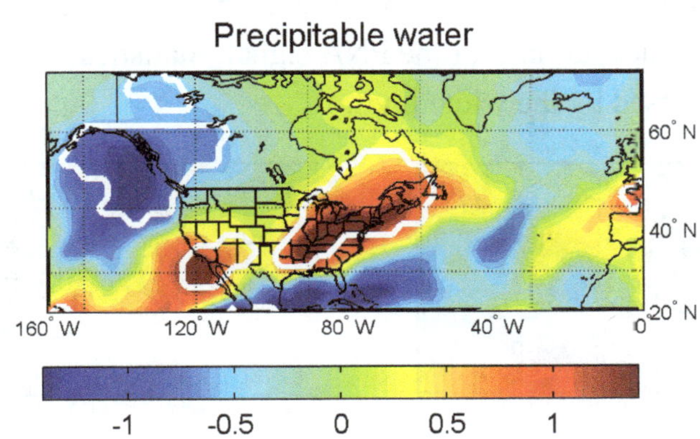

4.1.2. The Pacific/North American Index

The negative height anomaly located over the western U.S. and positive anomaly over the east-central North Pacific during the years of greatest wintertime lake-level increase (Figure 9d), is a pattern consistent with the negative phase of the Pacific/North American (PNA) index. In contrast, the anomalous high over the northeastern U.S. is positioned further north than that which is typically characterized by the PNA pattern. However, it is the western North American component of this pattern that has previously been linked to increased precipitation amounts during the winter months [24]. The negative PNA phase is characterized by a storm track that is more zonal than its positive counterpart, resulting in a more southerly storm track and enhanced moisture advection into the Great Lakes' region [4]. Indeed, the wintertime (November–January) Climate Prediction Center (CPC) PNA index [25] is well correlated with the lake-level changes (r = −0.50). In spite of this, the 8-y reconstructed lake-level time series dL_y^{rc8} is not well correlated with the PNA index (r = −0.06), indicating that the PNA pattern is not associated with 8-y lake-level cycle.

Another way to examine the PNA/lake-level connection is by computing the correlations between the 500-mb height anomalies and the wintertime lake-level changes dL_y^w. These correlations are illustrated in Figure 11a, where the white lines indicate areas that are significantly correlated at the 5% confidence level. The negative phase of the PNA pattern is indeed observed with a north-south dipole pattern in the North Pacific and an upper-level trough located in the northwest region of North America.

Because the PNA index is not well correlated with the 8-y cycle, we evaluated the residual lake-level signal, or the lake-level signal that is not associated with the PNA. To achieve this, we used the CPC PNA index as a predictor I and computed the best fit β from the linear model:

$$dL_y^w = I \times \beta + \varepsilon \quad (6)$$

where dL_y^w is the wintertime lake-level change. The residual lake-level change ε is significantly correlated with the 8-y reconstruction dL_y^{rc8} (r = 0.39), and the correlations between this residual signal ε and the 500-mb heights are illustrated in Figure 11b. It appears that while part of the wintertime lake-level signal is stemming from the North Pacific region, the 8-y cycle is instead rooted in the North Atlantic region. Statistically significant correlations between ε and SSTs were also identified in the North Atlantic (not shown).

Figure 11. Correlations between 500-mb height anomalies and (**a**) the actual Michigan-Huron wintertime lake-level changes (20-y moving average removed), and (**b**) the residual lake-level changes after accounting for the PNA signal. Statistically significant correlations ($p < 0.05$) are encompassed by the white lines.

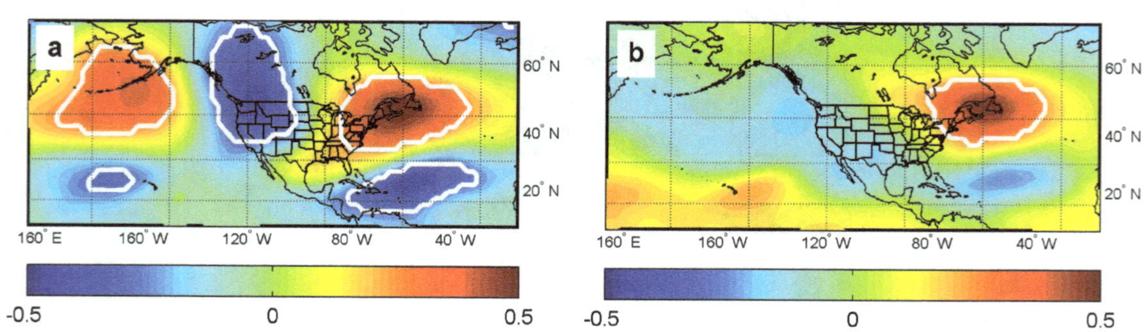

The North Atlantic SST's north-south dipole pattern is similar in structure to spatial patterns identified by Moron *et al.* [26] (Figure 10) and Da Costa and De Verdiere [27] (Figure 2), who analyzed SSTs and SLP fields and found a 7.7-y oscillation possibly rooted in coupled dynamics; the time scale and pattern of this oscillation (not shown) is remarkably similar to the ones associated with the ~8-y oscillation in the Michigan-Huron water levels, thus further substantiating our findings that the 8-y wintertime lake-level cycle is driven by processes in the North Atlantic region as opposed to those associated with the PNA in the North Pacific. We thus conclude that the 8-y cycle is initiated by temperature changes in the North Atlantic which modify long-wave synoptic patterns thus altering moisture transport and precipitation totals in the Great Lakes' region, thereby producing quasi-periodic lake-level changes during the winter months.

4.2. The 12-y Summertime Cycle

We concluded in Section 3 that the second near-decadal cycle, with a period of about 12 years, is associated with changes in precipitation frequency during the summer months of June–August. We thus compared summertime lake-level changes dL_y^s to summertime atmospheric fields and SSTs, analogous to the 8-y wintertime cycle analysis as described in Section 4.1. The resulting composite differences are illustrated in Figure 12, where the white lines again indicate significantly different means at the 5% level, between the years of most positive and most negative anomalous summertime lake-level changes.

Figure 12. Composite differences between the 12 years with the greatest increase and 12 years with the greatest decrease of summertime (JJA) lake-level change, for high-pass filtered seasonal anomalies of (**a**) SSTs; (**b**) 850-mb temperatures; (**c**) SLPs, and (**d**) 500-mb geopotential heights, for 1949–1998. Areas encompassed by white lines indicate significant differences at the 5% level.

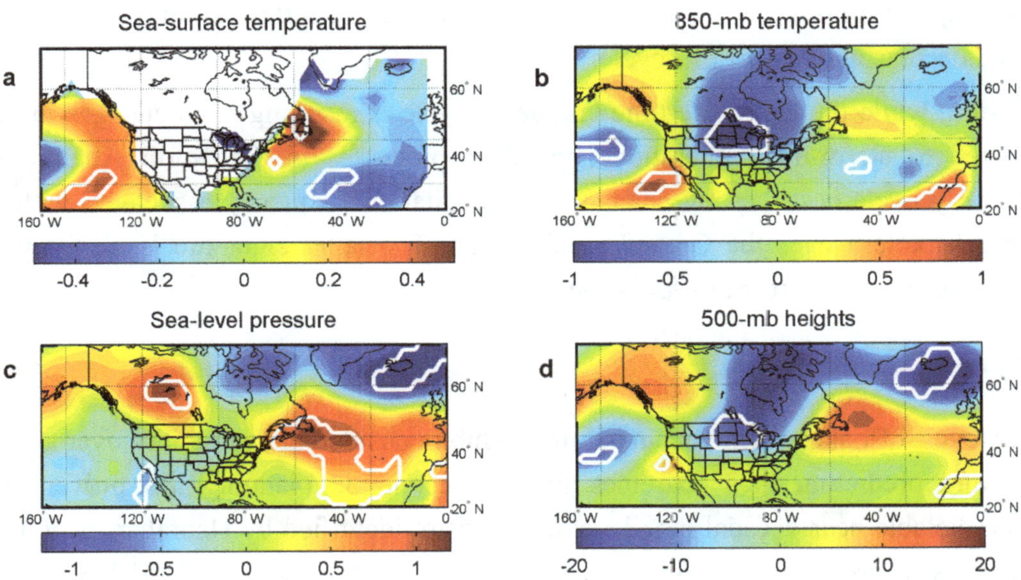

The years of most-increasing summertime lake-levels correspond to a large area of below-average 850-mb temperatures (Figure 12b) which underlie an upper-level trough over the Great Lakes region (Figure 12d). In agreement with Juckes and Smith [28], who examined the relationship between upper-level troughs and convective available potential energy (CAPE), Gold and Nielsen-Gammon [29] determined that the presence of an upper-level potential vorticity anomaly can increase CAPE in the region, thus triggering more frequent convection when conditions are favorable [30]. Thus, the existence of a quasi-periodic upper-level trough over the Great Lakes may alter the frequency of convective precipitation events during the summer months, producing lake-level changes which also exhibit this near-decadal cyclic behavior.

As with the wintertime composite plots, the SST anomalies (Figure 12a) largely appear to extend up to the 850-mb level, which is evident from the cold temperatures encompassed by the warm anomaly over the North Pacific. The warm SST anomalies near the Tropical Pacific region during the greatest summertime lake-level increases also agree with the findings of Wang *et al.* [31] who used El Niño events to skillfully forecast increased summertime precipitation in the Midwest. The North Atlantic SST pattern is similar to that of an identified ~13-y cycle by Moron *et al.* [26] (their Figure 9); the 13-y cycle SST pattern they described exhibits alternating SST signs between the tropical Atlantic off of Africa's western coast, and the eastern coast of the U.S. Hence, as with the 8-y lake-level cycle, it appears that the 12-y cycle can be associated with distinct differences in large-scale atmospheric and oceanic SST patterns.

We determined in Section 2 that the interannual lake-level changes are primarily precipitation driven, therefore indicating that both of the near-decadal cycles are precipitation driven. An evaluation of evaporative losses during different phases of the 12-y cycle, however, reveals that evaporation may also play a small role in the evolution of this periodicity through below-average evaporative losses when precipitation is high and above-average losses when precipitation is low (not shown). In Section 2.1, comparison of the interannual variability associated with both precipitation and evaporation indicated that the historic year-to-year changes in evaporation were too small to impact the lake levels in a significant way. While its historic impact may have been minimal, the existence of this cycle in the summertime evaporation time series substantiates our findings of differing atmospheric phases during increasing and decreasing years of this cycle.

In conclusion, while the 8-y wintertime lake-level cycle is resulting from changes in precipitation amounts related to storm-track alterations linked to processes in the North Atlantic region, the summertime 12-y cycle is driven by changes in precipitation frequency associated with a preferred upper-air trough pattern over the Great Lakes' region. The latter signature is also apparent in the historic time series of summertime evaporation.

5. Summary and Conclusions

The water levels of Lakes Michigan and Huron exhibit considerable variability over a wide range of time scales, and predicting their extremes for socioeconomic planning has proven problematic. Potentially predictable near-decadal lake-level cycles were identified by Hanrahan *et al.* [6] and were linked to changes in precipitation. However, while this and other studies [2–5] have suggested that regional precipitation changes are the primary interannual lake-level driver, no clear illustrations that connect precipitation variability to the lake-level time series have been made.

In this study, we used daily gridded precipitation data product to explicitly illustrate the impact of precipitation on the lake levels. By weighting the annual precipitation totals according to the historic GLERL component data, we isolated a precipitation-driven fraction of the total lake-level time series which closely resembles the full observed lake-level time series, thus confirming that precipitation has indeed been the primary lake-level driver (Section 2). Furthermore, we determined that regional precipitation effects are translated to the water levels essentially instantaneously, with a sub-monthly time lag.

Hanrahan [22] found that the AMO alters precipitation characteristics, and hence the lake levels, throughout the year. That is, lake levels tend to increase during the negative AMO phase and decrease during the positive AMO phase, during all seasons. In contrast, we concluded in Section 3 that the near-decadal lake-level cycles occur with distinct seasonality. This was verified by comparing the lake-level reconstructions to precipitation indices and reanalysis variables.

Our finding of two dominant, 8-y and 12-y signals in the lake-level and precipitation time series is consistent with previous observational studies based on the analysis of instrumental record [26,27]. Near-decadal spectral peaks were also found to be ubiquitous in tree-ring-based proxy reconstructions of the Great Lakes water levels [32–34], as well as in such reconstructions of the NAO index known to be correlated with temperature and precipitation conditions over the Eastern and Central U.S. [35–37].

We found that the 8-y cycle is occurring during winter months and is linked to variations in precipitation amount (Section 3). During the years of greatest wintertime lake-level changes, anomalously cold SSTs near the Gulf of Alaska coincide with locations of cold 850-mb temperatures, and below-average 500-mb heights, whereas anomalously warm SSTs off of the New England coast correspond to above average 850-mb temperatures and 500-mb heights (Section 4.1). This spatial pattern of an equatorward displaced storm track near the Rockies, and a poleward displaced track near the Great Lakes, is consistent with the timing of increased wintertime precipitation amounts. Rodionov [4] connected increased wintertime (December–February) Great Lakes' regional precipitation to an increased number of cyclones that originated from the south, which was attributed to a weakened 700-mb PNA teleconnection pattern. It was found here that while the wintertime lake levels are well correlated with the PNA index, the 8-y reconstructed cycle is not. After accounting for the PNA influence on the Michigan-Huron lake levels, we attribute variability in the residual lake-level time series, which contains the identified 8-y cycle, to processes in the North Atlantic region, also in agreement with Moron *et al.* [26] and Da Costa and De Verdiere [27]. The details pertaining to this connection are not yet entirely clear and are thus left for future work.

The 12-y cycle is primarily exhibited during the summer months and is driven by alternating frequencies in precipitation events (Section 3). During the years of largest summertime lake-level increases, we identified a 500-mb trough which blankets the Great Lakes region resulting in an increased number of precipitation days (Section 4). We further identified an 850-mb cold temperature anomaly that underlies the upper-level trough associated with this cycle. Our finding of increased precipitation frequency, which is characteristic of increased convective activity during the summer, is further supported by previous studies which have concluded that warm-season precipitation in the U.S. is primarily convectively driven [38,39].

In spite of these findings that largely consider precipitation effects, changes possibly associated with anthropogenic climate change may further complicate the connections between atmospheric patterns, precipitation, and the lake levels. Several studies have indicated that under global warming, the climatology of Northern Hemisphere cyclones will be altered in terms of both average storm track and intensity [40–44]. Furthermore, warmer temperatures may produce higher precipitation totals along storm tracks [45,46]. We hypothesize that changes such as these will ultimately modify the dynamics that have historically contributed to the 8-y lake-level cycle. In addition, increasing air temperatures may alter both rainfall frequency and intensity [47–51], therefore also modifying the behavior of the 12-y cycle. Watras *et al.* [7] already identified a recent shift in the amplitude of a similar cycle in small lake and groundwater levels. Hence, more work needs to be done to assess how the precipitation-driven near-decadal lake-level cycles may be altered in the presence of a changing climate.

While our findings indicate that precipitation has been the single most important variable when evaluating historical interannual lake-level fluctuations, recent research has identified changes in the way that the lake levels are responding to atmospheric processes. Specifically, some authors have found recent significant positive trends in evaporation [52,53], and although variability associated with evaporation has historically been too small to significantly affect the lake levels, a persistent positive decadal-scale evaporation anomaly is now resulting in significant cumulative lake-level changes [8]. As discussed in Section 2.1, the precipitation-driven lake-level components have explained a majority of the historic lake-level behavior; however, their time series are beginning to diverge, stemming from increasing evaporative losses over the past couple of decades. Thus, not only the precipitation patterns themselves are likely to be altered under climate change, the nature of the lake-level response to climate variability is itself changing. Therefore, while the discussed periodicities may still be useful in statistical predictions schemes for future lake-level variations, one must be cautious in interpreting these predictions (as with any statistical forecast scheme), due to possible non-stationarity of lake–climate connections.

Acknowledgments

The spectral analyses were performed using MTM-SSA toolkit developed by the Theoretical Climate Dynamics group at the University of California-Los Angeles [54]. This research was supported by the Office of Science (BER), U. S. Department of Energy (DOE) grant DE-FG02-07ER64428, NSF grant ATM-0852459, NSF grant ATM-1236620, UWM grant RGI05 (SK), UWM grant RGI05 (PR), and the Russian Department of Education and Science project 14.B25.31.0026 (SK).

Author Contributions

Janel Hanrahan conducted the data analyses and wrote the first draft of the manuscript. Paul Roebber and Sergey Kravtsov provided guidance throughout the research and writing stages. All authors reviewed the final manuscript.

References

1. World Lakes Website. Available online: http://www.worldlakes.org (accessed on 22 March 2014).
2. Changnon, S.A., Jr. Climate fluctuations and record-high levels of Lake Michigan. *Bull. Am. Meteorol. Soc.* **1987**, *68*, 1394–1402.
3. Changnon, S.A. Temporal behavior of levels of the Great Lakes and climate variability. *J. Gt. Lakes Res.* **2004**, *30*, 184–200.
4. Rodionov, S.N. Association between winter precipitation and water level fluctuations in the Great Lakes and atmospheric circulation patterns. *J. Clim.* **1994**, *7*, 1693–1706.

5. Polderman, N.J.; Pryor, S.C. Linking synoptic-scale climate phenomena to lake-level variability in the Lake Michigan-Huron Basin. *J. Gt. Lakes Res.* **2004**, *30*, 419–434.
6. Hanrahan, J.L.; Kravtsov, S.V.; Roebber, P.J. Quasi-periodic decadal cycles in levels of lakes Michigan and Huron. *J. Gt. Lakes Res.* **2009**, *35*, 30–35.
7. Watras, C.J.; Read, J.S.; Holman, K.D.; Liu, Z.; Song, Y.Y.; Watras, A.J.; Morgan, S.; Stanley, E.H. Decadal oscillation of lakes and aquifers in the upper Great Lakes region of North America: Hydroclimatic implications. *Geophys. Res. Lett.* **2014,** in press.
8. Hanrahan, J.L.; Kravtsov, S.V.; Roebber, P.J. Connecting past and present climate variability to the water levels of Lakes Michigan and Huron. *Geophys. Res. Lett.* **2010**, *37*, doi:10.1029/2009GL041707.
9. Delworth, T.L.; Mann, M.E. Observed and simulated multidecadal variability in the Northern Hemisphere. *Clim. Dyn.* **2000**, *16*, 661–676.
10. Gray, S.T.; Graumlich, L.J.; Betancourt, J.L.; Pederson, G.T. A tree-ring based reconstruction of the Atlantic Multidecadal Oscillation since 1567 AD. *Geophys. Res. Lett.* **2004**, *31*, doi:10.1029/2004GL019932
11. Knight, J.R.; Allan, R.J.; Folland, C.K.; Vellinga, M.; Mann, M.E. A signature of persistent natural thermohaline circulation cycles in observed climate. *Geophys. Res. Lett.* **2005**, *32*, doi:10.1029/2005GL024233
12. Enfield, D.B.; Mestas-Nuñez, A.M.; Trimble, P.J. The Atlantic multidecadal oscillation and its relation to rainfall and river flows in the continental US. *Geophys. Res. Lett.* **2001**, *28*, 2077–2080.
13. Rogers, J.C.; Coleman, J.S.M. Interactions between the Atlantic Multidecadal Oscillation, El Nino/La Nina, and the PNA in winter Mississippi valley stream flow. *Geophys. Res. Lett.* **2003**, *30*, doi:10.1029/2003GL017216.
14. McCabe, G.J.; Palecki, M.A.; Betancourt, J.L. Pacific and Atlantic Ocean influences on multidecadal drought frequency in the United States. *Proc. Natl. Acad. Sci. USA* **2004**, *101*, 4136–4141.
15. Sutton, R.T.; Hodson, D.L. Atlantic Ocean forcing of North American and European summer climate. *Science* **2005**, *309*, 115–118.
16. Ghil, M.; Allen, M.R.; Dettinger, M.D.; Ide, K.; Kondrashov, D.; Mann, M.E.; Robertson, A.W.; Saunders, A.; Tian, Y.; Varadi, F.; Yiou, P. Advanced spectral methods for climatic time series. *Rev. Geophys.* **2002**, *40*, 3:1–3:41.
17. Ghil, M.; Yiou, P.; Hallegatte, S.; Malamud, B.D.; Naveau, P.; Soloviev, A.; Friederichs, P.; Keilis-Borok, V.; Kondrashov, D.; Kossobokov, V.; *et al.* Extreme events: dynamics, statistics, and prediction. *Nonlin. Process. Geophys.* **2011**, *18*, 295–350.
18. GLERL data: Available online: http://www.glerl.noaa.gov/data/pgs/lake_levels.html (accessed on 1 June 2010).
19. Clites, A.H.; Quinn, F.H. The history of Lake Superior regulation: Implications for the future. *J. Gt. Lakes Res.* **2003**, *29*, 157–171.
20. International Research Institute/Lamont-Doherty Earth Observatory (IRI/LDEO) Climate Data Library. Available online: http://iridl.ldeo.columbia.edu/SOURCES/.NOAA/.NCEP/.CPC/.REGIONAL/.USA/.daily/.gridded/ (accessed on 1 March 2011).

21. Zolina, O.; Kapala, A.; Simmer, C.; Gulev, S.K. Analysis of extreme precipitation over Europe from different reanalyses: A comparative assessment. *Glob. Planet. Chang.* **2004**, *44*, 129–161.
22. Hanrahan, J.L. Connecting Past and Present Climate Variability to the Water Levels of Lakes Michigan and Huron. Ph.D. Thesis, Department of Mathematical Sciences, University of Wisconsin–Milwaukee, WI, USA, 2010.
23. Kaplan, A.; Cane, M.A.; Kushnir, Y.; Clement, A.C.; Blumenthal, M.B.; Rajagopalan, B. Analyses of global sea surface temperature 1856–1991. *J. Geophys. Res. Ocean.* **1998**, *103*, 18567–18589.
24. Leathers, D.J.; Yarnal, B.; Palecki, M.A. The Pacific/North American teleconnection pattern and United States climate. Part I: Regional temperature and precipitation associations. *J. Clim.* **1991**, *4*, 517–528.
25. National Weather Service, Climate Prediction Center. Available online: http://www.cpc.ncep.noaa.gov (accessed on 28 January 2011).
26. Moron, V.; Vautard, R.; Ghil, M. Trends, interdecadal and interannual oscillations in global sea-surface temperatures. *Clim. Dyn.* **1998**, *14*, 545–569.
27. Da Costa, E.D.; de Verdiere, A.C. The 7.7-year North Atlantic Oscillation. *Q. J. R. Meteorol. Soc.* **2002**, *128*, 797–817.
28. Juckes, M.; Smith, R.K. Convective destabilization by upper-level troughs. *Q. J. R. Meteorol. Soc.* **2000**, *126*, 111–123.
29. Gold, D.A.; Nielsen-Gammon, J.W. Potential vorticity diagnosis of the severe convective regime. Part III: The Hesston tornado outbreak. *Mon. Weather Rev.* **2008**, *136*, 1593–1611.
30. Van Klooster, S.L.; Roebber, P.J. Surface-based convective potential in the contiguous United States in a business-as-usual future climate. *J. Clim.* **2009**, *22*, 3317–3330.
31. Wang, H.; Ting, M.; Ji, M. Prediction of seasonal mean United States precipitation based on El Nino sea surface temperatures. *Geophys. Res. Lett.* **1999**, *26*, 1341–1344.
32. Brinkmann, W.A.R. Water supply to the Great Lakes reconstructed from tree-rings. *J. Clim. Appl. Meteorol.* **1987**, *26*, 530–538.
33. Quinn, F.H.; Sellinger, C.E. A reconstruction of Lake Michigan–Huron water levels derived from tree ring chronologies for the period 1600–1961. *J. Great Lakes Res.* **2006**, *32*, 29–39.
34. Wiles, G.C.; Krawiec, A.C.; D'Arrigo, R.D. A 265-year reconstruction of Lake Erie water levels based on North Pacific tree rings. *Geophys. Res. Lett.* **2009**, *36*, doi:10.1029/2009GL037164.
35. Cullen, H.M.; D'Arrigo, R.D.; Cook, E.R.; Mann, M.E. Multiproxy reconstructions of the North Atlantic Oscillation. *Paleoceanography* **2001**, *16*, 27–39.
36. Fye, F.K.; Stahle, D.W.; Cook, E.R.; Cleaveland, M.K. NAO influence on sub-decadal moisture variability over central North America. *Geophys. Res. Lett.* **2006**, *33*, doi:10.1029/2006GL026656.
37. D'Arrigo, R.D.; Anchukaitis, K.J.; Buckley, B.; Cook, E.; Wilson, R. Regional climatic and North Atlantic Oscillation signatures in West Virginia red cedar over the past millennium. *Glob. Planet. Change* **2012**, *84–85*, 8–13.
38. Fritsch, J.M.; Kane, R.J.; Chelius, C.R. The contribution of mesoscale convective weather systems to the warm-season precipitation in the United States. *J. Clim. Appl. Meteorol.* **1986** *25*, 1333–1345.
39. Heideman, K.F.; Michael Fritsch, J. Forcing mechanisms and other characteristics of significant summertime precipitation. *Weath. Forecast.* **1988**, *3*, 115–130.

40. Lambert, S.J. The effect of enhanced greenhouse warming on winter cyclone frequencies and strengths. *J. Clim.* **1995**, *8*, 1447–1452.
41. McCabe, G.J.; Clark, M.P.; Serreze, M.C. Trends in Northern Hemisphere surface cyclone frequency and intensity. *J. Clim.* **2001**, *14*, 2763–2768.
42. Zhang, X.; Zwiers, F.W.; Hegerl, G.C.; Lambert, F.H.; Gillett, N.P.; Solomon, S.; Stott, P.A.; Nozawa, T. Detection of human influence on twentieth-century precipitation trends. *Nature* **2007**, *448*, 461–465.
43. Catto, J.L.; Shaffrey, L.C.; Hodges, K.I. Northern Hemisphere Extratropical Cyclones in a Warming Climate in the HiGEM High-Resolution Climate Model. *J. Clim.* **2011**, *24*, 5336–5352.
44. McDonald, R.E. Understanding the impact of climate change on Northern Hemisphere extra-tropical cyclones. *Clim. Dyn.* **2011**, *37*, 1399–1425.
45. Bengtsson, L.; Hodges, K.I.; Keenlyside, N. Will extratropical storms intensify in a warmer climate? *J. Clim.* **2009**, *22*, 2276–2301.
46. Kunkel, K.E.; Karl, T.R.; Easterling, D.R.; Redmond, K.; Young, J.; Yin, X.; Hennon, P. Probable maximum precipitation and climate change. *Geophys. Res. Lett.* **2013**, *40*, 1402–1408.
47. Meehl, G.A.; Arblaster, J.M.; Tebaldi, C. Understanding future patterns of increased precipitation intensity in climate model simulations. *Geophys. Res. Lett.* **2005**, *32*, doi: 10.1029/2005GL023680
48. Sun, Y.; Solomon, S.; Dai, A.; Portmann, R.W. How often will it rain? *J. Clim.* **2007**, *20*, 4801–4818.
49. Allan, R.P; Soden, B.J. Atmospheric warming and the amplification of precipitation extremes. *Science* **2008**, *321*, 1481–1484.
50. Lenderink, G.; van Meijgaard, E. Increase in hourly precipitation extremes beyond expectations from temperature changes. *Nat. Geosci.* **2008**, *1*, 511–514.
51. Collins, M.; Knutti, R.; Arblaster, J.; Dufresne, J.L.; Fichefet, T.; Friedlingstein, P.; Gao, X.; Gutowski, W.J.; Johns, T.; Krinner, G.; *et al.* Long-term Climate Change: Projections, Commitments and Irreversibility. In *Climate Change 2013: The Physical Science Basis. Contribution of Working Group I to the Fifth Assessment Report of the Intergovernmental Panel on Climate Change*; Cambridge University Press: Cambridge, UK & New York, NY, USA, 2013.
52. Assel, R.A.; Quinn, F.H.; Sewnger, C.E. Hydroclimatic factors of the recent record drop in Laurentian Great Lakes water levels. *Bull. Am. Meteorol. Soc.* **2004**, *85*, 1143–1151.
53. Sellinger, C.E.; Stow, C.A.; Lamon, E.C.; Qian, S.S. Recent water level declines in the Lake Michigan−Huron System. *Environ. Sci. Technol.* **2007**, *42*, 367–373.
54. *MTM-SSA Toolkit*, software for spectral analysis; Theoretical Climate Dynamics group at the University of California-Los Angeles: Los Angeles, CA, USA, 2010; Available online: http://www.atmos.ucla.edu/tcd/ssa/ (accessed on 1 June 2010).

4

Farmers' Perceptions about Adaptation Practices to Climate Change and Barriers to Adaptation

Francis Ndamani [1,*] and Tsunemi Watanabe [2]

[1] Graduate School of Engineering, Kochi University of Technology, 2-22 Eikokuji, Kochi City, Kochi 780-8515, Japan

[2] School of Economics and Management, Kochi University of Technology, 2-22 Eikokuji, Kochi City, Kochi 780-8515, Japan; E-Mail: watanabe.tsunemi@kochi-tech.ac.jp

* Author to whom correspondence should be addressed; E-Mail: fndamani@gmail.com;

Academic Editor: Yingkui Li

Abstract: This study analyzed the farmer-perceived importance of adaptation practices to climate change and examined the barriers that impede adaptation. Perceptions about causes and effects of long-term changes in climatic variables were also investigated. A total of 100 farmer-households were randomly selected from four communities in the Lawra district of Ghana. Data was collected using semi-structured questionnaires and focus group discussions (FGDs). The results showed that 87% of respondents perceived a decrease in rainfall amount, while 82% perceived an increase in temperature over the past 10 years. The study revealed that adaptation was largely in response to dry spells and droughts (93.2%) rather than floods. About 67% of respondents have adjusted their farming activities in response to climate change. Empirical results of the weighted average index analysis showed that farmers ranked improved crop varieties and irrigation as the most important adaptation measures. It also revealed that farmers lacked the capacity to implement the highly ranked adaptation practices. The problem confrontation index analysis showed that unpredictable weather, high cost of farm inputs, limited access to weather information, and lack of water resources were the most critical barriers to adaptation. This analysis of adaptation practices and constraints at farmer level will help facilitate government policy formulation and implementation.

Keywords: climatic effects; agricultural production; farmer perceptions; adaptation constraints; Lawra district

1. Introduction

Climate change prediction models have indicated that the Sudan and Guinea Savanna zones of Ghana will continue to experience increasing temperature and decreasing precipitation trends [1]. This confirms previous findings that between 2030 and 2039 the rainy season might start in June or even later in Northern Ghana [2]. It is also projected that the standard deviation for the onset of the rainy season will increase [3], which suggests that not only will it shift but also it will become even more "erratic" [4]. The implications are that Northern Ghana would witness more extreme weather conditions such as droughts, dry spells, and floods. This situation will eventually affect agriculture, the environment, and human livelihoods. In particular, it is anticipated that adverse impacts on the agricultural sector will exacerbate the incidence of rural poverty [5]. Adaptation practices are therefore needed to help agrarian communities better face extreme weather conditions associated with climate variations [6].

Adaptations are adjustments or interventions that take place to manage the losses or take advantage of the opportunities presented by a changing climate. Adaptive capacity has been defined as the ability of a system to adjust to climate change (including climate variability and extremes), to moderate potential damages, to take advantage of opportunities, or to cope with the consequences [7]. Adaptation practices are therefore pre-emptive in nature. They are designed to mitigate potential adverse effects and take advantage of the potential benefits of an envisaged change in climatic variables.

Several studies in Ghana have reported adaptation practices in agriculture, including crop diversification, change of planting date, hybrid varieties, and soil moisture conservation techniques [8,9] In Uganda, income diversification, digging of drainage channels, and the use of drought-tolerant varieties have been reported [10]. In addition, mixed farming, mixed cropping, tree planting, use of different crop varieties, changing planting and harvesting dates, increased use of irrigation, increased use of water and soil conservation techniques, and diversifying from farm to non-farm activities have also been reported in Nigeria and in South Africa [11,12].

Globally, many studies have been used to understand farmers' perceptions about climate change and its associated effects on agriculture. Although perceptions are not necessarily consistent with reality, they must be considered to address socioeconomic challenges [13]. Perception has been defined as the process by which organisms interpret and organize sensation to produce a meaningful experience of the world [14]; and that a person's perceptions are based on experiences with natural and other environmental factors that vary in the extent to which such perceptions are enabled [15]. Previous studies have shown that the way in which people experience climate shocks varies across different social groups, geographic locations, and seasons of the year, with men, women, and children all experiencing different levels of hardship and opportunity in the face of climate change [16].

Discussions of adaptation practices and barriers to adoption need to be informed by empirical data from farmers. Adaptation practices in agriculture are generally location-specific [17]; hence, it is crucial to understand farmers' perceptions about the risks they face. To ensure farmers' readiness for extreme weather events and collaboratively learn about the evolution of weather patterns, efforts to focus on farmers and their current activities, knowledge, and perceptions are essential [18,19]. Farmers' willingness to accept and use prescribed measures could be enhanced if their perceptions and understanding are considered in designing such measures. By contrast, current models used in predictions of climate change and adaptation practices are at a global scale and need to be downscaled to accommodate realities at the community level [9].

In the Lawra district of Ghana, agriculture production is the dominant source of food and household incomes for the vast majority of rural households. Agriculture production is largely rain-fed. Farmers' dependence on an annual mono-modal rainfall pattern coupled with farm resource constraints make agriculture very vulnerable to the impacts of climate change. Results of previous studies have revealed a negative correlation between seasonal rainfall and volume of staple crops *(i.e.*, sorghum, millet, and groundnut) produced annually in the Lawra district over the past 20 years [20]. This study explored farmers' perceptions regarding long-term changes in climatic variables and the associated effects on farming. It also identified and prioritized adaptation practices based on farmers' perceived importance. Constraints on the use of adaptation measures were also identified and ranked. This study will help government policy decisions about suitable adaptation practices that are applicable and most preferred by farmers. It will also ensure that critical barriers to adoption are effectively addressed.

2. Materials and Methods

2.1. Survey Design and Study Area

This study is based on a cross-sectional survey data from farming households across four communities (*i.e.*, Brifo-chaa, Methuo, Kalsagri, and Oribili) in the Lawra district of Ghana, located at longitude 10°30′ N and latitude 2°35′ W. The district lies within the Guinea Savanna Zone, with mean annual rainfall ranging from 900 to 1200 mm. It has two seasons: the dry season (November–April) and the rainy season (May–October). The vegetation is guinea savanna grassland characterized by shrubs and medium-sized trees, such as shea-tree, dawadawa, baobab, and acacia. The soils are mainly laterite soils developed from birimian and granite rocks. These soils are shallow sandy loam with medium coarse quartz stones. Recurrent droughts, dry spells, and floods tend to have adverse effects on crop production. The major crops produced include maize, sorghum, millet, and groundnut. Crop production activities take place within the rainy season. Eighty percent of the district's total population of 100,929 is engaged in rain-fed subsistence agriculture [21] The district was chosen because, based on historical data from the Ghana Meteorological Agency, it is more prone to extreme weather conditions. According to the [21], Lawra is the poorest district in the upper west region of Ghana.

A total of 100 farming households were randomly selected for the interviews. Semi-structured questionnaires were used to investigate farmers' perceived changes in temperature and rainfall, causes and effects of climate change, and adaptation practices being used by farmers. Four focus group discussions (FGDs) were conducted to double check the survey data. The household survey and FGDs were conducted between February and November 2014 with the assistance of three regional and four district agricultural officers. The selection of communities was based on the accessibility and knowledge of agricultural officers.

2.2. Statistical Analysis

Data were entered and analyzed using statistical package for the social sciences (SPSS). Frequencies, percentages, and means are the basic descriptive statistical tools used to represent farmers' perceptions about long-term changes in climatic variables and the associated causes (Table 1). In determining farmers' perceived importance of adaptation practices, respondents were requested to score selected practices based on a 0–3 scale, where 0 is the least important practice and 3 is the most important practice. The adaptation practices were then ranked using the weighted average index (WAI):

$$\text{WAI} = \frac{\sum Fi\,Wi}{\sum Fi} \tag{1}$$

where F = frequency of response; W = weight of each score; and i = score (3 = highly important; 2 = moderately important; 1 = less important; 0 = not important).

Table 1. Description of data variables.

Variables	Mean	Standard Deviation
Age (15–34 years = 1; 35–54 years = 2; above 55 years = 3)	2.26	0.73
Education level (literate = 1; illiterate = 0)	0.21	0.41
Farm size (continuous)	4.95	1.70
Household size (continuous)	8.20	5.12
Family labor (continuous)	4.65	3.04
Annual farm income—Ghana cedi (continuous)	949.42	1909.55
Annual off-farm income—Ghana cedi (continuous)	797.20	2459.92
Farmer's adaptation (adapted = 1; not adapted = 0)	0.67	0.47

Previous studies have also applied the weighted average index (WAI) to assess farmers' perceived important adaptation strategies in Bangladesh and barriers of adaptation to climate change in Nepal [22,23].

To identify the critical constraints that hinder farmers from using adaptation practices, a ranking was conducted using the Problem Confrontation Index (PCI). Respondents were asked to grade their perceived barriers based on a 0–3 Likert scale (*i.e.*, ranging from "not a problem" to "highly problematic"). The PCI value was estimated using the formula below:

$$\text{PCI} = P_n \times 0 + P_l \times 1 + P_m \times 2 + P_h \times 3 \tag{2}$$

where:

PCI = Problem Confrontation Index;
P_n = Number of respondents who graded the constraint as no problem;
P_l = Number of respondents who graded the constraint as low;
P_m = Number of respondents who graded the constraint as moderate;
P_h = Number of respondents who graded the constraint as high.

3. Results

3.1. Farmers' Perceptions of Long-Term Temperature and Rainfall Changes

The majority of farmers (82%) perceived an increase in temperature over the past 10 years (Figure 1). About 9% of respondents perceived no change, 6% perceived a decreasing change in temperature, and 3% did not know if there was a long-term change in temperature (Figure 1). Similar results were obtained from the focus group discussion. Generally, farmers believe that the increasing temperature trend was associated with the changes in precipitation. A total of 87% of respondents claimed that the rainfall amount has been decreasing over the past 10 years, 6% perceived no change in precipitation, and 7% gave other responses. Results obtained from FGDs proved that this perception was unanimous among farmers.

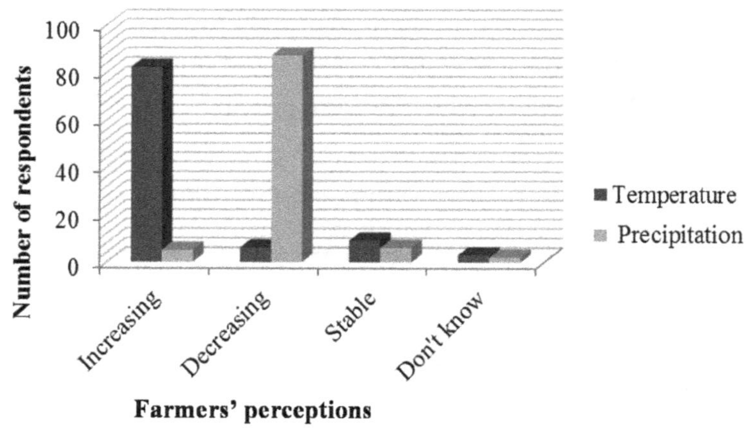

Figure 1. Farmers' perceptions of long-term changes in temperature and precipitation in the Lawra district of Ghana.

To verify farmers' perceptions regarding the precipitation trend, available historical annual rainfall data from 1980 to 2012 were obtained from the upper west regional weather station of the Ghana Meteorological Agency. The results indicated high variability rather than a clear decreasing trend in precipitation (Figure 2).

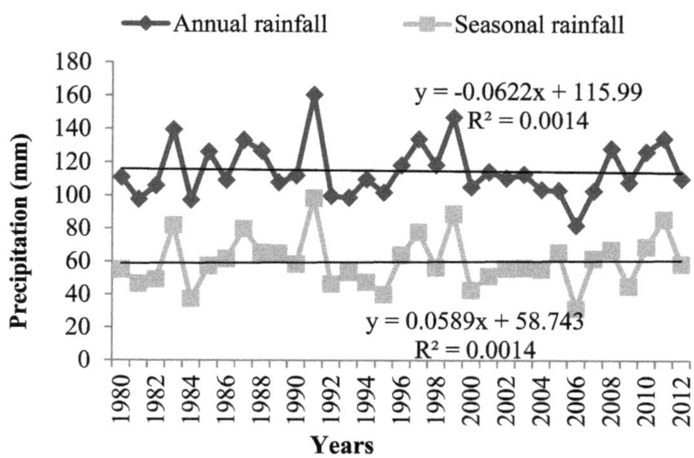

Figure 2. Annual and seasonal rainfall (mm) in the Lawra district of Ghana (The rainfall data was collected directly from the office of the meteorological agency: Ghana Meteorological Agency, 2014).

In addition, the results of discussions with the district agricultural officers confirmed the results of statistical analysis on the long-term trend in precipitation. Thus, the difference between farmers' perceptions and statistical results is due to the fact that farmers' responses are based solely on recall. The high illiteracy rate among farmers in Lawra district hinders their ability to keep formal records, and so accurately recalling long-term trends of rainfall could be difficult.

3.2. Farmers' Perceived Causes of Climatic Variability on Agriculture

Most farmers attributed climate change to human-related causes such as bush fires (51%) and deforestation (14%). While deforestation is largely perceived as being for the purposes of fuel wood, charcoal, and farm expansion, bush fires are believed to be caused by the 'negligence' of hunters and cigarette smokers. About 9.3% of respondents also claimed that traditional gods and ancestral spirits were responsible for the perceived changes in rainfall and temperature trends. During the FGDs, farmers indicated that the gods/ancestors were angry because many taboos have been broken by people (e.g., destroying sacred groves or woods, catching of sacred fish, *etc.*). Additionally, 23.3% of respondents claimed that climate change is caused by many factors, while 2.4% could not give any cause (Table 2).

Table 2. Farmers' perceptions about the causes of climate change in the Lawra district of Ghana (*Number of respondents* = 100).

Cause Variable	Percentage of Respondents
Deforestation	14.0
Bush fires	51.0
More than one cause	23.3
Gods/ancestral spirits	9.3
Do not know	2.4
Total	100

3.3. Farmers' Adaptation to Climate Change

The results revealed that farmers' adaptations are largely in response to dry spells (73%). However, 20% and 7% of respondents indicated that they used adaptation options in response to droughts and floods, respectively. Although an overwhelming majority of farmers recognized climate change, 33% of respondents still do not use any adaptation practices (Table 3).

Table 3. Proportion of farmers by adaptation classification and reasons for adaptation in the Lawra district of Ghana (*Number of respondents* = 100).

Variable	Percentage of Respondents
a. Adaptation classification	
Adapted	67
Not adapted	33
b. Reasons for adaptation	
Reduce effects of flood	7
Reduce effects of drought	20
Reduce effects of dry spell	73

3.4. Farmer-Perceived Importance of Adaptation Practices

The ranking of adaptation practices based on farmers' perceived importance is presented in Table 4. Among the seven adaptation practices, improved crop varieties and irrigation practice ranked first and second with a WAI of 2.15 and 2.09, respectively. The increasing incidence of drought and dry spells makes drought-tolerant crop varieties and irrigation preferable to farmers. On the other hand, income-generating activities and agroforestry practice were ranked the least important with a WAI of 0.77 and 0.74, respectively. Results of FGDs showed that farmers considered trading and agroforestry as capital-intensive activities. Crop diversification, farm diversification, and change of planting date were ranked as moderately important.

Table 4. Farmers' ranking of adaptation practices in the Lawra district of Ghana (Number of respondents = 100).

Adaptation Practice	Frequency by Each Level of Importance				WAI	Rank
	Highly Important	Moderately Important	Less Important	Not Important		
Improved crop varieties	35	48	14	3	2.15	1
Irrigation	30	51	17	8	2.09	2
Crop diversification	14	76	8	2	2.02	3
Farm diversification	7	67	23	3	1.78	4
Change of planting date	10	44	26	20	1.44	5
Income generating activities	3	20	28	49	0.77	6
Agroforestry practice	0	9	56	35	0.74	7

The results of actual adaptation measures being implemented by farmers are presented in Figure 3. The majority of farmers use crop diversification practices such as mixed cropping (41%) and crop rotation (10%). About 12% of the respondents use improved crop varieties (*i.e.*, drought-tolerant and early maturing varieties), while 23% adopted change of planting date. Other identified adaptation practices being implemented are off-farm jobs (6%), composting and mulching (3%), reduction in farm size (3%), and dry season gardening (2%).

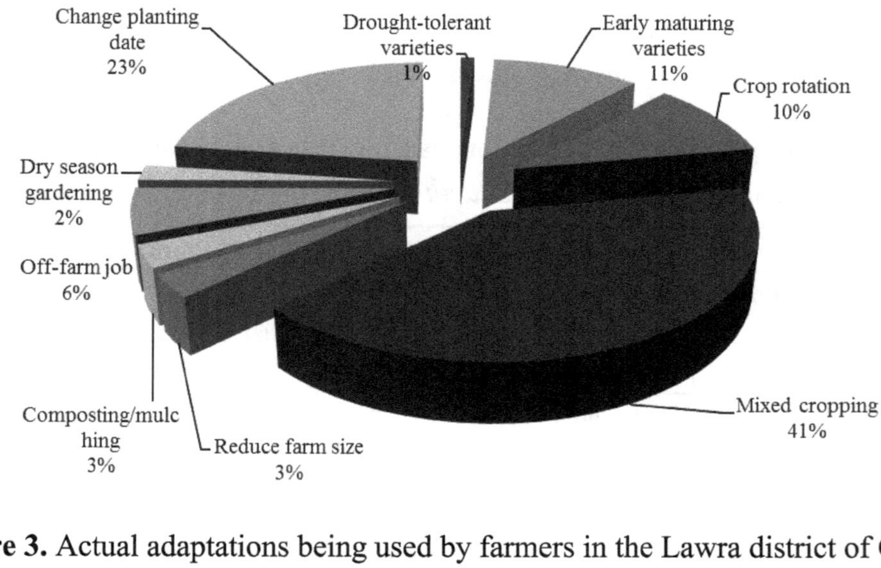

Figure 3. Actual adaptations being used by farmers in the Lawra district of Ghana.

3.5. Perceived Constraints to Adaptation to Climate Change

Results on barriers to use of adaptation practices are presented in Table 5. With a PCI value of 215, unpredictable weather was ranked the most critical impediment to use of adaptation options. High cost of farm inputs, lack of access to timely weather information, and lack of water resources were ranked the second, third, and fourth most pressing problems, respectively. The FGDs showed that farmers' main source of weather information was colleagues who visited peri-urban towns during market days. The focus group discussions also confirmed that farmers considered lack of credit facilities, agricultural subsidies, and poor soil fertility as moderate constraints, while lack of access to agricultural extension officers, agricultural markets, farm labor, and farm size are the lowest constraints.

Table 5. Problems affecting implementation of adaptation practices in the Lawra district of Ghana (Number of respondents = 100).

Constraints to Adoption	Degree of Constraint				PCI	Rank
	High	Moderate	Low	No Problem		
Unpredictable weather	35	48	14	3	215	1
High cost of farm inputs	14	76	8	2	202	2
Lack of access to timely weather information	7	67	23	3	178	3
Lack of access to water resources (e.g., dams)	10	44	26	20	144	4
Lack of access to credit facilities	2	32	11	55	81	5
Lack of access to agricultural subsidies	3	20	28	49	77	6
Poor soil fertility	0	9	56	35	74	7
Limited access to agricultural extension officers	3	19	7	71	54	8
Limited access to agricultural markets	0	0	24	76	24	9
Inadequate farm labor	0	1	19	80	21	10
Limited farm size	0	0	13	87	13	11

4. Discussion

Generally, farmers are aware of climate change, since more than 80% of the surveyed respondents have perceived long-term changes in temperature and precipitation trends. In Sub-Saharan Africa, similar findings have been reported in the Sekyedumase and Wa West districts of Ghana [8,13], Uganda [10], and Senegal [24]. Other studies have also shown that, in the last 100 years, there has been an average global temperature increase of 0.74 °C [25].

Climate change model predictions for the Guinea Savannah Zone of Ghana revealed that the increasing temperature and decreasing precipitation trends will continue [1]. This implies that agricultural stakeholders should identify relevant and applicable adaptation practices to mitigate the effects of the impending change in climatic variables.

The study showed that farmers' perceptions about the causes of climate change are mostly centered on human factors (*i.e.*, deforestation and bushfires) and gods/ancestral curses. Similar findings have been reported in the Wa West district of Ghana [13] and in Northern Nigeria [5].

This study showed that some farmers are already adjusting their farming activities in response to droughts, dry spells, and floods. The FGDs revealed that increased access to agricultural extension officers has impacted positively (67%) on farmers' implementation of adaptation options. Similar findings were reported in Bangladesh, where more than 75% of respondents were using adaptation practices [23]. However, a previous study conducted in the Sekyedumase district of Ghana showed that less than 44% of farmers use adaptation measures due to lack of funds [8].

This study also revealed that farmer-perceived important adaptation practices were different from the actual practices being implemented. Although farmers ranked improved crop varieties (e.g., drought-tolerant and early maturing crops) and irrigation as the most important adaptation strategies, only 14% actually implemented measures in these categories. The majority of respondents (51%) used crop diversification activities (*i.e.*, mixed cropping and crop rotation). Similar findings were reported in Northern Nigeria [5]. Feedback from the group discussions showed that most farmers did not have access to improved crop varieties; hence, they could not implement their most preferred measure. Results of the group discussion showed that farmers are generally aware of the annual recurrent dry spells and droughts. Also, although they view irrigation as the most important solution to these extreme climatic events, they failed to rank it as such. This is because, according to farmers, water resources such as dams and dugouts are very limited in the district. Field observation showed that most of the available water bodies for irrigation are broken down.

Also, in this study, unpredictable weather, high farm input cost, and limited access to timely weather information and water resources were identified as the most critical barriers to adoption. This is likely the case, because in Ghana, the main sources of weather information are television and radio broadcasts. The majority of farmers surveyed did not have electronic gadgets and hence could not readily access weather information. Also, the FGDs revealed that farmers in the Lawra district operate under limited resources due to limited agricultural credit and subsidies. Field observations revealed that the limited number of irrigation facilities (*i.e.*, dams and dugouts) were either broken down or dried out. Similar barriers to adoption have been reported in South Africa [12] and Nigeria [26].

5. Conclusions

With properly tailored policies, smallholder farmers can adjust to climate change and improve their crop production. To do this, climate change policies need to factor in farmers' understanding of the risks they face and potential adaptations to climate change. In this regard, interventions of the Ghanaian government should focus on the development of improved crop varieties and irrigation facilities. More specifically, the development of drought-tolerant crop varieties and the construction of dams and dugouts need to be prioritized in the list of climate change adaptation practices in the Lawra district. Also, there is a need for stakeholders to adhere to proper management and maintenance of existing irrigation facilities.

The perception that climate change is also caused by traditional gods and ancestral curses implies that scientists and development experts should consider the cultural and traditional beliefs of farmers when designing adaptation practices. As such, a bottom-up approach must be used to ensure that farmers' beliefs and understanding are a crucial part of the design and dissemination of adaptation practices.

Farmers' access to timely weather information also needs to be prioritized to help farmers in their production decision-making processes (e.g., selection of adaptation options). The Ghana Meteorological Agency and agricultural staff need to be properly trained and resourced to collect, collate, and disseminate accurate weather information timely and widely. Also, the government should boost the capacity of scientists and agricultural staff to develop and promote appropriate and effective technologies to help farmers adapt to climate change. In addition, the prevailing high cost of farm inputs and lack of credit facilities and subsidies require the government to ensure that agricultural loans with flexible terms are made available to farmers to boost their capacity to adapt to the changing climate.

Finally, further research is recommended to assess the feasibility of farm-level adaptation practices to climate change. This will help governments, researchers, non-governmental organizations (NGOs), and farmers to develop and implement adaptation measures that are sustainable, resilient, and reliable.

Acknowledgments

This study was funded by the Kochi University of Technology, Kochi, Japan. The authors gratefully acknowledge the valuable support of the farmers, agricultural officers, and local government staff of the Lawra district of Ghana.

Author Contributions

Francis Ndamani conceived the idea of the study, carried out data collection and data analysis. Tsunemi Watanabe supervised data collection and data analysis. Both authors drafted the manuscript, read and approved the final manuscript.

References

1. USAID. Ghana Climate Change Vulnerability and Adaptation Assessment. Available online: http://www.encapafrica.org/documents/biofor/Climate%20Change%20Assessment_Ghana_%20FINAL.pdf (accessed on 23 April 2015).
2. Jung, G.; Kunstmann, H. High-resolution regional climate modelling for the Volta Basin of West Africa. *J. Geophys. Res.* **2007**, *112*, doi:10.1029/2006JD007951.
3. Laux, P.; Kunstmann, H.; Bárdossy, A. Predicting the regional onset of the rainy season in West Africa. *Int. J. Climatol.* **2008**, *28*, 329–342.
4. Laube, W.; Schraven, B.; Awo, M. Smallholder adaptation to climate change: Dynamics and limits in Northern Ghana. *Clim. Chang.* **2009**, *111*, 753–774.
5. Farauta, B.K.; Egbule, C.L.; Idrisa, Y.L.; Agu, V.C. Farmers' Perceptions of Climate Change and Adaptation Strategies in Northern Nigeria: An Empirical Assessment. Available online: http://www.atpsnet.org/Files/rps15.pdf (accessed on 15 May 2015).
6. Adger, W.N.; Huq, S.; Brown, K.; Conway, D.; Hulme, M. Adaptation to Climate Change in the Developing World. Available online: http://www.glerl.noaa.gov/seagrant/ClimateChangeWhiteboard/Resources/Uncertainty/climatech/adger03PR.pdf (accessed on 9 May 2015).
7. Intergovernmental Panel on Climate Change (IPCC). *Climate Change 2001: Impacts, Adaptation, and Vulnerability. Intergovernmental Panel on Climate Change*; Cambridge University Press: Cambridge, UK, 2001.
8. Fosu-Mensah, B.Y.; Vlek, P.L.G.; MacCarthy, D.S. Farmers' perceptions and adaptation to climate change: A case study of Sekyeredumase district in Ghana. *Environ. Dev. Sustain.* **2012**, *14*, 495–505.

9. Nhamo, N.; Daniel, M.; Fritz, O.T. Adaptation strategies to climate extremes among smallholder farmers: A case of cropping practices in the Volta Region of Ghana. *Br. J. Appl. Sci. Technol.* **2014**, *4*, 198–213.
10. Okonya, J.S.; Syndikus, K.; Kroschel, J. Farmers' perceptions of and copping strategies to climate change: Evidence from six agro-ecological zones of Uganda. *J. Agric. Sci.* **2013**, *5*, doi:10.5539/jas.v5n8p252.
11. Apata, T.G. Factors Influencing the Perception and Choice of Adaptation Measures to Climate Change among Farmers in Nigeria. Evidence From farm Households in Southwest Nigeria. Available online: http://businessperspectives.org/journals_free/ee/2011/ee_2011_04_Apata.pdf (accessed on 9 May 2015).
12. Nhemachena, C.; Hassan, R. Micro-Level Analysis of Farmers' Adaptation to Climate Change in Southern Africa. Available online: http://cdm15738.contentdm.oclc.org/utils/getfile/collection/p15738coll2/id/39726/filename/39727.pdf (accessed on 23 May 2015).
13. Kusakari, Y.; Asubonteng, K.O.; Jasaw, G.S.; Dayour, F.; Dzivenu, T.; Lolig, V.; Donkoh, S.A.; Obeng, F.K.; Gandaa, B.; Kranjac-Berisavljevic, G. Farmer-perceived effects of climate change on livelihoods in Wa West District, Upper West Region of Ghana. *J. Disaster Res.* **2014**, *9*, 516–528.
14. Lindsay, P.H.; Norman, D.A. Human Information Processing: An Introduction to Psychology. Available online: http://www.phon.ucl.ac.uk/courses/spsci/audper/SDT%20Lindsay%20&%20Norman%20App%20B.pdf (accessed on 4 August 2015).
15. Harig, T.; Kaiser, F.G.; Bowler, P.A. Psychological restoration in nature as a positive motivation for ecological behavior. *Environ. Behav.* **2001**, *33*, 590–607.
16. Ministry of Environment, Science and Technology (MEST) of Ghana. Ghana Goes for Green Growth: National Engagement on Climate Change. Available online: http://prod-http-80-800498448.us-east-1.elb.amazonaws.com/w/images/2/29/GhanaGreen.pdf (accessed on 4 August 2015).
17. Luni, P.; Maharjan, K.L.; Joshi, N.P. Perceptions and realities of climate change among the Chepong Communities in rural mid-hills of Nepal. *J. Contemp. India Stud. Space Soc.* **2012**, *2*, 35–50.
18. Di Falco, S.; Yesuf, M.; Kohlin, G.; Ringler, C. Estimating the impact of climate change on agriculture in low-income countries: Household level evidence from the Nile Basin, Ethiopia. *Environ. Resour. Econ.* **2012**, *52*, 457–478.
19. Jackson, L.E.; Pascual, U.; Hodgkin, T. Utilizing and conserving agrobiodiversity in agricultural landscapes. *Agric. Ecosyst. Environ.* **2007**, *121*, 196–210.
20. Ndamani, F.; Watanabe, T. Influences of rainfall on crop production and suggestions for adaptation. *Int. J. Agric. Sci.* **2015**, *5*, 367–374.
21. Ghana Statistical Service (GSS). Population by Region, District, Locality of Residence, Age Groups and Sex, 2010. Available online: http://www.statsghana.gov.gh/docfiles/population_by_region_district_locality_of_residence_age_groups_and_sex,_2010.pdf (accessed on 15 May 2015).
22. Uddin, M.N.; Bokelmann, W.; Entsminger, J.S. Factors affecting farmers' adaptation strategies to environmental degradation and climate change effects: A farm level study in Bangladesh. *Climate* **2014**, *2*, 223–241.
23. Devkota, R.P.; Cockfield, G.; Maraseni, T.N. Perceived community-based flood adaptation strategies under climate change in Nepal. *Int. J. Glob. Warm.* **2014**, *6*, 113–124.
24. Mertz, O.; Mbow, C.; Reenberg, A.; Diouf, A. Farmers' perceptions of climate change and agricultural adaptation strategies in rural Sahel. *Environ. Manag.* **2009**, *43*, 804–816.

25. Intergovernmental Panel on Climate Change (IPCC). *Contribution of Working Group II to the Fourth Assessment Report of the Intergovernmental Panel on Climate Change, 2007*; Parry, M.L., Canziani, O.F., Palutik, J.P., Linden, P.J. Hanson, C.E., Eds.; Cambridge University Press: Cambridge, UK, 2007.
26. Kandlinkar, M.; Risbey, J. Agricultural impacts of climate change: If adaptation is the answer, what is the question? *Clim. Chang.* **2000**, *45*, 529–539.

5

Assessment of Short Term Rainfall and Stream Flows in South Australia

Mohammad Kamruzzaman [1,*], Md Sumon Shahriar [2] and Simon Beecham [1]

[1] Centre for Water Management and Reuse (CWMR), School of Natural and Built Environments, University of South Australia, Mawson Lakes, SA 5095, Australia; E-Mail: simon.beecham@unisa.edu.au

[2] CSIRO Computational Informatics (CCI), Hobart, TAS 7001, Australia; E-Mail: mdsumon.shahriar@csiro.au

* Author to whom correspondence should be addressed; E-Mail: Mohammad.kamruzzaman@unisa.edu.au.

External Editor: Athanasios Loukas

Abstract: The aim of this study is to assess the relationship between rainfall and stream flow at Broughton River in Mooroola, Torrance River in Mount Pleasant, and Wakefield River near Rhyine, in South Australia, from 1990 to 2010. Initially, we present a short term relationship between rainfall and stream flow, in terms of correlations, lagged correlations, and estimated variability between wavelet coefficients at each level. A deterministic regression based response model is used to detect linear, quadratic and polynomial trends, while allowing for seasonality effects. Antecedent rainfall data were considered to predict stream flow. The best fitting model was selected based on maximum adjusted R^2 values (R^2_{adj}), minimum sigma square (σ^2), and a minimum Akaike Information Criterion (AIC). The best performance in the response model is lag rainfall, which indicates at least one day and up to 7 days (past) difference in rainfall, including offset cross products of lag rainfall. With the inclusion of antecedent stream flow as an input with one day time lag, the result shows a significant improvement of the R^2_{adj} values from 0.18, 0.26 and 0.14 to 0.35, 0.42 and 0.21 at Broughton River, Torrance River and Wakefield River, respectively. A benchmark comparison was made with an Artificial Neural Network analysis. The optimization strategy involved adopting a minimum mean absolute error (MAE).

Keywords: Artificial Neural Network; lag response model; rainfall and stream flow; wavelet coefficient; Akaike Information Criterion

1. Introduction

A review of rainfall-runoff modeling has been given by [1]. Rainfall and stream flow models can be applied to a diverse range of purposes including daily control of reservoirs, projecting future stream flows and flood management. Rainfall and stream-flow models can be classified as physically based, conceptual and empirical. Physically-based models include the Système Hydrologique Européan with sediment and solute transport [2] and Gridded Surface Subsurface Hydrologic Analysis [3] both of which require extensive spatial and temporal data and typically are used for small catchments. An example of a conceptual based model is the Modèle du Génie Rural à 4 paramètres Journalier (GR4J), which has been developed for understanding catchment hydrological behavior [4]. Other examples of conceptual rainfall-runoff models are the Sacramento Soil Moisture Accounting Model [5] and the SIMulation and HYDrologic model (SIMHYD) [6], which can be applied either as a lumped or gridded application. SIMHYD estimates daily stream flows from daily rainfall and areal potential evapotranspiration data. The class of empirical models includes time series models [7–13]. An advantage of an empirical model is that it can be fitted to situations where the hydrological data are restricted to rainfall and stream flow time series. A further advantage is that in a parametric test, a distribution can be fitted for assessing the hydrological behavior for any time period in any region. In addition, they can represent either linear or non-linear relationships. Time series models perform as well as physically-based alternatives [14]. Combined a conceptual model with an artificial neural network (ANN) for forecasting inflow into the Daecheong Dam in Korea [15]. Compared the wavelet decompositions of rainfall and runoff at four sites in the Tianshan Mountains [16]. They aimed to distinguish between errors in timing and errors in magnitude of hydrograph peaks. They used a cross-wavelet technique to quantify timing errors and hence provided an empirical adjustment to model predictions of stream flow.

In this study, we have proposed a novel method for assessing short-term rainfall and stream flow models. The travel time between rainfall and stream flow gauges using cross-correlation functions [10,17]. They reported that the travel time was less than one day for the Onkaparinga catchment in South Australia. In this paper, we presume that there is a higher order relationship between rainfall gauge and stream flow data. It is, therefore, important in this study to construct the correlation structure. Linear regression models are commonly used for time series analysis [18], particularly for assessing evidence of trends, higher order changes and variability, including allowing for seasonality. We developed deseasonalized and detrended time series rainfall and stream flow models from deterministic regression models including linear, quadratic and cubic terms. These models take account of both lag rainfall and the influence of stream flow. The results of this study will be useful for water managers and policy makers involved in sustainable water resource management and climate change adaptation for the catchments used in this study. The approach is capable of modeling the non-linear relationships between inputs and outputs using ANNs [19]. The first advantage of ANN is that it only requires a small number of parameters and learns through a number of training iterations involving adjusting the parameters (weights) of the network [20]. A second advantage is that it is useful in situations where it is complex to build a physical or conceptual model, such as hydrological modeling of rainfall- stream flow processes [21–25]. ANN models were useful to find the relationships between rainfall and river flow data in a river basin in India [26]. We present a statistical approach that uses the deterministic features of a regression model to build many neural networks with a combination of different lagged input patterns. A wavelet based regression model for stream flow using the discrete wavelet transform (DWT) of the entire time series [27]. They also provided a comparison of their model performance with ANN. A chaotic stream flow model using an ensemble wavelet network [28]. Used wavelet analyses of rainfall and runoff and wavelet rainfall–runoff cross-analyses to investigate the temporal variability of the rainfall-runoff relationship [17,29]. They found that wavelet transforms provide a physical explanation of the temporal structure of the catchment response.

2. Data Collection and Preparation

The analysis is based on data from three rainfall and stream flow stations in South Australia, as presented in Figure 1. The Broughton River (BR) station is at Mooroola, which is located approximately 40 km north of Port Broughton and 20 km south west of Port Pirie. Torrance River (TR) station is located at Mount Pleasant, and its rivers and tributaries are highly variable in flow and together drain an area of 508 km^2. Wakefield River (WR) is an ephemeral river near Rhynie, with a catchment area of approximately 1913 km^2.

Figure 1. Location of Broughton River (BR), Torrance River (TR), and Wakefield River (WR).

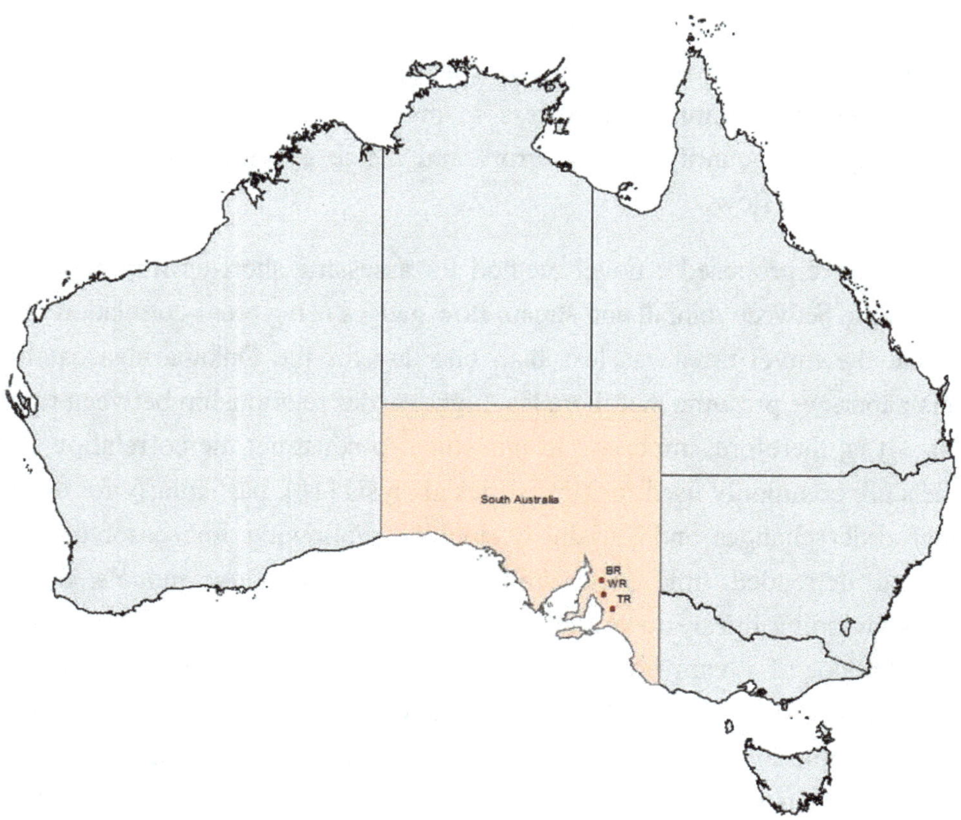

The elevation of the river may indicate the hydrological feature, presented in Table 1, Column 4. These stations were selected because they had long records of rainfall and stream flow and the highest quality control in terms of Australian Bureau of Meteorology, [30] and the Department for Environment, Water and Natural Resources [31] quality designations for rainfall and stream flow records. Information on these stations and data quality are presented in Table 1.

Table 1. Weather stations information, data quality and observations.

Stations name	ID	Location		Elevation	Variables	Data period		% of
		Latitude	Longitude			Start	End	Missing
Broughton River at Mooroola	A5070503	−33.53	138.51	196 m	Rainfall	Jun. 1989	Dec. 2011	0.1
					Stream flow	Jun. 1972	Dec. 2011	0.7
Torrance at Mount Pleasant	A5040512	−34.78	139.02	414.7 m	Rainfall	Jun. 1989	Dec. 2011	0.6
					Stream flow	May 1973	Dec. 2011	0.1
Wakefield river near Rhyine	A5060500	−34.13	138.63	202 m	Rainfall	Sep. 1985	Dec. 2011	0.9
					Stream flow	Jun. 1971	Dec. 2011	0.2

In this paper, there was less than 1% missing data and these were replaced by the mean of the series of rainfall and stream flow, to give an unbroken time series for analysis. Methods for replacing periods of missing values are discussed [18,32]. In this paper, we propose a dyadic signal time period (*i.e.*, 2^n where n is an integer and $n \geq 0$, for assessing the relationship between daily rainfall and stream flow during the period 1990–2012. We observe the discrete sequence of time series $\{y_t\}$ where $\{y_t\}$ is an integer ranging in length. We extract multi-level information of observed rainfall and stream flow series in three catchments in South Australia using the Haar wavelet decomposition. We split $\{y_t\}$ into 10 sub-time series of length power two *i.e.*, 2^n, where n is the level of the time series, starting from 0. We also investigate the correlation between rainfall and stream flow patterns for each sub-series from levels 0 to 8.

3. Statistical Analysis

3.1. Assessing the Relationship between Rainfall and Stream Flow

The open source software R [33] was used for the analyses in this paper. We calculate 10 subseries of rainfall and stream flow from 1990 to 2012 using the "wavethresh" R routine packages [34,35] for assessing the relationship between rainfall and stream flow. The length of time taken into account in 10 subseries for rainfall and stream flow is a period of 512 days.

The relationship between rainfall and stream flow within 10 subseries is presented in Figure 2. The maximum correlation coefficients are 0.08, 0.23 and 0.31 at Broughton River, Torrance River and Wakefield River, respectively. These values are between −1 and +1 in all cases, indicating the degree of linear dependence between rainfall and stream flow. For assessing short term spatial variability, a correlation coefficient of the sub-series of rainfall and stream flow less than 0.4 indicates a significant difference from 0 at each station. For example, in sub-series 2, the correlation coefficient was 0.04, 0.15 and 0.28 which indicates the independence of rainfall and stream flow at Broughton River, Torrance River and Wakefield River, respectively.

Figure 2. Correlation pattern subseries of rainfall and stream flow time series.

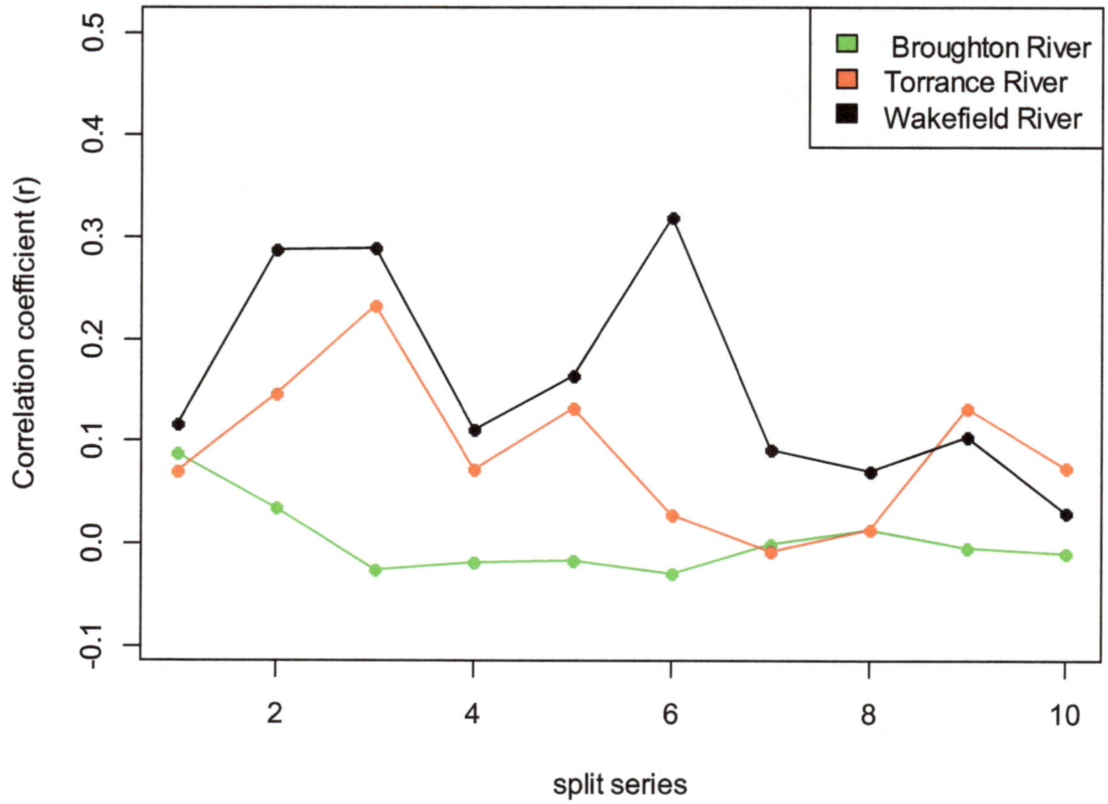

In order to understand stream flow availability under the climatic conditions in South Australia, we investigated the characteristics of rainfall and stream flow patterns, as categorized by climatic phenomena. A statistical measure of the dispersion of rainfall and stream flow patterns around the mean is defined as follows:

$$CV = \frac{S_x}{\mu_x} \tag{1}$$

where CV is defined as the coefficient of variation and is represented by the ratio of the standard deviation (S_x) to the mean (μ_x). Table 2 shows the degree of variation in rainfall and stream flow patterns.

Table 2. Rainfall and stream flow variability at Broughton River, Torrance River and Wakefield River in South Australia (SA) from 1990 to 2011.

Statistics	Broughton River		Torrance River		Wakefield River	
	Rainfall	Stream flow	Rainfall	Stream flow	Rainfall	Stream flow
Mean	1.653	9.817	1.530	5.396	1.282	25.333
Estimated standard deviation	0.385	4.075	0.296	4.201	0.223	21.300
Coefficient of variation (CV)	23.31%	41.51%	19.36%	77.85%	17.40%	84.07%

In Table 2, the CV for stream flow patterns indicates higher variability than for the rainfall series.

Figure 3 shows the variability of the wavelet coefficients from levels 0 to 8. The evidence of association between the rainfall and stream flow coefficient is strongly correlated at the 5% significance level in Table 1.

Figure 3. Standard deviations of wavelet coefficients of rainfall and stream flow from level 0 to 8. (**a**) Rainfall; (**b**) Stream flow.

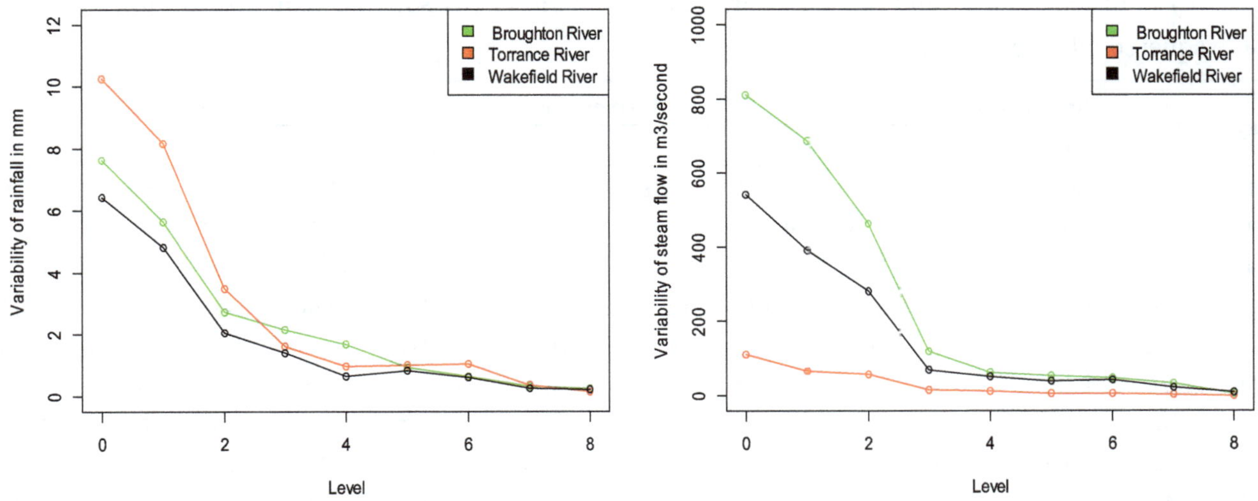

(**a**) (**b**)

3.2. Correlation Structures between Rainfall and Stream Flow

In the previous sections, we calculated wavelet coefficients for each subset of the rainfall and stream flow series. In order to filter each of those series, we applied Haar wavelets.

The constructed correlation pattern for each rainfall and stream flow sub-series for levels 0 to 8 is given by:

$$r_k = \sum_{k=0}^{8} \frac{\sum_{j=1}^{10} (X_{wd_j k} - \overline{X}_{wd,k})(Y_{wd_j k} - \overline{Y}_{wd,k})}{S_{xwd} S_{ywd}} \tag{2}$$

and

$$\overline{X}_{wd} = \sum_{j=1}^{10} X_{wd_j} / 10,\ \overline{Y}_{wd} = \sum_{j=1}^{10} Y_{wd_j} / 10,\ (j = 1, 2, 3, \ldots 10, \text{ and } k = 0, 1, 2, \ldots 8) \tag{3}$$

where r_k is the constructed correlation with level n from 0 to 8 and X_{wd_j}, Y_{wd_j} is the *j*th sub-series of the rainfall and stream flow wavelet decomposed with the Haar procedure. The results are presented in Table 3.

The evidence of significant correlation ($r \geq 0.50$) between rainfall and stream flow wavelet coefficient series with at least a 5% significance level is shown in Table 3. Furthermore, to avoid co-linearity problems, the squared rainfall and stream flow wavelet coefficient series are also included. We found that a correlation structure ($r = 0.56$) such as stream flow is determined by rainfall on at least 4 days with 5% level at the Broughton River Basin and Torrance River Basin, as shown in Table 3. The adjusted squared stream flow and rainfall has a little evidence of correlations (*i.e.*, at 5% level) up to 64 days at Torrance River at, also a marginal correlation ($r = 0.51$) up to 128 days within squared adjusted rainfall and adjusted stream flow at Wakefield River. The rainfall and stream flow relationship was used to develop a response model for predicting stream flow.

Table 3. Constructed correlation pattern for different levels between (a) adjusted rainfall and adjusted stream flow; (b) squared adjusted rainfall and adjusted stream flow; (c) adjusted rainfall and squared adjusted stream flow; (d) squared adjusted rainfall and squared adjusted stream flow.

Days	Broughton River				Torrance River				Wakefield River			
	a	b	c	d	a	b	c	d	a	b	c	d
1	0.71 **	0.89 ***	0.70 **	0.86 ***	0.76 **	−0.184	−0.513	0.447	0.76 **	−0.53 *	−0.59 *	0.86 ***
2	0.65 *	0.264	0.469	0.449	0.72 **	0.57	0.158	0.265	0.384	0.413	0.201	0.038
4	0.56 *	−0.189	0.257	0.146	0.63 *	0.061	−0.27	0.125	0.324	−0.28	−0.341	0.115
8	0.087	0.369	−0.296	0.103	0.032	0.55 *	0.173	−0.51 *	−0.369	0.483	0.191	−0.418
16	0.233	0.08	−0.009	−0.326	0.275	−0.15	0.009	−0.311	0.306	−0.652	0.081	−0.393
32	0.094	0.055	−0.059	−0.248	0.68 *	−0.84 **	−0.81 ***	0.97 ***	−0.116	0.002	−0.382	−0.121
64	0.411	0.036	0.292	0.238	0.488	0.67 *	0.456	0.71 **	−0.091	0.166	−0.005	−0.301
128	0.423	−0.409	0.604 *	0.68 *	0.299	−0.162	0.186	−0.223	0.279	−0.575	0.128	−0.51 *
512	0.456	0.354	−0.292	0.343	−0.218	−0.405	0.007	−0.405	0.094	0.117	0.098	−0.163

Notes: * Coefficients are statistically significant at 5%; ** Coefficients are statistically significant at 1%; *** Coefficients are statistically significant at 0.1%

3.3. Rainfall-Stream Flow Response Modeling

The constructed correlations described in the previous section may be partly due to common seasonal variations and trends, so a first step is to estimate these deterministic features with regression models for entire period from 1990 to 2010. The residuals from these regressions are reformed to the deseasonalized and detrended (dsdt) time series. For all three stations, a cubic trend gave a statistically improved fit over a linear or quadratic trend over the study period. The seasonal variation was reasonably modelled by a sinusoidal curve. Therefore, the regression models are of the form:

$$T_i = \beta_0 + \beta_1 \times time + \beta_2 \times time^2 + \beta_3 \times time^3 + \beta_4 \times C + \beta_5 \times S + \varepsilon_t \tag{4}$$

where, T_i represents either rainfall or stream flow; time is the mean adjusted time, that is $(t-\bar{t})$ where t is the number of days from the start of the record and $\bar{t}$ is the mean of t, $time^2$ and $time^3$, which allows for possible quadratic and cubic trends; C is cos(2πt/365.25) and S is sin(2πt/365.25) and together these allow for seasonal variation of period one cycle per year; β_j are the unknown coefficients to be estimated; and ε_t are random variations with mean 0 and constant standard derivation.

For the estimated coefficients, only a few values are significantly different from 0 even at the 5% significance level, as shown in Table 4. There is evidence of significantly different trends in rainfall at Wakefield River, which may have corresponded to increased stream flows if rainfall is increased. We have predicted the stream flow (Y_t) on day t from rainfall (X_t) with corresponding lags k. This is referred to as a Response Model (RM). The regression is defined as:

$$Y_t = \beta_0 + \beta_1 \times X_1 + \beta_2 \times X_2 + \ldots\ldots\ldots\ldots \beta_{128} \times X_{128} + \varepsilon_t \tag{5}$$

We assess stream flow in response to rainfall at lags 0 to 128. The best fitted model is selected based on the adjusted coefficient of determination; (R^2_{adj}); minimum sigma squared (σ^2) and the Akaike Criterion Information (AIC); The AIC is defined as:

$$\text{AIC} = 2 \times \text{number of parameters} - 2\ \text{Log(L)} \quad (6)$$

where L is the maximized value of the likelihood function for the estimated model. Comparisons of the AIC for different model is as shown in Table 5. The R^2_{adj} value significantly reduces and the estimated stream flow influence is close to zero after the exogenous rainfall at lag 7. Therefore, we reduced the exogenous rainfall at lags from 128 to 7 in the response model; referred to as RM0 in Table 5. This strategy is sub-optimal inasmuch as rejected terms might meet the retention criterion if added back individually. However; any small improvement in R^2_{adj} would be balanced by increased complexity in the model; which is undesirable if interaction and squared terms are added. The regression model is defined as RM:

$$Y_t = \beta_0 + \beta_1 \times X_1 + \beta_2 \times X_2 + \beta_3 \times X_3 + \beta_4 \times X_4 + \beta_5 \times X_5 + \beta_6 \times X_6 + \beta_7 \times X_7 \quad (7)$$

In the second model, we add deterministic features to the regression model including linear, quadratic and cubic terms of t, allowing for seasonality effects. This model is defined as RM_D:

$$Y_t = \beta_0 + L + \sum_{l=1}^{7} \beta_{5+l} \times X_l \quad (8)$$

where $L = \beta_1 \times time + \beta_2 \times time^2 + \beta_3 \times time^3 + \beta_4 \times C + \beta_5 \times S_t$.

The third model is defined as RMD_AR[1] and is an autoregressive model of order 1 (AR[1]) with RM_D. It can be written in the form:

$$Y_t = \beta_0 + L + \sum_{l=1}^{7} \beta_{5+l} \times X_l + \beta_{13} \times Y_{t-1} \quad (9)$$

The fourth model is defined as RMD_AR[2], and is an autoregressive model of order 2 (AR[2]) with RMD_AR[1]. It can be written in the form:

$$Y_t = \beta_0 + L + \sum_{l=1}^{7} \beta_{5+l} \times X_l + \beta_{13} \times Y_{t-1} + \beta_{14} \times Y_{t-2} \quad (10)$$

Table 4. Estimated coefficients of rainfall and stream flow variability from 1990 to 2012.

Station	Statistical Summary	Intercept (β_0)	Linear Term t	Quadratic Term t	Cubic Term t
Broughton River	Estimated rainfall	1.58	−0.000042	−0.000000001	−0.000000000003
	Variability of rainfall	0.106	0.00008	0.000000017	0.000000000008
	Estimated stream flow	52.08	−0.01244 *	0.0000031 *	−0.000000000258
	Variability of stream flow	6.18	0.004661	0.0000009	0.000000000485
Torrance River	Estimated rainfall	1.424	−0.00007	0.000000019	0.000000000004
	Variability of rainfall	0.077	0.00006	0.000000012	0.000000000006
	Estimated stream flow	3.47	−0.00174 *	0.0000003 *	0.000000000149 *
	Variability of stream flow	0.607	0.0004573	0.000000092	0.000000000048
Wakefield River	Estimated rainfall	1.226	−0.000123 *	0.0000000037	0.00000000001 *
	Variability of rainfall	0.067	0.000051	0.00000001	0.000000000005
	Estimated stream flow	15.95	−0.01144 *	0.0000007	0.0000000008 *
	Variability of stream flow	3.576	0.002694	0.0000005	0.000000000280

Note: * statistical significance at 5%.

Table 5. Fitted regression model for Broughton River, Torrance River and Wakefield River.

Model	Broughton River				Torrance River				Wakefield River			
	R^2_{adj}	Std. Error	AIC	RMSE *	R^2_{adj}	Std. Error	AIC	RMSE *	R^2_{adj}	Std. Error	AIC	RMSE *
RMO	0.16	333.6	1104.9	3.5107	0.24	31.19	742.77	5.074	0.13	195.8	1023.53	1.1777
RM_D	0.18	331.3	1103.9	3.1507	0.26	31.02	741.9	5.012	0.14	195.3	1023.1	1.1777
RMD_AR[1]	0.35	292.9	1085.1	0.0353	0.42	27.35	722.7	0.052	0.21	187.4	1016.9	0.11777
RMD_AR[2]	0.36	291.7	1084.5	0.0313	0.43	27.35	722.5	0.0452	0.22	187.4	1016.8	0.10777
RMD_tau	0.39	285.8	1081.4	0.0035	0.42	27.32	722.1	0.0411	0.23	187.3	1016.1	0.10178

Note: Asterisk (*) units are in m^3s^{-1}.

Finally, we develop a model for a benchmark comparison of stream flow on day t based on the entire previous period of stream flow and their influence (τ) adding with model RM_D. This model is defined as RMD_tau. Tau (τ) is 0 if there is no stream flow influence from the previous day's rainfall. We have demonstrated an example of count stream flow influence in Table 6.

Table 6. An example of count tau and stream flow influence rainfall over time.

Stream flow	**y1**	**y2**	**y3**	**y4**	**y5**	**y6**	**y7**	**y8**	**y9**	**y10**	**y11**	**y12**	**y13**
	8	9	0	0	0	2	9	22	3	5	8	8	6
Rainfall	**x1**	**x2**	**x3**	**x4**	**x5**	**x6**	**x7**	**x8**	**x9**	**x10**	**x11**	**x12**	**x13**
	3	2	5	3.2	3	2.8	2.6	2.4	2.2	2	1.8	1.6	1.4

In the Table 6, when the day t = 6, Y6 = 2, then we count tau = 3 (number of 0), and Y6-3-1 = 9, can be applied in the referred model RMD_tau.

The model RMD_tau can be written in the form:

$$Y_t = \beta_0 + L + \sum_{l=1}^{7} \beta_{5+l} \times X_l + \beta_{13} \times \tau + \beta_{14} \times Y_{t-\tau-1} \tag{11}$$

The fitted model for predicted stream flow in response to exogenous rainfall, deterministic features of the regression model, and previous stream flow influence, is presented in Table 5. The best fitting model selection was based on minimum AIC and minimum root mean square Error (RMSE). The RMSE is defined as:

$$RMSE = \sqrt{E(\hat{Y}_t - Y_t)^2} \tag{12}$$

where, $\hat{Y}_t$ is defined as the estimated stream flow and Y_t is the observed stream flow, respectively.

The response model RM0 has 128 predictor variables namely the rainfall lags at 0 to 128. Therefore, there are 129 parameters to estimate including the intercept. The estimated rainfall effects belong to 0 up to 7 days lag, therefore we reduced the rainfall lags from 128 to 7 days and the optimized R^2_{adj} values for this model are 0.16, 0.24 and 0.13 for Broughton River, Torrance River and Wakefield River, respectively, as presented in Table 5. We also offset the cross product term of lags to further reduce the complexity of this model. The second model included linear quadratic and cubic terms, and this model is denoted as RM_D. The number of parameters to be estimated is therefore 8 + 3 = 11 and the R^2_{adj} increased to 0.18, 0.26 and 0.14 for Broughton River, Torrance River and Wakefield River, respectively, which is a practical and statistically significant improvement. We then added a first order autoregressive term, referred to as a RMD_AR[1] model, and a second order autoregressive term referred to as a RMD_AR[2] model. We also made a benchmark comparison by using the entire stream flow record and this model is denoted RMD_tau, as presented in Table 5.

In Table 5, there is evidence of improvement of R^2_{adj} values, RMSE in m^3s^{-1} from RM to RM_D. Adding autoregressive order 1 (AR[1]) with RM_D results in substantially improved R^2_{adj} values (from 0.18, 0.26, and 0.14 to 0.35, 0.42 and 0.21 for Broughton River, Torrance River and Wakefield River, respectively. Furthermore, when adding autoregressive order 1 (AR[1]) with RM_D, there is evidence of improvement but this may be offset by the increasing number of parameters that affect the complexity of the model. In addition, the RMD_tau model represents a small improvement for two of the three river basins. The best fitted models are RMD_tau for Broughton River, RMD_AR[2] for Torrance River and RMD_tau for Wakefield River, were selected based on the minimum Akaike Information Criterion (AIC) and minimum root mean square error (RMSE) in m^3s^{-1}. The residuals from the best fitted models were transformed to normalized form by factor multiplication. A factor was calculated, which allows for the fact that the mean of a non-linear function of a random variable is not equal to that function of the mean. The transform series follow an identically normalized form with mean (μ) of zero, standard deviation (σ^2) of 1 and a random disturbance term (ε_t) which is uncorrelated. The transformed series were used to predict the stream flow on day t based on the predicted stream flow influence over the short term, as shown in Figure 4.

Figure 4. Predicted stream flow based on dsdt rainfall for (**a**) Broughton River; (**b**) Torrance River; and (**c**) Wakefield River from 1990 to 2010.

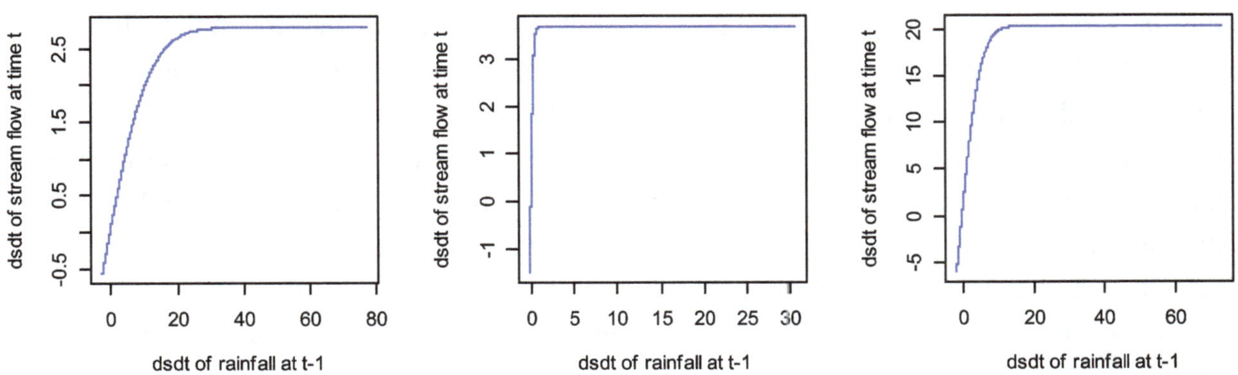

(**a**) (**b**) (**c**)

In Figure 4, we demonstrate the versatility of stream flow prediction. It can be seen that this is a non-linear relationship when expressed in terms of the physical interpretation of stream flow based on rainfall.

3.4. Modeling Stream Flow Using an Artificial Neural Network

Artificial neural network (ANN) techniques are motivated by the principles of biological nervous systems [36]. Although there are different types of ANN, the multilayer feed forward network is the most commonly used technique. For example, a common approaches of training using back-propagation in a multi-layer feed forward network [23]. The network consists of input, hidden and output layers. Each layer is fully connected with the proceeding layer with weights in each connection, as shown in Figure 5.

Figure 5. A schematic ANN including input, hidden and output layers.

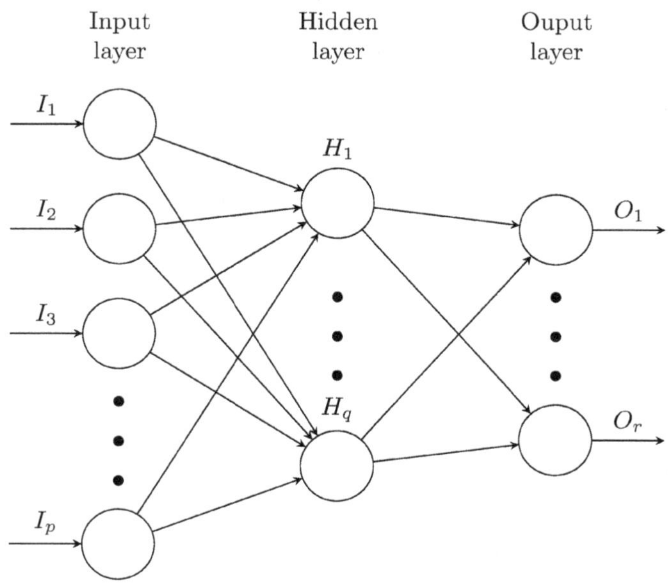

In Figure 5, the number of nodes in the input layer is p, the number of nodes in the hidden layer is q and the number of nodes in the output layer is r. The initial assigned random weights are updated during the training process by comparing the predicted output and the known output for errors. Errors are then back-propagated to adjust the weights. The dsdt of daily rainfall and stream flow data from the regression model developed in the previous section are considered for developing a prediction model for each of the three river basins for the years 1990 to 2010. A certain methods proposed such as input selection, model architecture selection, model calibration (training) and validation (testing) [37]. In addition, we emphasize the fact that ANN set-up has to be carefully achieved and described to get the reliable results. This study described the steps in building the prediction models for stream flow. We consider the prediction function as: $S_{t+1} = f(S_t, S_{t-1}, S_{t-2}, \ldots, S_{t-m}, R_t, R_{t-1}, R_{t-2}, \ldots, R_{t-n})$ where S represents stream flow, R represents rainfall, t is the current day, m = {3,...,8}, n = {3,...,8} and f represents the ANN as a regression function. We investigate necessary lagged inputs of rainfall and river flow for modeling the river flows at three locations in South Australia. We apply an artificial neural network (ANN) technique for modeling river flow. ANN models are developed with all combinations of rainfall and river flow input ranges. In addition, a standard range of nodes in the hidden layer are also considered. Among all models based on inputs and hidden nodes, the best model is selected based on mean absolute error criteria. This entire process is applied to all three locations. ANN models capture the non-linear relationships of rainfall and river flow patterns in modeling river flows from large time series data. For example, if we consider 3 days lag of stream flow and 5 days lag of rainfall, then the total number of input nodes in the ANN structure will be 8 and we consider the number of nodes in the hidden layers ranging from 1 to 10. To achieve the best model using ANN for each location, all inputs not only apply in combination, but we also consider setting a range of parameters, such as different number of nodes in the hidden layer, for each combination of inputs.

In predicting stream flow one day ahead as output, we consider stream flow and rainfall with combinations of consecutive lags where the minimum lag is 3 days and the maximum lag is 8 days. Thus, for each location, the total number of models to be trained becomes 36. As the data set is large, one year of data is considered initially for testing. For training ANN models at each location, we consider stream flow and rainfall data for the period 1990 to 2009. The remaining data for the year 2010 is used for testing the best model found in the training phase.

For the Multilayer Perceptron (MLP) function, the ANN stream flow prediction model was built using the RWeka package in R Language [38]. One of the important parameters to specify is the number of nodes in the hidden layer, which may vary for time series modeling in different locations. Using trial and error, the number of nodes in the hidden layer is considered from 1 to 10. This range is widely used in hydrological time series modeling [21]. We consider the learning rate (the amount the weights are updated) to be 0.3, momentum is 0.2 and the number of epochs to train is 500.

Application of back propagation in ANN with a sigmoidal function was used to set the normalized data in the MLP function. Furthermore, the mean absolute error (MAE) in m^3s^{-1} was minimized through an iteration process that varied the number of nodes in the hidden layer.

The best lag combination at each location is presented in Figure 6.

Figure 6. MAE for training data (1990–2009) using ANN with best lag combinations at each location, units in m^3s^{-1}.

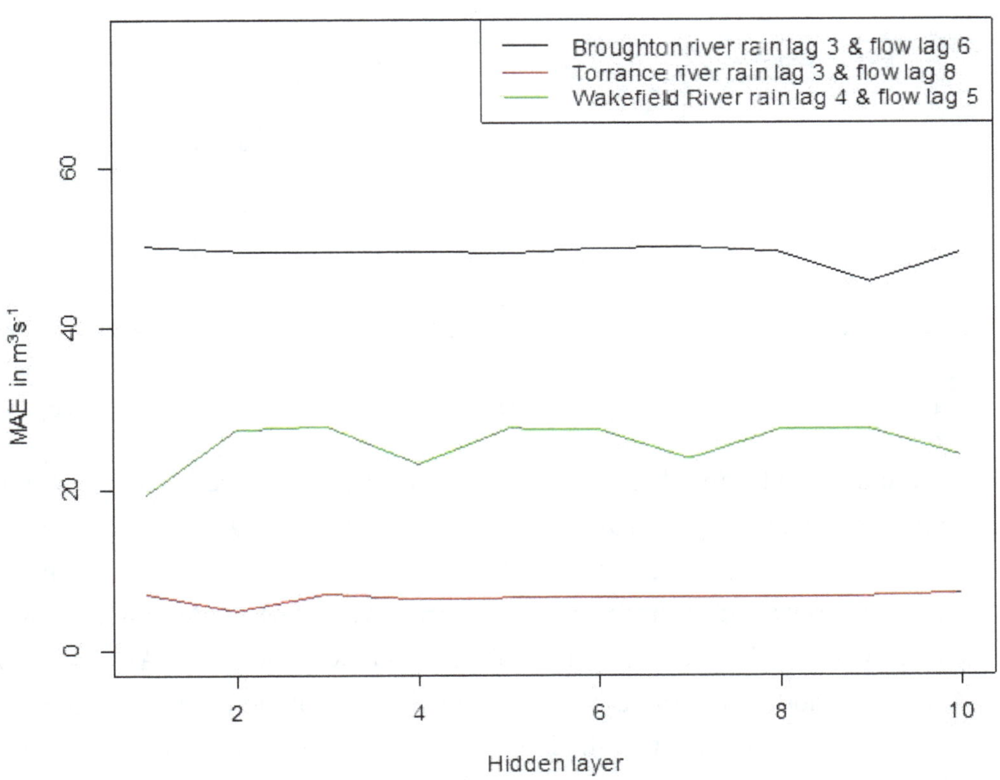

We find that both input lags and nodes in the hidden layer are different for each location. The best model based on correlation coefficient (R^2_{adj}) and the lowest root mean square error (RMSE) and mean absolute error (MAE) for each location is presented in Table 7. For Broughton River, 3 days rainfall and 6 days stream flow as lagged inputs with 9 nodes in the hidden layers produces the lowest MSE.

At Torrance River, 3 days rainfall and 8 days stream flow as lagged inputs with 2 nodes in the hidden layers produces the lowest MSE. For Wakefield River, 4 days rainfall and 5 days stream flow as lagged inputs with only one node in the hidden layer produces the lowest MSE. This indicates the variability in the ANN models for different locations.

When the best model is identified based on the training data for each location, we use this model on testing data prediction. This study show the prediction results for the testing data for each location. Figure 7 shows the predicted and observed stream flows using testing data for the locations Broughton River, Torrance River and Wakefield River, respectively.

Table 7. Best prediction model based on R^2_{adj}, lowest RMSE and MAE are in m^3s^{-1} on the training data.

Location	Input Lags	Nodes in Hidden Layer in ANN(H)	R^2_{adj}	RMSE *	MAE *
Broughton River	3 days rain, 6 days stream flow	9	0.68	270.33	45.53
Torrance River	3 days rain, 8 days stream flow	2	0.71	24.54	4.89
Wakefield River	4 days rain, 5 days stream flow	1	0.45	179.42	19.28

Note: Asterisk (*) units are in m^3s^{-1}.

Figure 7. Observed and predicted stream flow for **(a)** Broughton River; **(b)** Torrance River; and **(c)** Wakefield River for the year 2010.

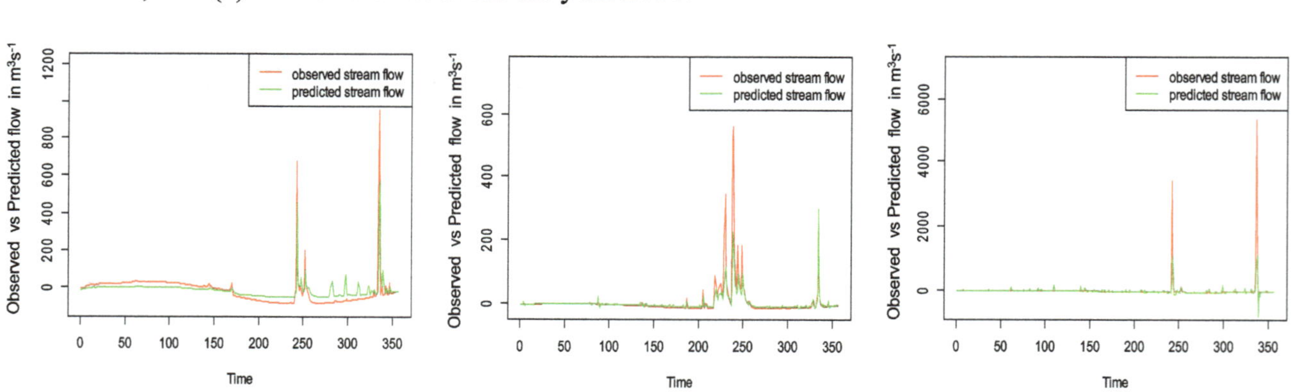

(a) (b) (c)

The MAE for training and testing data is shown in Figure 8 for all three locations. We observed that the MAE for the training and testing data at Broughton and Torrance Rivers do not vary significantly.

For Broughton, in training, the best ANN model structure includes 3 days lagged rainfall and 6 days lagged stream flow as inputs with 9 nodes in the hidden layer. This model has the lowest MAE, at 45.53 m^3s^{-1}. We further use this best model for testing and we find the MAE of 32.43 m^3s^{-1}. For Torrance, the ANN best model in training has 3 days lagged rainfall and 8 days lagged stream flow as inputs with 2 nodes in the hidden layer achieving the MAE of 4.89 m^3s^{-1}. For testing data, this model gives a MAE of 9.27 m^3s^{-1}. In case of Wakefield, the best ANN model has 4 days lagged rainfall and 5 days lagged stream flow as inputs with 1 node in the hidden layer achieving the MAE of 19.28 m^3s^{-1}. For the testing data, this model achieves an MAE of 42.88 m^3s^{-1}. The reason for the difference in MAE between the training and testing phases could be due to this river's ephemeral nature, and its substantial dependence on rainfall.

Figure 8. Comparison of MAE for training and testing data, units are in m^3s^{-1}.

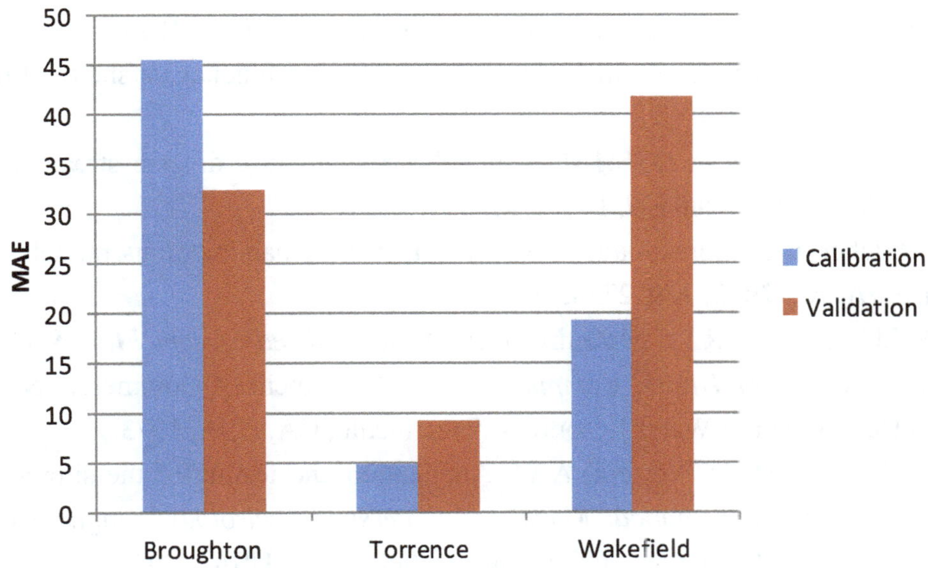

4. Conclusions

Initially, we split the whole series with a dyadic signal process for assessing the short term relationship between rainfall and stream flow including correlation using Haar wavelets. We have presented an innovative idea for the hydrological community for assessing stream flow for any catchment. In particular, the end user could assess the variability of changes and construct higher order correlations from 2 days up to as long as required. In addition, this study would be helpful for predicting stream flows using deterministic regression techniques, particularly where there is evidence of changes of statistical distribution characteristics, which is important for Water Sensitive Urban Design, as clearly demonstrated [39]. Using a deterministic regression based response model we found an increasing trend in stream flow when rainfall increased significantly. Predicted stream flow was more influenced by the previous few days' stream flows than when considering the entire previous period of stream flow. We also developed artificial neural network models for three locations. The results show that the influence of lagged rainfall and stream flow lies within a short temporal window. The results demonstrate that the ANN models perform better for Broughton and Torrance River in capturing the rainfall and stream flow relationships.

Acknowledgments

This study was funded by the Goyder Institute for Water Research under their Climate Change program. The researchers are grateful to the developers of the R project for software code and to the Australian Bureau of Meteorology for providing meteorological data.

Author Contributions

This manuscript, draft design by Mohammad Kamruzzaman and Md Sumon Shahriar, complete revised by Simon Beecham.

References

1. Moradkhani, H.; Sorooshian, S. General review of rainfall-runoff modeling: Model calibration, data assimilation, and uncertainty analysis. *Water Sci. Technol. Libr.* **2008**, *63*, 1–23.
2. Birkinshaw, S.J. Introduction. Available online: http://research.ncl.ac.uk/shetran/Introduction.htm (accessed on 26 July 2013).
3. Downer, C.W.; Ogden, F.L. GSSHA: A model for simulating diverse stream flow generating processes. *J. Hydrol. Eng.* **2004**, *9*, 161–174.
4. Perrin, C.; Michel, C.; Andreassian, V. Improvement of a parsimonious model for stream flow simulation. *J. Hydrol.* **2003**, *279*, 275–289.
5. Burnash, R.J.C.; Ferreal, R.L.; McGuire, R.A. *A Generalized Stream Flow Simulation System: Conceptual Modeling for Digital Computers*; U.S. Department of Commerce, National Weather Service and Department of Water Resources: Sacramento, CA, USA, 1973.
6. Chiew, F.H.S.; Peel, M.C.; Western, A.W. Application and testing of the simple rainfall-runoff model SIMHYD. In *Mathematical Models of Watershed Hydrology*; Singh, V.P., Frevert, D., Meyer, S., Eds.; Water Resources Publications: Littleton, CO, USA, 2002.
7. Beven, K.J. *Rainfall-Runoff Modelling: The Primer*; Wiley: Chichester, UK, 2011; pp. 1–360.
8. Castellano-Méndez, M.; Gonzàlez-Manteigo, W.; Febrero-Bande, M.; Prada-Sànchz, J.M.; Lozano-Calderon, R. Modelling of the monthly and daily behaviour of the runoff of the Xallas rivers using Box-Jenkins and neural networks methods. *J. Hydrol.* **2004**, *296*, 38–58.
9. Hipel, K.W. (Ed.) *Stochastic and Statistical Methods in Hydrology and Environmental Engineering: Time Series Analysis in Hydrology and Environmental Engineering, Water Science and Technology*; Springer: New York, NY, USA, 2010.
10. Kamruzzaman, M.; Metcalfe, A.; Beecham, S. Wavelet based rainfall-stream flow models for the South-East Murray Darling Basin. *ASCE J. Hydrol. Eng.* **2014**, *19*, 1283–1293.
11. Kamruzzaman, M.; Beecham, S.; Metcalfe, A. Climatic influence on rainfall and runoff variability in the South-East region of the Murray Darling Basin. *Int. J. Climatol.* **2013**, *33*, 291–311.
12. Kamruzzaman, M.; Beecham, S.; Metcalfe, A. Evidence for Changes in Daily Rainfall Extremes in South Australia. In Proceedings of 9th International Workshop on Precipitation in Urban Areas, IWA/IAHR, St. Moritz, Switzerland, 6–9 December 2009.
13. Solomatine, D.; See, L.; Abrahart, R. Data-driven modelling: Concepts, approaches and experiences. In *Practical Hydroinformatics*; Abrahart, R., See, L., Solomatine, D., Eds.; Springer Berlin: Heidelberg, Germany, 2008; Volume 68, pp. 17–30.
14. McIntyre, N.; Al-Qurashi, A.; Wheater, H. Regression analysis of rainfall-runoff events from an arid catchment in Oman. *Hydrol. Sci. J.* **2007**, *52*, 1103–1118.
15. Kim, Y.; Jeong, D.; Ko, I. Combining rainfall-runoff model outputs for improving ensemble stream flow prediction. *J. Hydrol. Eng.* **2006**, *11*, 578–588.
16. Liu, H.L.; Bao, A.; Chen, X.; Wang, L.; Pan, X.L. Response analysis of rainfall-runoff processes using wavelet transform: A case study of the alpine meadow belt. *Hydrol. Processes* **2011**, *25*, 2179–2187.
17. Kamruzzaman, M.; Beecham, S.; Metcalfe, A. Wavelet Based Assessment of Relationship between Hydrological Time Series in South East Australia. In Proceedings of 2nd Conference on Practical Responses to Climate Change, Engineers Australia, Canberra, Australia, 1–3 May 2012.

18. Kamruzzaman, M.; Beecham, S.; Metcalfe, A. Non-stationarity in rainfall and temperature in the Murray Darling Basin. *Hydrol. Process* **2011**, *25*, 1659–1675.
19. Hsu, K.-L.; Gupta, H.V.; Sorooshian, S. Artificial neural network modeling of the rainfall-runoff process. *Water Resour. Res.* **1995**, *31*, 2517–2530.
20. Sajikumar, N.; Thandaveswara, B.S. A non-linear rainfall-runoff model using an artificial neural network. *J. Hydrol.* **1999**, *216*, 32–55.
21. Kisi, O.; Shiri, J.; Tombul, M. Modelling rainfall-runoff process using soft computing techniques. *Comput. Geosci.* **2013**, *51*, 108–117.
22. Kisi, O. River flow forecasting and estimation using different artificial neural network techniques. *Hydrol. Res.* **2008**, *39*, 27–40.
23. Kisi, O. Stream flow forecasting using different artificial neural network algorithms. *J. Hydrol. Eng.* **2007**, *12*, 532–539.
24. Rezaeian Zadeh, M.; Amin, S.; Khalili, D.; Singh, V.P. Daily outflow prediction by Multi Layer perceptron with logistic sigmoid and tangent sigmoid activation functions. *Water Resour. Manag.* **2010**, *24*, 2673–2688.
25. Rezaeianzadeh, M.; Stein, A.; Tabari, H.; Abghari, H.; Jalalkamali, N.; Hosseinipour, E.Z.; Singh, V.P. Assessment of a conceptual hydrological model and Artificial Neural Networks for daily outflows forecasting. *Int. J. Environ. Sci. Technol.* **2013**, *10*, 1181–1192.
26. Sudheer, K.P.; Gosain, A.K.; Ramasastri, K.S. A data-driven algorithm for constructing artificial neural network rainfall-runoff models. *Hydrol. Processes* **2002**, *16*, 1325–1330.
27. Kisi, O. Wavelet regression model as an alternative to neural networks for river stage forecasting. *Water Resour. Manag.* **2011**, *25*, 579–600.
28. Dhanya, C.T.; Kumar, D.N. Predictive uncertainty of chaotic daily stream flow using ensemble wavelet networks approach. *Water Resour. Res.* **2011**, *47*, doi:10.1029/2010WR010173.
29. Labat, D.; Ababou, R.; Mangin, A. Rainfall-runoff relations for karstic springs. Part II: Continuous wavelet and discrete orthogonal multi resolution analyses. *J. Hydrol.* **2000**, *238*, 149–178.
30. Australian Bureau of Meteorology (BoM). Climate Data Online. Available online: http://www.bom.gov.au/climate/data/index.shtml (accessed on 12 January 2013).
31. Department for Environment, Water and Natural Resources (DEWNR) 2013. Available online: https://www.waterconnect.sa.gov.au/Systems (accessed on 12 March 2013).
32. Weerasinghe, S. A missing values imputation method for time series data: An efficient method to investigate the health effects of sulphur dioxide levels. *Environmetrics* **2010**, *21*, 162–172.
33. R Development Core Team. *Language and Environment for Statistical Computing*; R Foundation for Statistical Computing: Vienna, Austria, 2011.
34. Nason, G.P. *Wavelet Methods in Statistics with R*; Springer: New York, NY, USA, 2008.
35. R routine in open access. Wavethresh package, 2013. Available online: http://cran.ms.unimelb.edu.au/web/packages/wavethresh/index.html (accessed on 10 November 2013).
36. Haykin, S. *Neural Networks—A Comprehensive Foundation*, 2nd ed; Prentice-Hall: Upper Saddle River, NJ, USA, 1998; pp. 26–32.
37. Maier, R.H.; Jain, A.; Graeme, C.D.; Sudheer, K.P. Methods used for the development of neural networks for the prediction of water resource variables in river systems: Current status and future directions. *Environ. Model. Softw.* **2010**, *25*, 891–909.

38. R routine in open access. RWeka package, 2014. Available online: http://cran.r-project.org/web/packages/RWeka/index.html (accessed on 28 April 2014).
39. Beecham, S.; Chowdhury, R. Effects of changing rainfall patterns on WSUD in Australia. *Proc. ICE Water Manag.* **2012**, *165*, 285–298.

6

Trends in Levels of Allochthonous Dissolved Organic Carbon in Natural Water: A Review of Potential Mechanisms under a Changing Climate

Todd Pagano [1,*], Morgan Bida [1] and Jonathan E. Kenny [2]

[1] Laboratory Science Technology Program, Department of Science & Mathematics, Rochester Institute of Technology, Rochester, NY 14623, USA; E-Mail: mrbnts@rit.edu

[2] Department of Chemistry, Tufts University, Medford, MA 02155, USA; E-Mail: jonathan.kenny@tufts.edu

* Author to whom correspondence should be addressed; E-Mail: tepnts@rit.edu;

External Editor: Julia Piantadosi

Abstract: Over the past several decades, dissolved organic carbon (DOC) in inland natural water systems has been a popular research topic to a variety of scientific disciplines. Part of the attention has been due to observed changes in DOC concentrations in many of the water systems of the Northern Hemisphere. Shifts in DOC levels, and changes in its composition, are of concern due to its significance in aquatic ecosystem functioning and its potential and realized negative effects on waters that might be treated for drinking purposes. While it may not be possible to establish sound cause and effect relationships using a limited number of drivers, through long-term DOC monitoring studies and a variety of laboratory/field experiments, several explanations for increasing DOC trends have been proposed, including two key mechanisms: decreased atmospheric acid deposition and the increasing impact of climate change agents. The purpose of this review is three-fold: to outline frequently discussed conceptual mechanisms used to explain DOC increases (especially under a changing climate), to discuss the structure of DOC and the impact of higher levels of DOC on drinking water resources, and to provide renewed/sustained interest in DOC research that can encourage interdisciplinary collaboration. Understanding the cycling of carbon from terrestrial ecosystems into natural waters is necessary in the face of a variable and changing climate, as climate change-related mechanisms may become increasingly responsible for variations in the inputs of allochthonous DOC concentrations in water.

Keywords: allochthonous dissolved organic carbon; atmospheric acid deposition; climate change; phenolic compounds; drinking water

---

1. Introduction

Dissolved organic carbon (DOC) is a complex mixture of aromatic and aliphatic carbon-rich compounds that are important natural components of aquatic ecosystems, modulating many basic biogeochemical and ecological processes [1]. Over the past few decades, several long-term studies have reported an apparent trend of increasing DOC concentrations in inland surface waters over large areas of the Northern Hemisphere. According to a comprehensive review, there remains some uncertainty as to the ubiquity of the trend in increasing DOC levels [2]. However, the review reports that it is clear that a number of studies purport to show an increasing trend in DOC levels over large regions located in the Northern Hemisphere [2], and under a changing climate, it is feasible that DOC changes may become more pronounced—making examination of the potential drivers worthy of attention. At sites where DOC has increased, waters have often become "colored" (brownish tint) or "darker" (more absorbing of radiant light) [3], resulting in changes in light penetration and availability to plants in aquatic ecosystems. DOC concentrations also impact the transport of nutrients as well as pollutants, like heavy metals, in ecosystems. Further, since natural water supplies are sometimes treated and used for drinking water, higher levels of DOC have been shown to contribute to an increased formation of potentially dangerous chemical by-products (the most publicized being the trihalomethanes (THMs)) upon traditional chlorine disinfection treatment. High DOC can also dictate an increased demand for pretreatment steps in the water disinfection process, effectively making water more costly to treat [4,5].

Several factors have been considered to explain the observed increases in surface water DOC, including changes in soil and surface water acidity (e.g., [6,7]) as a result of declining sulfur deposition (e.g., [8,9]), continued nitrogen deposition (e.g., [10]), climate change (e.g., [11,12]), changes in hydrology and precipitation (e.g., [13,14]), changes in land-use patterns (e.g., [15]), as well as combinations of some of these factors. Despite intensive research over a few decades, much remains unknown in regard to our understanding of the terrestrial cycling of carbon. While many studies suggest that changes in DOC are occurring over large regional areas and the potential mechanisms discussed here have the capability to influence these changes to varying degrees, it should be recognized that it may not be possible to establish sound cause and effect relationships using a limited number of drivers. Many drivers, in fact, can work to produce a variety of effects on DOC release from the terrestrial landscape to natural waters and these effects can even work in opposition to each other. Further, many drivers may be covariate and it may, therefore, be difficult to isolate specific effects on DOC, especially in ecosystems where small changes in DOC or in a particular driver are observed. Nonetheless, the drivers outlined in this review are important to consider in areas experiencing changes in allochthonous DOC inputs, but further research should take a critical approach when attempting to ascribe significant statistical relationships between a reduced number of drivers and changes in DOC.

It is also recognized that changes in DOC can vary significantly according to geographic region [16]. For example, arctic and subarctic regions will respond differently to the complex drivers of DOC change when compared to temperate and tropic regions. It is not within the scope of this review to cover in detail how each ecoregion will respond to changes in the carbon cycle and many of the published research articles we reviewed were performed in temperate and subarctic regions. To understand more about carbon cycling in the arctic and changes in arctic carbon stocks, the authors direct readers to a review by McGuire *et al.* [17]. A highly cited review in regard to carbon cycling in tropical forests is also available [18], as well as several articles highlighting organic-matter [19,20] and climate change [21] in the tropics.

Despite the complexity of this issue, much of the attention devoted to explaining the observed long-term changes in DOC has sometimes been reduced to two key hypotheses. The first attributes the increases in DOC in surface waters to a reduction in acidic deposition as a result of tougher air standards internationally [8]. Proponents of this hypothesis argue that the changes in DOC are representative of a return to pre-industrial DOC levels, prior to the addition of sulfate pollution in the atmosphere [8]. There appears to be supporting evidence in the trends of reduced acidity in many surface waters in the U.K. in conjunction with an increase in DOC levels [22], but in some regions not subject to intense acid-deposition, such as northern, high latitude regions, the sulfate reduction hypotheses alone may not be able to completely explain increasing levels of DOC [23].

The second hypothesis offered to explain the DOC increases in water observed in boreal regions and other regions with low acid-deposition, is climate change and all of its inherent consequences (e.g., increasing temperatures, unpredictable weather, increasing atmospheric CO_2, *etc.*) as the main driver for increases in DOC [11]. Increasing temperature and atmospheric CO_2 concentrations can result in greater primary production and the accumulation of degrading biomass to contribute to the ecosystem's DOC pool. Researchers have indicated this observed increase in DOC may be a result of a change in the way in which soils and inland waters store and respire carbon [11,24], potentially representing a systemic environmental change; essentially a response in the carbon cycle to enhanced atmospheric CO_2 concentrations. Other factors of climate change, like varied precipitation and associated droughts and hydrology changes can also affect DOC levels found in water, but their implications are complex and varied in the ways that they may impact DOC trends.

A purpose of this review is to highlight the atmospheric deposition and climate change paradigms used to explain observations of increasing DOC and to briefly discuss other possible drivers discussed in the literature, such as land-use. The topic of DOC increases has garnered much research over the last few decades, perhaps due to the implication of increasing costs for treating water, but also as a means to better understand the processes governing the terrestrial transport of carbon to natural waters [24] and the associated ecological consequences of increasing DOC. Understanding the mechanisms that govern DOC release to natural waters is important, as it can enhance the ability to model future changes in the chemistry and ecosystem functioning of aquatic environments, especially in the face of impending climate change.

2. Characterization of Natural Organic Matter in Water Systems

2.1. Natural Organic Matter Definition and Composition

There is a range of terminology that is used to describe organic matter found in water systems and the terms to discuss its different fractions are sometimes used ambiguously, while accounting for the variety of methods used in different studies to measure organic matter portions (for a quality review of this issue and a comprehensive study of terminology/methodology used in the literature, please see ref. [25]). For the purposes of this review, the following terminology will be used for the various forms of organic matter found in natural waters. The Venn diagram displayed in Figure 1 can help to categorize and visualize the forms of organic matter and related terminology in a simplified way.

Total (or Natural) Organic Matter (TOM) has approximately 50% carbon by weight, and the organic carbon species found in natural water are often referred to as Total Organic Carbon (TOC) [26]. TOC encompasses all molecular organic carbon species found in water (from small molecules like methane to macro-molecular structures such as lignins and proteins) [26]. Dissolved Organic Matter (DOM) is a term used generally when discussing dissolved organic substances, but the term is often used interchangeably with DOC in the literature, despite the fact that DOC makes up a fraction of the DOM profile. Physically, DOC is categorized as organic compounds that can pass through a 0.45 μm filter, though this is merely an operational definition [27]. Any particulates that do not filter through are designated particulate organic carbon (POC) [27].

DOC can be further divided, based on composition, into humic and non-humic fractions. Humic material contains both aromatic and aliphatic components with amide, carboxyl, ketone, and other functional groups [28]. The humic fraction can be further categorized into humic acids, fulvic acids, and humin. The criteria for inclusion into each of these three categories are based on solubility properties at specific pH levels [27]. Humic acids are soluble in water until $pH < 2$, and contain the highest molecular weight samples of the humics—in the range of 1500–5000 Da in natural waters [29]. Fulvic acids are soluble at all pH levels found in nature, and range in molecular weight between 600 and 1000 Da in natural waters [29], while humin is not soluble in water at any pH [27]. Further, fulvic acid characteristically has more carboxylic groups and oxygen atoms, while humic acid has more phenolic and aromatic groups with longer aliphatic chains [30]. Because humic acids have longer aliphatic chains, they tend to be more nonpolar than fulvic acids, making them less soluble in water [30]. Humics tend to have fairly broad absorption spectra, absorbing most intensely toward the blue region of the visible spectrum [31]. Though models of the molecular makeup of humic compounds are still debated, a representative model structure of a humic acid has been suggested in the literature (see ref. [32] for a representative image). More recent studies have increased the understanding of the composition of DOC beyond the classic categories of humic and fulvic acids using techniques such as nuclear magnetic resonance (NMR) [33], high performance liquid chromatography (HPLC) [34], and Fourier transform ion cyclotron resonance mass spectrometry (FTICRMS) [35]. FTICRMS, coupled with electrospray ionization (ESI), appears to be a promising method for resolving humic and fulvic acids at a molecular level. ESI-FTICRMS has been used, for example, to differentiate between autochthonous and allochthonous dissolved organic matter in marine environments [36], to assess

changes in DOM brought about by its interaction with sunlight [37], and to compare and contrast humic and fulvic acids at nearly the molecular level [38,39].

Figure 1. Simplified Venn representation of the various forms of organic matter found in natural waters. Total Organic Matter (TOM), Total Organic Carbon (TOC), Dissolved Organic Matter (DOM), Dissolved Organic Carbon (DOC), Particulate Organic Carbon (POC), Dissolved Organic Nitrogen (DON), and Dissolved Organic Phosphorus (DOP) are represented. DOC can be further broken down to its humic (humic acid, fulvic acid, and humin) and non-humic material, while new analytical methods continue to reveal more molecular-level detail.

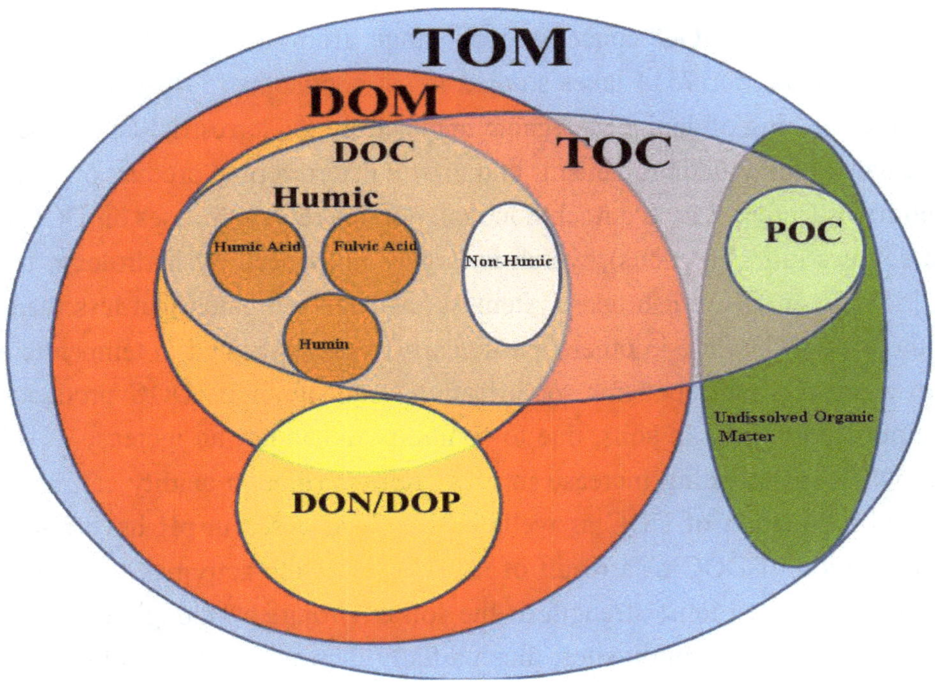

DOC is a complex material containing many molecules with a variety of functional groups—including phenolic compounds (which are of particular interest in this review). The phenolic components of DOC tend to be more recalcitrant to biodegradation [40], especially when compared to aliphatic compounds, and only certain microorganisms, including specialized fungi [41], seem to be able to decompose them [42]. As such, phenolic compounds have the potential to remain in the environment longer and some suggest the breakdown of lignin phenolic compounds may be a rate limiting step in the terrestrial carbon cycle [42,43]. In contrast, lignin phenols can be highly susceptible to photodegradation, and in areas that receive high levels of solar radiation, phenolic compounds have been shown to have faster rates of degradation [40]. Humic material can arise from the microbial breakdown of plants and organisms during the process of humification. The exact molecular mechanism of the humification process, though not fully understood, has been explained by lignin-based and polymerization/condensation-based processes [27]. Humic material can also be completely decomposed over time and ultimately returned to the atmosphere as CO_2.

The varied sources of humic material give rise to different spectroscopic properties. As an example, researchers have noted that humics from microbial-derived sources have less aromaticity than humics from terrestrial origins [44–46]. McKnight *et al.*, have proposed that samples with a Fluorescence index (FI) (fluorescence intensity at 450 nm/intensity at 500 nm, both measured at 370 nm excitation) of ~1.9 can be attributed to fulvic acid components of microbial origin [45]. In fact, SanClements *et al.*, showed that increasing DOC in U.S. lakes, which accompanied decreasing sulfate deposition, correlated with a decrease in FI [47]. Similar characteristic spectral responses are the basis for the classic fluorescence spectral classification schemes for DOC proposed by Coble [48,49]. Some studies have further used multidimensional fluorescence spectroscopy coupled with chemometric analyses to categorize modeled components with various types/origins of humic material [50–53].

2.2. DOC Sources and Production

Sobek *et al.*, report that DOC concentrations range from 0.1 to 322 mg/L, with a median concentration of 5.71 mg/L, for 7514 lakes studied in six continents [54]. A main source of DOC in natural waters is the leaching of degraded organic matter from the terrestrial landscape into freshwater systems. The breakdown of aquatic organisms and *in-situ* heterotrophic production can also contribute to DOC concentrations (referred to as autochthonous), but this review focuses on DOC that is exported from terrestrial sources (allochthonous), which is largely influenced by hydrology, temperature, and land-use/cover [55–58]. In a Swedish lake system, it has been estimated that less than 5% of aquatic DOC is autochthonous [59], while in others, particularly systems located in temperate ecoregions and those subject to extensive anthropogenic modification, autochthonous DOC production can exceed allochthonous inputs [55,60,61]. In soils, due to solubility dynamics, an increase of 0.5 pH units can lead to about a 50% corresponding increase in DOC release (though acidity can have an impact on microbial mediated production of DOC as well) [62] and increases in pH have also been shown to enhance the mobilization of DOC as a result of increased enzyme activities [63]. DOC mobilization can be reduced with increased ionic strength of the soil environment, though anions like sulfate, can compete with DOC for soil adsorption sites, thus partially increasing DOC mobility [64] Acidity and ionic strength are important players in the atmospheric chemical deposition mechanism for explaining DOC increases.

With terrestrial origins, allochthonous DOC production is principally described as the result of carbon assimilation via primary production in higher order plants, and is largely a function of the subsequent microbial degradation (or the leaching of plant litter in soils, which may be partially broken down) [64]. Greater amounts of primary production yield a higher accumulation of plant biomass that can become DOC. The plant material can introduce organic matter to its surroundings directly through root exudates, which can be further broken down to become DOC [64]. In addition to the mentioned biotic pathways (enzyme-catalyzed reactions), biomass can also be degraded abiotically (largely through photochemical processes), but often to a lesser extent than biotic processes [65], depending on regional climatic conditions. With plants, a portion of their biomass will be consumed by herbivores [66] whose subsequent death and degradation can be a secondary contribution to the net DOC. Another source of DOC production is through the release of organic matter via the breakdown of the microbial biomass itself [64]. The type of plant species producing the DOC is important in terms of the ultimate quantity that is produced; for example, coniferous dominated ecosystems produce about 50% more DOC than hardwood dominated ecosystems [64]. Likewise, the availability of inorganic nutrients during degradation, microbial environment acidity, and soil moisture (and its related conditions for dictating aerobic or anaerobic microbial degradation) can dictate the amount of DOC produced [67].

It is worth noting the importance of high latitude ecoregions on the bulk production of DOC. The subarctic and boreal regions have been subject to much attention in regard to DOC mobility/runoff into waterways in part because they are responsible for an estimated one-third (390–455 Gt C) of global soil carbon [68], representing 10%–15% of global terrestrial carbon [69], and the arctic delivers 25–36 $TgCyr^{-1}$ [70] to the Arctic Ocean, a quantity that is predicted to increase with a warmer climate. The subarctic is home to large stocks of peat, produced mainly by various species of *Sphagnum*, making it an important region for studies of global carbon sequestration and DOC production. Ultimately, peatlands can add to the global carbon budget in that they are areas where primary production can be greater than decomposition, resulting in a net carbon sink [68] and are therefore of particular significance in the broader context of climate change.

3. Changing DOC Concentrations in Natural Waters

DOC has shown changes in measured concentrations in natural water systems over the past few decades, with a large number of reports coming from North America and northern Europe [22,31,47]. At these sites, over the monitored timeframe (typically one to three decades), the DOC levels have largely been shown to be increasing, and in some cases, may have more than doubled. While the ubiquity of the observed changes in DOC is still up for debate, the vast majority of sites claimed data that show an agreement with the stated trend, though a few geographical areas report no increase and sometimes even a decrease in DOC concentration. Below we will discuss several notable examples of DOC trends in the literature. For a critical review with comprehensive analyses of analytical and data treatment methods of long-term (>10 years) DOC monitoring studies in the literature, readers are encouraged to see ref. [2]. In this review, our intent is to highlight the primary drivers responsible for changes in DOC and to emphasize the complexity of these drivers as they relate to changes in DOC.

A review of the literature indicates that the systematic monitoring of DOC in natural water systems was not pervasive much before the 1980s, though a few studies go back as far as the 1970s—and even the 1960s. DOC in natural waters has been measured in several countries, including parts of Northern Europe, United States, Canada, Korea [71] and Africa [20] and almost all studies report DOC to be increasing [8,9,11,12,14,31,72–95]. Natural waters in the U.K. have been among the most thoroughly studied, in part because of the monitoring conducted by the U.K. Acid Waters Monitoring Network (UKAWMN), which has been measuring the chemical, physical, and biological traits of 11 streams and 11 lakes since 1988 [22,31]. All of the 22 UKAWMN sites showed DOC concentration increases, with an average increase of 91% over a 15 year period [9]. In the U.S., general increases in DOC levels have been reported in the waters of the Adirondack and Catskill Mountain regions of New York State [8,75,81,84,85,96]. Monitored sites in Vermont and Maine showed a few areas of small to moderate rises in DOC levels, while Pennsylvania did not show general increases [8,84]. A study in Canada showed DOC levels increasing 30%–80% in parts of that country during a 20 year period [97]. Areas of Central Europe [98,99] and Siberia [100–102] have also shown marked increases in DOC levels of their natural waters. Models have also predicted DOC level increases in the natural waters of Japan, New Zealand, and portions of China [82].

As one of the most comprehensive intergovernmental studies [84] of DOC, another report of natural waters over a dispersed area (studying 189 water sites in six regions of Europe and six regions of the U.S.) showed some interesting long term trends. All of the European, and all but one of the U.S., regions showed increasing DOC trends in monitored natural water over a twelve year period [84]. The one area that did not show the increasing DOC trend was the Blue Ridge Mountain region of the U.S. [84]. And in another significant study, 68% of 315 varying size natural water sites in the U.K. displayed significant increases in DOC levels over one to four decades [91]. In an even larger study of 522 streams and lakes in North American and Europe, 88% of the surveyed sites showed increases in DOC concentration between the years 1990 and 2004 [8].

4. Mechanisms for Changing DOC Levels

There are several mechanisms that have been used to explain variations in DOC availability and production in freshwaters. Readers are encouraged to see the reviews in refs. [64,103] for discussions of the biogeochemistry and mechanisms potentially responsible for DOC concentration variations in water. As stated, a focus of this review is to explore the widely discussed mechanisms of decreased atmospheric acid deposition and climate change agents, specifically elevated atmospheric CO_2 and temperature, but related precipitation/hydrology and land-use will be introduced as well. Of particular interest here are the mechanisms that are, or will be, factors under a changing climate. Caution should be applied, however, when attempting to explain DOC changes using a reduced number of drivers as changes in DOC are controlled by a variety of complex, interrelated processes, many of which are not fully understood. Fitting schematic diagrams and tabulated information illustrating the complexity of these drivers at work in the terrestrial transport of carbon to inland waters can be found in two quality reviews [104,105].

4.1. Atmospheric Chemical Deposition-Related Increases in DOC

Though the impact of climate change agents are intriguing and are a focus here in subsequent sections of this review, some researchers believe that the impacts of climate change cannot explain the magnitude of reported DOC increases [8] at their study sites. Atmospheric chemical deposition (often referred to as "acid deposition") has been postulated as the main driver responsible for the observed increase in DOC concentrations in natural waters at specific study sites [8,9,31,73–75,83,84,88,94,106–108]. One of the more cited papers on the topic reports that the observed increase in DOC levels is the result of decreases in anthropogenic sulfate, sea salt, and other chemical species in the atmosphere that deposit in the soil [8]. This inverse relationship is believed potentially to be able to account for the magnitude of DOC increase that elevated temperatures and increased atmospheric CO_2 levels alone are seemingly unable to approach, but due to the fact these drivers vary contemporaneously, it is difficult to rule out one driver over another.

As previously highlighted, measurements from 522 lakes and streams in the U.S., Canada, and northern Europe were used to show how increases in DOC concentrations during the time span of 1994–2004 correlate with reductions in atmospheric chemical deposition [8]. In fact, the argument could be made that the observed DOC increases are merely steps toward the DOC levels returning to pre-industrial levels, before anthropogenic sources began adding chemical pollutants to the atmosphere (beginning in about 1750, peaking in the 1970s, and decreasing ever since) [8]. The major chemical in the atmospheric chemical deposition model, sulfate deposition from air pollution, has decreased by about 50% between the years 1986 and 2001 in the U.K. [109] due to notably successful international air regulations. As a consequence, large-scale reductions in sulfate deposition have been observed in many countries.

Simultaneously, increases in nitrogen emissions over the past 150 years have resulted in increases in deposited atmospheric nitrogen (N), which has resulted in an accumulation of N in soils and greater fluxes of N in rivers and streams (in addition to N from agricultural runoff) [110]. It has been hypothesized that increased soil N will stimulate the microbial processing of DOM, resulting in a net decrease in DOC that is available for export to the fluvial network as it is potentially outgassed as CO_2 [10], though experiments designed to test this hypothesis have produced conflicting results [111–114], possibly indicating that N deposition alone is not responsible for changes in DOC [103]. One study points to the significance of nitrogen deposition in bogs, because bogs receive their nutrient supply almost exclusively through atmospheric deposition [114]. The authors of this study show that increased N deposition may result in increased DOC release and an enhancement of phenol oxidase activity in *Sphagnum* litter, promoting peatland carbon release through DOC export and increased outgassing to CO_2. More research into the interplay between C and N cycling is warranted, as much remains not fully understood, particularly the extent to which high N deposition rates will influence plant productivity and litter decomposition. Further, for the purposes of this review we chose to focus primarily on sulfur deposition as it is the more abundant atmospheric deposition chemical referenced in the literature for explaining DOC increases in freshwaters.

It has been observed in the aforementioned studies that sulfate deposition has become significantly reduced [84], while coastal areas in northern Europe recorded reductions in sea salt (a proposed secondary contributor to DOC increases) [115]. And though most of the 522 studied sites showed the inverse relationship between DOC and sulfate concentrations, some coastal areas showed increasing DOC that was not accompanied by significant decreases in sulfate—but these areas showed significant decreases in chlorine concentrations [8]. Models showed that more than 85% of the total anion effect on DOC was a function of anthropogenic sulfate reductions in most of the studied geographic sites [8]. In some areas, the total anion effect was dictated by chlorine to about the same extent as sulfate, and the two anions likely impact DOC concentrations in water by similar mechanisms [8].

The solubility of the bulk of humic matter in soils is largely dependent on acidity [116] and/or ionic strength [117], as increases in either will decrease DOC solubility in laboratory experiments [64]. Acid deposition, specifically by sulfuric acid will increase the acidity as well as increase the total anionic strength (predominantly, increased $[SO_4^{2-}]$) of the soil environment, while sea salt deposition also increases the total anionic strength (increased $[Cl^-]$). A net result of the greater deposition of these chemical species is the decreased solubility of DOC in soils, yielding less mobility to water catchments and resulting in lower DOC levels. Conversely, the observed trends in the reduction of atmospheric chemical deposition should lower soil acidity and anionic strength, allowing for greater solubility of DOC and accounting for the observed increases in DOC concentration in natural waters. Indeed, in a recent study using U.K.-wide upland soil pH datasets, a correlated relationship between DOC and acidity changes was observed [118], lending further support to the chemical deposition paradigm for explaining DOC increases in these systems.

Increasing DOC levels in natural waters can actually be accompanied by increasing organic acidity, as many of the components of DOC are acidic (like humic and fulvic acids). In fact, the presence of these organic acids may be counterbalancing the contributions of declining acidity due to acid deposition—slowing (buffering) recovery from the acidification of the natural waters [8].

Though the trend toward increases in DOC levels, and specifically the link with decreasing sulfate deposition, is compelling, there are a few datasets that show regional areas where DOC is not increasing, despite declining sulfate deposition [76,90,91]. Some argue that DOC concentrations were increasing in the 1960s (or earlier) when sulfate deposition was still increasing and that laboratory studies on the quantitative role of acidification on the mobility of DOC can be ambiguous [119]. In all, it is apparent that a causal relationship between DOC levels and reduced acid deposition may be difficult to establish, especially in areas where the magnitude of DOC changes are small, and that the exact mechanism for changing DOC may include a more complex combination of drivers.

4.2. Climate Change Driven Increases in DOC

Due largely to anthropogenic contributions resulting from the combustion of fossil fuels, the atmospheric concentration of CO_2 has increased by over 60 ppm since 1960 and could rise to about double its current concentration of about 400 ppm by the year 2050 according to IPCC estimates and interpretation of their various models [120,121]. Likewise, an average increase in the global temperature of 0.74 °C has been observed over the past 100 years, and the IPCC estimates that the global temperature could rise by about another 3 °C by the year 2100 (depending on the prediction model chosen) [121]. Many detrimental effects are predicted as a result of climate change, however, the impact of increasing levels of DOC in natural waters has not received as much attention in the IPCC report, perhaps in part because the complex dynamics of terrestrial carbon export are not yet fully understood [121] and available for prediction. Furthermore, net temperature increases are expected to be greatest and most rapid at mid and high latitudes, where the majority of the global peat stock is contained [122].

The various agents associated with climate change have been reported as possible causes for the observed increased DOC levels in several natural water systems [11,12,14,68,69,76,77,79,89,97,98, 100,101,123–129]. As precipitation and its associated hydrology changes are functions of our natural climate, they too have been proposed as dynamics that will change with climate change [121]. Further, hydrologic changes have also been postulated as mechanisms for the observed increases in DOC [92,130,131].

Much of the primary literature related to climate change and its presumed impact on DOC concentration in natural waters deals with increased levels of atmospheric CO_2 and temperature warming. As will be discussed in relation to DOC increases, temperature can play a role in microbial degradation rates and atmospheric CO_2 can increase primary production and root exudates. A secondary outcome of elevated levels of atmospheric CO_2 can be the shift in the population of one plant species to another, where the species taking over the environmental area might assimilate and turnover biomass to DOC more readily. Other drivers are significant in terms of regional observations in DOC, including land-use change and modified hydraulic flow-paths, which will be discussed in subsequent sections.

4.2.1. Increasing Atmospheric CO_2

Increases in plant biomass can accompany elevated atmospheric CO_2 concentrations above ambient levels. This increase is largely due to enhanced levels of primary productivity by plants, and related increases in root exudates [12,69]. In fact, CO_2 "enrichment"/"fertilization" is a technique that has been used by farmers/gardeners in the U.S. to increase their crop yields (strategic biomass) since the mid-1960s [132]. Several controlled laboratory experiments have demonstrated increased biomass accumulation and increased primary productivity under increased atmospheric CO_2. Since 1999, free-air CO_2 enrichment (FACE) technology has been applied to a variety of ecosystem studies; including deciduous, alpine, and evergreen forests; deserts; grasslands; and bogs [133]. FACE experiments have been crucial to our understanding and defining of the ecological processes and responses that occur under elevated atmospheric CO_2 concentrations, especially as it pertains to refining quantitative modeling, but FACE experiments are not always conducive to application to global ecosystem processes [133]. One of the major advantages of FACE experimentation, over laboratory microcosm experimentation, has been the ability to experiment in a larger temporal and spatial context, allowing for the creation of improved models of real-world ecosystems.

Through empirical studies, researchers have predicted an increase of about 20% DOC production when atmospheric CO_2 levels are at twice their recent ambient levels [134]. During a three year experiment of peat monolith samples, the concentration of CO_2 in solardomes was held constant at 372 ppm (matching that of the ambient air at the time) in control samples and at 607 ppm (ambient + 235 ppm) in elevated CO_2 samples [69]. At the end of the three year experiment, pulse isotope labeling studies were conducted on the control and elevated CO_2 samples by exposing them to $^{13}CO_2$ for five hours and monitoring the assimilation of the ^{13}C at various time intervals using collected leachate and plant tissue for analyses [69]. Under conditions of elevated CO_2 levels, measurements of DOC in leachate collections resulted in a 66% increase in concentration over measurements taken from the control samples, while the above and below ground biomass increased by 115% and 96%, respectively [69]. Likewise, $^{13}CO_2$ pulse labeling showed about a ten times increase in "new" (produced within 24 hours of the label exposure) labeled DOC (^{13}DOC, in this case) [69]. The pulse labeling experiments indicate increased exudation (from the roots of increased biomass) as the dominant driver of DOC production under elevated atmospheric CO_2, while microbial degradation still plays a role over the longer term of the isolated ecosystem. Also, the *Sphagnum* moss dominated peat samples shifted greatly toward vascular (lignified tissue) plants, which assimilated 49% more ^{13}C during isotope pulse labeling and led to higher overall DOC production [69].

In terms of allochthonous DOC, some is further broken down and fractions are even returned to the environment as outgassed CO_2, contributing to atmospheric CO_2 concentrations. If further decomposition of organic matter is inhibited, greater quantities of DOC could be released to runoff. In the case of experiments under elevated CO_2, a presence of greater amounts of phenolic compounds (which may further inhibit the breakdown of DOC) was found [89]. The phenolic content of DOC can be an important indicator in the climate change-related mechanisms and its presence is also of concern when water is chlorinated for drinking purposes (discussed in subsequent sections).

One of the more cited articles on the subject of the role of elevated atmospheric CO_2 levels reports increases in DOC concentrations of 14%, 49%, and 61% for bog, fen, and riparian peatlands samples, respectively, at an elevated atmospheric CO_2 concentration of (235 ppm + ambient = 607 ppm) over three years of the controlled study [12]. For clarification, this study [12] was conducted by the same researchers who performed the above experiment [69], and in a parallel timeframe, but on two additional peat types (one of the three peat samples discussed here was the sole sample type reported in the study above). Of course, variations in both estimated and measured DOC levels are expected with the different ecosystems that researchers might be studying (e.g., different forms of peatlands, pine *vs.* deciduous), nutrient availability needed to assist humification, soil moisture, and other variables. This study also showed the similar trend in the shifting of species composition of the isolated environment toward more vascular plants [12]. Fen and riparian peatland soils tend to be rich in the nutrients needed for humification, while bogs tend to be deficient in nutrients, including phosphorus and nitrogen [135]. This nutrient dependence points toward a mechanism of primary production enhancement accompanying elevated atmospheric CO_2 levels for the production of the DOC end product, as the nutrients are needed for primary production of biomass [12]. In soil types where nutrients are less available (like in bogs) [135], the amount of nutrients will be the limiting factor in primary production—showing less sensitivity to environmental stimuli (like increases in atmospheric CO_2 levels) [136].

To support the hypothesis that primary production is responsible for the DOC increases under increased atmospheric CO_2 levels, $^{13}CO_2$ was again used as a marker when it was exposed for five hours to the contained atmosphere of peat monoliths that had been under ambient and elevated (235 ppm + ambient) atmospheric CO_2 levels for about three years [12]. Through photosynthesis and translocation processes, the ^{13}C was traced to the DOC leachate from the monoliths [12]. Not only was more ^{13}C translocated into the sample's biomass (56% more in tissue studies) under elevated atmospheric CO_2 conditions, but the amount of DOC in the soil that was attributed to the "newer" assimilated ^{13}C throughput was an order of magnitude higher [12]. Again, the authors attributed this to the fact that primary production (and associated root exudates) appears to be more sensitive to elevated atmospheric CO_2 levels than microbial degradation [12], which will be discussed further in relation to temperature sensitivity.

Despite the presented data, some believe that elevated atmospheric CO_2 levels cannot account for the magnitude of the observed global trend in DOC concentration increases. The elevated atmospheric CO_2 level (about 607 ppm) used in the two related studies outlined above is much higher than the approximate 20 ppm in atmospheric CO_2 increase that has been measured during the timeframe in which the observed global DOC concentrations were increasing. This poses a challenge to arguments surrounding elevated atmospheric CO_2 levels being solely responsible for the magnitude of observed global DOC concentration increases in freshwaters. Opponents of the elevated atmospheric CO_2 mechanism state that, assuming a linear relationship between DOC increases and elevated atmospheric CO_2 levels, that the group's [12] data for bog, fen, and riparian peat sample would represent only 1.2%, 4.2%, and 5.2% DOC concentration increases, respectively—at the approximate 20 ppm increase in CO_2 concentration over the 20 years of the UKAWMN measurements [9]. It should be noted that bog, fen, and riparian peatlands dominate the U.K. ecosystems of the UKAWMN monitored waters [9]. This is perhaps one reason why the atmospheric chemical deposition theory has been mentioned as an additional or alternative theory to this atmospheric CO_2-controlled increase.

Many FACE experiments have been conducted in a variety of ecosystems. For example, a group at Duke University's Forest FACE experimental station has established an experiment that subjected an actual pine plantation (*Pinus taeda)* in North Carolina to elevated atmospheric CO_2 levels equal to some climate model projections for the year 2050 (about 550 ppm) for twelve years [127]. In a more recent example, a FACE study in the Swiss Alps looked at the effect of elevated atmospheric CO_2 concentrations on a forest near the timberline consisting mainly of a pine species (*Pinus uncinata)* and a deciduous tree species (*Larix decidua*) [137]. More than eight different hypotheses were tested during the most intensive period of FACE experimentation at various sites (1999–2011) and many proved difficult to support [133]. The most well supported hypotheses appear to be in relation to (1) an observed increase in net primary production per unit of leaf-area index (LAI); (2) an apparent closure of leaf stoma in response to elevated CO_2; (3) elevated CO_2 can apparently alter intra- and inter-specific competition for soil resources amongst forest trees, potentially signifying a change in species composition with elevated atmospheric carbon; (4) a weak increase in soil microbial activity under elevated CO_2; and (5) that elevated atmospheric CO_2 does not necessarily imply an increased ecosystem C storage [133], which may have implications for DOC levels within catchments. While FACE experiments have increased our understanding of some ecosystem responses to increased CO_2, many questions remain, but FACE experimentation has demonstrated how complex the interactions of C in an ecosystem are, how they may vary regionally, and how a multidisciplinary approach to DOC research is further warranted to fully understand the dynamics of carbon through ecosystems [133].

4.2.2. Increasing Temperature

Temperature rises are believed to increase DOC levels in natural waters, though the increase has been primarily attributed to improved activity by microorganisms, and specifically by enzyme activity (especially by phenol oxidase) [11]. Phenol oxidase is an enzyme that catalyzes the oxidation of phenolic compounds to quinines [138]—which may occur by extracellular enzyme hydrolysis or by microbial metabolism [12]. Controlled studies on peat samples showed direct increases in DOC production with warmer temperatures [11,129]. In one study, a 10 °C increase in sample environment yielded a 36% increase in phenol oxidase activity, a 33% increase in DOC concentration found in leachate, and a 72% increase in phenolic compounds found in leachate [11]. The presence of higher levels of phenolic compounds is significant, as they are known to inhibit the further breakdown of DOC [139]—making them more available for transport to natural waters [140], including those used as drinking water sources. Additionally, in a study of peat porewater DOC concentrations, warming with infrared lamps (temperature increase of 1.9 °C) produced DOC concentrations about 15% higher than those in controlled, non-heated peatland plots [129]. Further, DOC from the warmed plots degraded faster in lability experiments than the control plots and showed a decreased aromaticity using "specific UV absorbance" ($SUVA_{254}$ = Absorbance @ 254 nm/[DOC]), which suggests an associated increase in microbial activity [129].

Despite the above preliminary findings, the effect on DOC levels using the net global temperature increase estimated by IPCC (0.5–0.7 °C) over the past several decades falls short of explaining the observed DOC trends that have been monitored globally—and a 10 °C increase would likely be a more reasonable temperature rise required to explain the observed global DOC trends [12]. It should be noted that a 10 °C increase in the next century is not projected by any of the current IPCC models. For clarification, the 10 °C rise in temperature chosen in the experimental method of the early study [11] was likely chosen to correlate with the common biological activity coefficient, Q10 (the measure of biological activity over an applied 10 °C increase), not to necessarily match IPCC climate change scenarios. Also, these earlier temperature increase studies were performed on very nutrient rich peat samples that might be expected to be more sensitive to warming (with sufficient nutrient availability to accommodate enhanced microbial activity), but might not be fully representative of all the geographical areas that are producing the observed DOC increases [9]. Nevertheless, other peat soil sample types have been found to be even more sensitive to DOC production due to warming, with even greater increases in aerobic microbial conditions [130]. Though water saturated peat samples are known to produce DOC largely by anaerobic means, there are layers that degrade aerobically due to seasonal drying in the soil [9]. Aerobic degradation and possible future droughts predicted by climate change scenarios could have a synergistic impact on increases in DOC production in soils.

It is estimated that the global 0.5–0.7 °C temperature increase during the time when many DOC levels were actively being measured could potentially be responsible for a 10%–20% increase in DOC concentration in the areas and soil types that were monitored [9]. Still, researchers who make these estimates maintain that they measured a much larger average DOC increase (91%) over about 20 years on a large and geographically diverse data set from the U.K. [9]. Another study cites temperature as potentially being responsible for about 12% of the 78% total DOC increase in a U.K. peat catchment [79]. So, even though rising temperatures appear to be a plausible player in explaining DOC increases, scientists are looking for other mechanisms to account for the larger magnitude of observed increases.

4.2.3. Combined Effect of Increased Atmospheric CO_2 Concentration and Temperature

The combination of elevated CO_2 and temperatures, in conjunction with longer growing seasons, might show additive or synergistic increases in the DOC levels found in natural waters. Experimental investigations on the separate roles of these climate change agents show that they can have at least a partial effect on DOC concentrations in natural waters. Inasmuch as climate change predicted scenarios strongly indicate increases in both atmospheric CO_2 levels and temperature, the interaction of these two factors on natural water DOC concentrations is also of interest. Perhaps, the combined effect can account for both individual models' shortcomings in regard to the magnitude of observed DOC levels that they appear to entail.

In an extension of the studies discussed above (separate/isolated warming and elevated atmospheric CO_2 experiments), the same research group performed a parallel study, again on peat monoliths of riparian peatlands in solardomes and over a three year period [89]. Their data showed a 119% increase in DOC concentration in collected leachate when under a combination of elevated atmospheric CO_2 (again, ambient + 235 ppm CO_2) and temperature (ambient + 3 °C, for this experiment—as opposed to 10 °C discussed for the above experiments) conditions compared to samples kept under ambient (average 372 ppm CO_2 and seasonal average temperatures) conditions [89]. Separate samples kept in elevated atmospheric CO_2 and elevated temperature conditions alone produced 36% and 22% more DOC, respectively, than ambient control samples in collected leachate [89]. It is important to note that the additive effect of the isolated elevated conditions accounts for less than half of the DOC production under the combined condition. The synergistic increase in DOC concentrations for this study is indicative of the potential impact of climate change on DOC flux in natural waters worldwide.

In this same study, both above and below ground DOC concentrations were shown to increase synergistically with the combined treatment conditions [89]; above ground biomass increased by 284% and below ground biomass increased by 407% [89]. Peat Poly-β-Hydroxyalkanoate (PHA) is a microbial nutrient stress indicator that is used to assess competition for inorganic nutrients between the plants and the decomposition microbial agents [128], and was shown to increase 30% under elevated CO_2, 19% under elevated temperature, and 51% under the combined conditions [89]. This was suggestive of an approximate additive relationship in the combined conditions. Phenol oxidase is an enzyme often accredited with the storage and translocation of organic carbon [11]. In this experiment, phenol oxidase decreased by 58% under combined conditions in an apparent synergistic relationship [89]. β-Glucosidase plays a role in the breaking down of general organic substances [89,141] and was shown to decrease by 27% in a slightly synergistic relationship between increased CO_2 and temperature [89]. Similar to the rationale for monitoring PHA, phosphatase can be used to determine whether there has been competition for specific phosphorous nutrients [89]. Phosphatase was shown to increase by 24% under combined conditions, but decrease 9% under elevated temperature and increase 33% under increased CO_2, suggesting neither an additive or synergistic relationship [89]. Phenolic compounds in collected leachate almost doubled in separate elevated atmospheric CO_2 and temperature conditions, and nearly quadrupled in the combined conditions [89]. This finding again indicates that these compounds are more prevalent because of the synergistic reduction of phenol oxidase activity (which breaks down phenolic compounds like those found in lignins, humic acid, and fulvic acid) in the combined conditions [89]. The combined treatment also displayed an additive increase in general microbial nutrient stress, showing microbial competition with the plants, straining the microbes to find available inorganic nutrients—which can be a limiting factor to net DOC production [89]. The results of these chemical measurements led to the development of mechanisms that explain the increased DOC measurements in terms of additive and synergistic effects of elevated CO_2 and elevated temperatures.

The enzyme, phenol oxidase, is an interesting single molecule in the global carbon cycle and has been examined in the literature in detail in regard to its mechanism of carbon assimilation. Phenol oxidase has been labeled the delicate "enzyme latch to the global carbon stock" as it regulates the stability of some 455 Gt C supply in peatlands, and is largely responsible for preventing the release of large amounts of CO_2 back into the environment and could have serious impacts on climate change [142]. Peatlands hold 20%–30% of the world's soil carbon supply, and in anaerobic conditions (true of most water saturated peatlands), the microbial breakdown of phenolic compounds is restrained [142,143]. Under these conditions, phenolic compounds remain in the soil and become available for transport, as part of the net DOC profile, to aquatic systems. Consequently, phenolic compounds are known to inhibit other enzymes from further breaking down organic matter, adding to the net availability of DOC for transport [142]. Some researchers believe that this "enzyme latch" mechanism could be responsible for a portion of observed DOC increases [42,79,142,143]. However, in the presence of droughts (that could accompany climate change), the microbial conditions could become more aerobic, increasing the activity of phenol oxidase, and encouraging the further breakdown of phenolic compounds (and subsequently, other organic compounds) [142,143]. Under these more aerobic conditions, total DOC available for transport may be diminished, but the process could release vast amounts of CO_2 back into the atmosphere [143] with potentially harmful consequences. It is interesting to note that biosensors based on the enzyme interaction of phenol oxidase have been studied for use as phenol sensors in water treatment processes [144].

Along with greater biomass production that is seen with the climate change agents, the suppression of the further breakdown of DOC by phenolic compounds might be a strong accelerator of DOC levels toward current trends. In the study detailed above, there were also changes in plant species composition of the peat samples in the elevated atmospheric CO_2 (slight shift), temperature (moderate shift), and combined (extreme shift) conditions [89]. Some of these potential climate change induced shifts in species composition of an environmental area might also add to the observed global increases of DOC levels in natural waters, as some species exude DOC more readily [89].

Even though it is still a point of debate, while warming and elevated atmospheric CO_2 mechanisms alone appear to fall short in explaining observed DOC level trends, it is conceivable that the synergistic effect of the two conditions combined might approach quantitative explanations of DOC trends. It is also noteworthy that the experimental conditions used in some studies—namely, combined elevated temperature and atmospheric CO_2 levels—are within feasible ranges of IPCC estimates for climate change conditions that could be experienced within the next 50 years or so. Therefore, even opponents of this mechanism for explaining recently observed DOC concentration trends in natural waters should be cognizant of the notion that these could well be the factors driving DOC trends under a changing climate in the near future and beyond.

4.2.4. Changing Hydrology and Its Effect on DOC Concentrations

Many changes to the water cycle are predicted as global temperatures rise, as the water cycle is particularly sensitive to changes in the climate [24]. It is predicted that there will be alterations to precipitation (flooding and drought), discharge, modifications to terrestrial flow-paths of water, and transformations to the hydrologic connectively of the world's water catchments with a warmer climate [121]. Variations in the hydrology of catchments will result in changes to the way DOC is transported, affecting the quantity and composition upon delivery to inland and marine waters. Tranvik and Jasson [145] discussed the significance of hydrology on the transport of DOC from streams to the ocean and stated the ways in which warming might have affected DOC export over a decade ago. Here, we intend to provide several examples of how hydrology affects allochthonous DOC quantity and composition in natural waters. Readers seeking more information, particularly in regard to autochthonous DOC and hydrologic changes, should also consider the comprehensive review of hydrologic changes and DOC associated with climate change provided by Porcal *et al.* [103].

Precipitation is expected to increase, decrease, and/or become more variable with climate change. Increases in the frequency and severity of droughts have been predicted in some geographic regions, while in others, increases in precipitation are likely, with more severe storm events and floods expected [121]. Changes in the amount, frequency, and seasonal timing of hydrologic events are significant considerations in the prediction of future DOC, as precipitation can increase terrestrial primary production and enhance the export of DOC to natural waters [104,146]. In a study with experimentally controlled drought and rainfall, drought was accompanied by decreases in DOC and its phenolic components, and increased precipitation was accompanied by significant increases in allochthonous DOC and an even more pronounced increase in the phenolic content of allochthonous

DOC [42]. Similarly, DOC concentrations decreased in response to a simulated drought that was shown to impact the metabolic activity of biofilms in waters from an experimental wetland [147]. In a study of the effect of drought on litter and peat, drought was shown to be the dominant factor to explain decreases in DOC releases from both sources and the DOC that was released was less hydrophilic in character, suggesting more difficulty in its removal for water treatment purposes [148]. Additionally, in a study in Australia, a severe, decade-long drought was followed by extreme flood events that effected large areas and persisted for several months [149]. During the drought conditions of this study, organic material accumulated on land and in dry fluvial channels, but once this material was again inundated, DOC rapidly leached into the water column of a multitude of streams, rivers, and lakes, resulting in conditions favorable for rapid microbial metabolism and hypoxia [149]. These events, commonly referred to as blackwater events, have also been reported in Brazil [150] and the southern United States [151] and often result in large fish kills. Blackwater events provide insight into the potential water quality challenges under more variable precipitation conditions in the future [149].

Soils that are well-drained, a characteristic of many upland watersheds, tend to have greater proportions of degraded, microbially-derived organic matter than soils located in lowland areas, such as swamps and marshes, where the soil is poorly-drained [152]. This suggests that microorganisms that breakdown DOC and phenolic compounds do so most efficiently in aerobic (water table drawdown) conditions. Under aerobic conditions, phenol oxidase activity increases and organic matter and its constituent phenolic compounds are more readily degraded [128,143]. This hypothesis has been further tested in peatlands, where drought was shown to increase phenol oxidase activity and bacterial growth in a series of *in vitro* and mesocosm experiments, as well as field observations [143]. Conversely, in high flow or flood conditions (anaerobic), not only might there be higher levels of phenolic compounds [63], but they could be more readily swept away by access to water flow. This was true in peatlands previously experiencing drought, where the effect of high flow conditions resulted in a re-wetting of the peat and carried markedly high concentrations of carbon away, in addition to re-establishing the anoxic pre-drought conditions [143].

The amount of water discharged from fluvial networks is an important consideration, as it can alter both vertical and lateral hydraulic flow-paths in catchments. Multiple studies have indicated that in temperate streams, for example, much of the DOC that is exported occurs during high-flow conditions [153–155] and is largely governed by intra-annual variation in hydrologic regimes [61,156]. Additionally, the composition of DOC exported during high-flow or low-flow conditions has been shown to change and is thought to represent alterations to the flow-path of water due to precipitation. For example, under high-flow conditions, stream waters were shown to contain DOM with large C-to-N ratios, increased aromaticity, and ^{14}C values which indicated that DOM had a less degraded nature, all variables that suggest DOM originating from the upper soil horizons, indicative of a surficial flow-path [152,157]. In contrast, streams under low-flow conditions contained DOM with much older ^{14}C ages, small C-to-N ratios, and a decrease in aromaticity—all indicative of a microbially-processed DOM from lower soil horizons, suggestive of a deeper, ground-water flow-path. Additionally, watersheds with mixed land-use, different land cover types, and anthropogenically modified hydraulic flow-paths can contribute to the DOC quantity and quality in rivers and streams (land-use will be discussed further in a subsequent section).

Jencso *et al.*, highlight the importance of hydrologic connectivity of a watershed in the transport of solutes [158] such as DOC. Hydrologic connectivity is subject to seasonal variation, for example, low connectivity occurs during the winter, when flows over land, to streams, and to rivers are often frozen or otherwise disconnected [158]. In the spring and summer seasons, DOC found in rivers shows signs of more terrestrial inputs (*i.e.*, increased lignin phenols) [159], suggesting a more connected landscape where water carries solutes from all regions of a watershed. Further, the IPCC has predicted that runoff resulting from the thawing of permafrost in the arctic and subarctic areas and subsequent formation of new ponds and lakes will increase the availability of stored carbon stocks [121] through increased hydrologic connectivity [160]. As permafrost thaws, the hydrologic connectivity of arctic and subarctic regions will increase, resulting in greater export of allochthonous DOC to fluvial networks, and ultimately, to the ocean.

Changes in the hydrology experienced by various watersheds will be an important predictor of DOC composition and quantity in the future, especially in areas of high and mid latitudes [161], on both regional and global spatial scales and short and long-term temporal scales. Additionally, increases or decreases in the abundance of ponds, lakes, and man-made reservoirs/impoundments are expected to change the way DOC is exported [104] and processed on land differently in different geographic regions, so additional studies to understand the effect of precipitation, discharge, hydraulic flow-path and the hydrologic connectivity of a catchment are warranted, especially under a changing climate.

4.3. Land-Use and Its Effect on DOC Levels

Analysis of satellite imagery reveals that human land-use activities, such as clearing of forested areas for agriculture and residential land-use, are transforming a large proportion of the earth's surface [162]. Human land-use, including modifications to hydraulic flow-paths, alters biogeochemical cycling, directly affecting water quality, including DOC input to aquatic ecosystems [163–165]. For example, Yallop *et al.*, found that land management may be the most significant driver of humic DOC fluctuations for their study conducted in the U.K. [166] and Gough *et al.*, showed that significant differences in DOC concentration, quality, and THM formation potential are all influenced by land cover type [165]. Further, nutrient enrichment of agricultural land is currently a leading cause of degraded water quality in U.S. coastal waters [167]. Waters with excess nitrogen and phosphorus are impaired by direct effects, like eutrophication, as well as indirect effects, such as changes in temperature, pH, and light attenuation [163,168]. But much less attention has been given to drivers of terrestrial carbon cycling in regard to land-use, especially in terms of land management decisions, where DOC is often overlooked [58]. Further, little is understood about the impact climate change will have on the drivers of DOC in varied land-use catchments.

Researchers have learned about the influence of land-use and land management on stream biochemistry in regard to carbon by analyzing the fluorescence characteristics of DOM [169]. In a recent study, the character of riverine DOM in 34 watersheds in south-central Ontario, Canada was examined in relation to catchment land-use, with findings showing a relationship between agricultural land cover and the composition of DOM in water draining these areas [170]. Values for FI and β/α (a ratio used to estimate the contribution of recently produced DOC, β, to its more degraded form, α) increased as the amount of continuous crop cover increased in the different watersheds [170]. The authors suggest that this indicates a more microbially-derived DOM character [170]. Additionally, they found that there was a strong correlation between total dissolved nitrogen and the DOM composition [170]. It seems that the relationship between DOM composition and the high nitrogen availability associated with agricultural land-use disproportionately increases microbial respiration and changes the composition of DOM, such that it is composed of more microbially-derived moieties. This implies that there may be decreased DOM availability in systems downstream from agriculturally dominated watersheds and that DOM transport distances are decreased [170].

In a similar study, the hypothesis that the optical characteristics of DOM are influenced by microbial activity and land-use, specifically that increased agricultural land-use leads to higher microbial signature in the optical properties of DOM flowing downstream, was tested [55]. They found that the microbial activity increased for DOM in streams with a higher proportion of anthropogenic land-use and that the fluorescence characteristics associated with more labile components in DOM increased with more agricultural land-use [55]. The authors mentioned that their study period was particularly wet and that the increased soil moisture actually made their correlation between more labile, microbially derived DOM stronger [55], pointing, again, to the significance of hydrologic regimes in regional DOC studies.

The main points from these spectroscopic studies are that land-use altered the quantity and quality of DOM exported from human-influenced streams when compared to more natural streams and that DOM from agriculturally dominated watersheds was more labile and supported higher microbial activity when compared to DOM from forested watersheds [55,170]. It is fairly unclear whether land-use will magnify or buffer changes to DOC concentrations and composition in regard to predicted climate change variables, but it is clear that land-use can contribute to changes in DOC levels in natural waters. Therefore, continued studies from varied ecosystem types can contribute to the overall understanding of terrestrial carbon cycling and its prediction in the future.

5. The Impact of Increased DOC on Drinking Water Supply and Treatment

Regardless of the exact driver for the observed increases in water DOC levels, there are several environmental and health concerns associated with the trends, especially in regard to its impact on drinking water. Because increases in aquatic DOC concentration can intensify the "color" of natural waters, higher levels can block the sun's radiation from penetrating to reach deeper ecosystems, even though some DOC is beneficial to blocking solar ultraviolet (UV) radiation from damaging aquatic life. DOC also plays a role in natural nutrient transport [171], which in the case of excessive concentrations of DOC, can also increase the bioavailability of pollutants like mercury in natural waters [172]. As discussed previously, higher levels of DOC, and thus, organic acidity, can also make natural waters more acidic.

Higher concentrations of DOC in water systems that are used for drinking water supplies are a considerable concern for the treatment and disinfection of drinking water [107,173–179]. When treated by the most common chlorination practices, drinking water produced from sources with elevated levels of DOC can produce potentially dangerous by-products; including chloroform, haloacetonitriles, and chloral hydrate—along with the infamous trihalomethane by-products (U.S. EPA maximum drinking water level of THM = 80 ppb [176,180]). Alternatives to traditional chlorination disinfection techniques, like ozonation, and pretreatment strategies, like coagulation/flocculation or filtering, might be necessary for the treatment of natural water that is high in DOC concentration. Since 2006, the U.S. EPA has imposed more rigid regulations on water treatment disinfection by-products through "Stage 2" of their "Disinfection and Disinfection By-product Rule" [181].

In addition to the EPA, disinfection by-products are regulated by the World Health Organization and the European Union [176]. The problem of DOC levels and drinking water treatment have been reported in the U.S. [175,176] Norway [14,173,182], Sweden [178], Australia [173], and the U.K. [31,179], among other locations. In the Central Valley of California, where 23 million people receive their drinking water from the Sacramento-San Joaquin River watershed, DOC levels are a major concern for their drinking water supply [176]. As such, Sacramento's drinking water treatment has been the source of much study [175–177,183].

Studies that are aimed at characterizing the chemical structure of DOC in different input water systems, in order to select the best treatment strategy, are important. In one study, natural water sources of drinking water supplies for Norwegian and Australian systems found differing potentials toward the formation of disinfection by-products upon chlorination related to the chemical characterization of their DOC profiles [173]. In particular, the phenolic/aromatic portion of some DOC profiles are believed to be reactive sites for by-product formation [173]. In fact, a phenolic component of natural DOC, meta-dihydroxybenzene (resorcinol), is thought to be one of the major precursors to the formation of THMs [4]. As well, β-diketones found in DOC are also believed to be THM precursors [184]. Also, methoxyphenol is believed to be one of the most chlorine reactive components of DOC [185]. The phenolic content of DOC may increase under predicted climate change conditions, making DOC levels an even greater concern for water resource managers and treatment facilities.

Using multidimensional fluorescence with parallel factor analysis (PARAFAC) and ^{13}C-NMR studies, quinones (diketones being oxidized-quinones), were shown to represent significant portions of the DOC fluorescence profile [33]. Likewise, phenolic portions of humic acids from the International Humic Substance Society (IHSS) were identified with fluorescence/PARAFAC and reagent-based phenol assays [53] and lignin phenols found in several large artic rivers have been attributed to a PARAFAC-derived component [160]. In kinetic studies of THM formation upon chlorination of natural waters, it was shown that resorcinol components make up 15%–30% of THM precursors in natural water and are likely fast reacting (formed within first hours) THM precursors [4], while other phenolic compounds are likely slow reacting (from hours to weeks) THM precursors [4]. A mechanism has been suggested that involves chlorination by electrophilic aromatic substitutions of phenol that causes ring cleavage, followed by an addition/elimination pathway [186].

Aromatic molecules absorb strongly at 254 nm, thus UV_{254} (observed absorbance at 254 nm) has been suggested as a standard for measuring DOC [20,187]. Specific UV_{254} ($SUVA_{254}$), is the absorbance at 254 nm, measured in inverse meters, divided by an average DOC concentration in milligrams per liter. $SUVA_{254}$ has been proposed as a predicting tool for the evaluation of the suitability of waters containing DOC for the different water treatment methods [188,189]. The variability in "color" of DOC-laden water from different sources is often linked to its aromatic/phenolic content [26]. In the disinfection study mentioned above, phenolic portions of DOC were correlated to THM formation—and the relationship between UV_{254} readings and measured THM concentrations were examined [4]. The relationship between phenolic components of DOC in natural waters and climate change agents has been discussed here in detail, and the inherent importance in monitoring drinking water sources for such components appears to be an area of importance for future studies.

The different sources of natural waters within a system will contain different concentrations and compositions of DOC and inherent phenolic moieties. In effect, two different sampled locations could have the same net quantity of DOC, but because of differences in the composition of the native DOC, they will have different potentials for forming disinfection by-products. Hydrologic factors, including water flow, mixing, and transport, are crucial to understanding the fate of DOC (and its phenolic content) from its source location to its end water supply location. Land-use, anthropogenic sources, and seasonal variations can all impact the composition of DOC and phenolic compounds in water. Additionally, whether a result of climate change or a natural climate function, changes in the water table levels, including severe occurrences of droughts and floods, have an impact on DOC and inherent phenolic compound levels.

In the presence of higher levels of DOC, water treatment facilities may be forced to pre-treat their water with filtration, carbon adsorption, coagulation, or chemical techniques prior to disinfection by chlorination. These costly pretreatment steps often remove much, but not all, of the THM precursors and one study suggests some pretreatment steps may even double the proportion of brominated THMs, a disinfection by-product that is considered to be more carcinogenic than chlorine-derived THMs [179]. Furthermore, less time consuming and more cost effective methods will be required to assist water treatment plants and regulating agencies in monitoring drinking water in regard to potential for forming disinfection by-products. It has been reported that coagulation can reduce THM formation by 40%–50%, depending on the water source [190,191] and different portions of the DOC profile are removed more efficiently than others [50,192]. Pre-oxidation with ozone or chlorine dioxide leads to lower yields of THM upon chlorination [4]. Some scientists suggest that controlling DOC at the geographic source may even be the most economical way to control disinfection by-product formation [176,193]. Sometimes, drinking water supplies, primarily from surface waters, are mixed with groundwater reserves in order to dilute DOC concentrations; however, this practice can be counterproductive if the groundwater also contains DOC (or even a high fraction of phenolic compounds). In groundwater mixing of water supplies, bromine can also be introduced to the system, thereby introducing the potential for formation of brominated disinfection by-products [191]. Readers are also encouraged to see the review by Ritson *et al.*, in regard to changing DOM with climate change and the potential effects on water treatment processes [105].

6. Conclusions

The intent here was to review the issues and research related to observed increases in allochthonous DOC concentrations in several natural water systems, with particular attention to the impact on water resources under a changing climate, in the hope of providing renewed/sustained interest in the study of carbon cycling across the terrestrial landscape and to encourage a multidisciplinary approach toward current research in regard to the drivers of increasing allochthonous DOC levels. Climate change agents such as increases in temperature, variations in precipitation, drought, and related hydrology issues are complex in the way that they impact DOC levels, and may lead to either increases or decreases in aquatic DOC. The experimental evidence that relates elevated CO_2 and temperatures associated with climate change to allochthonous DOC increases in water provides the basis of a plausible mechanism for the increased production, and decreased net degradation, of DOC that is subsequently available for transport to area water catchments. Though it can be inferred that climate change has contributed to the observed increases in the DOC levels of natural water systems over the past few decades, the contribution might be less substantial thus far, as estimates show that elevated temperature and atmospheric CO_2 levels could account for only 10%–20% and 1%–5% of the observed DOC concentration increases, respectively [9]. The experimental controls in the climate change-DOC studies [11,12,69,89,129] often exceed the levels of warming and atmospheric CO_2 coinciding with the current increases in DOC, which makes it difficult to translate the results on a broader scale. In fact, the experimental conditions of most of these experiments exceed the IPCC predicted model for the next 50–100 years. Further, it is possible that climate change-related mechanisms might become increasingly responsible for continued variations in the net DOC concentrations in freshwaters—even before the climatic levels reach those of the experimental conditions.

The atmospheric chemical deposition paradigm predicts that DOC levels might stabilize in the future as sulfate deposition trends reach a lowered steady-state. In short, the current trend of increasing DOC potentially represents a return to DOC levels similar to those of the pre-industrial environment [8,118]. If atmospheric chemical deposition is a key driver responsible for the observed increased DOC levels over recent decades, the prevalence of its role will likely wane with decreasing deposition. Under these conditions, climate change agents may become a dominant mechanism in regulating DOC production and flux. Regardless, the role of DOC levels in selecting appropriate drinking water treatment techniques is an area of paramount importance.

Though not as heavily examined in the primary literature, it is likely that a combination of the major conceptual mechanisms could be occurring. Such a combination has been cited as possible causes for the observed trends in DOC [79,98,130,182,194]. Some studies indirectly state that more than 85% of DOC increases in different geographical areas can be explained by decreases in anthropogenic sulfate deposition, leaving room for other factors (like those of climate change) to contribute [8]. At the core of differentiating the two key conceptual models (or accepting a combination of the two) is whether primary production of DOC and microbial activity (and perhaps the related role of phenolic compounds) or solubility dynamics are driving the observed DOC changes in natural waters. As stated in the introduction, we have simplified our interpretation of current research to a reduced number of drivers of changes in DOC concentration by placing emphasis on two main paradigms presented commonly in the literature. While these drivers may correlate well at specific study sites, the complexity of the processes that govern the release of allochthonous DOC to inland natural waters is immense and may not lend itself to the establishment of distinct cause and effect relationships between a specific driver and specific instances of changing DOC over the spatial and temporal scales included in this review. Future studies should take a critical approach when assigning causal relationships to correlations, especially to explain DOC changes of a lesser magnitude.

It is proposed that future work on this topic should focus, in part, on the analysis of the phenolic composition of DOC. Not only could such a focus potentially assist in deciphering the two main mechanisms for DOC increases, but because of increasing projected climate change conditions (as well as trends of reducing atmospheric chemical deposition), the climate change model (and its inherent projected increases on the phenolic content of DOC) could become the primary driver of DOC levels in the future, especially in areas not subjected to historic acid deposition, areas undergoing land-use change, areas subject to a changing hydrologic regime, and areas situated in mid-high latitudes [118,161]. Studies of the phenolic composition would also be prudent due to concerns related to the role of phenolic compounds in the current and future treatment of drinking water.

Understanding carbon transfer across the terrestrial landscape and into aquatic ecosystems is important, especially considering the unprecedented anthropogenic environmental changes we are experiencing in the biosphere, lithosphere, atmosphere, and hydrosphere and the threat to drinking water sources [152]. To understand the effect of a changing climate on DOC quantity and quality in natural waters and to be better able to predict future DOC changes, it is apparent that work from multiple scientific disciplines is required to unravel the complexity of this issue. The multidisciplinary approach requires a commitment of collaboration from different communities of scientific research that often operate in isolation from each other [152]. Training new scientists to think in a more multidisciplinary way toward the subject of carbon cycling across the terrestrial environment will require exposure to a broad field of subjects in an educational setting. Ultimately, multidisciplinary collaboration will require effective communication of experimental findings and proposed paradigms, as well as an increased exchange between soil and aquatic disciplines [152].

Acknowledgments

Funding to support this work was provided by the National Technical Institute for the Deaf (NTID), through their *Innovation Fund.* The authors would like to thank Susan B. Smith for her helpful comments and technical assistance in preparation of this manuscript. The authors would also like to thank the anonymous reviewers who offered excellent insight and feedback to the manuscript.

References

1. Sinsabaugh, R.L.; Findlay, S. Dissolved organic matter: Out of the black box into the mainstream. In *Aquatic Ecosystems: Interactivity of Dissolved Organic Matter*; Elsevier Science: San Diego, CA, USA, 2003; pp. 479–496.
2. Filella, M.; Rodriguez-Murillo, J.C. Long-Term trends of organic carbon concentrations in freshwaters: Strengths and weaknesses of existing evidence. *Water* **2014**, *6*, 1360–1418.
3. Worrall, F.; Burt, T.P. Has the composition of fluvial DOC changed? Spatiotemporal patterns in the DOC-color relationship. *Glob. Biogeochem. Cycl.* **2010**, *24*, doi:10.1029/2008GB003445.
4. Gallard, H.; von Gunten, U. Chlorination of natural organic matter: Kinetics of chlorination and of THM formation. *Water Res.* **2002**, *36*, 65–74.
5. Sharp, E.L.; Parsons, S.A.; Jefferson, B. Seasonal variations in natural organic matter and its impact on coagulation in water treatment. *Sci. Total Environ.* **2006**, *363*, 183–194.
6. Reynolds, B.; Chamberlain, P.M.; Poskitt, J.; Woods, C.; Scott, W.A.; Rowe, E.C.; Robinson, D.A.; Frogbrook, Z.L.; Keith, A.M.; Henrys, P.A.; *et al.* Countryside survey: National "Soil change"

1978–2007 for topsoils in great britain-acidity, carbon, and total nitrogen status. *Vadose Zone J.* **2013**, *12*, doi:10.2136/vzj2012.0114.

7. Evans, C.D.; Monteith, D.T.; Fowler, D.; Cape, J.N.; Brayshaw, S. Hydrochloric acid: An overlooked driver of environmental change. *Environ. Sci. Technol.* **2011**, *45*, 1887–1894.
8. Monteith, D.T.; Stoddard, J.L.; Evans, C.D.; de Wit, H.A.; Forsius, M.; Hogasen, T.; Wilander, A.; Skjelkvale, B.L.; Jeffries, D.S.; Vuorenmaa, J.; *et al.* Dissolved organic carbon trends resulting from changes in atmospheric deposition chemistry. *Nature* **2007**, *450*, 537–540.
9. Evans, C.D.; Chapman, P.J.; Clark, J.M.; Monteith, D.T.; Cresser, M.S. Alternative explanations for rising dissolved organic carbon export from organic soils. *Glob. Chang. Biol.* **2006**, *12*, 2044–2053.
10. Findlay, S.E.G. Increased carbon transport in the hudson river: Unexpected consequence of nitrogen deposition? *Front. Ecol. Environ.* **2005**, *3*, 133–137.
11. Freeman, C.; Evans, C.D.; Monteith, D.T.; Reynolds, B.; Fenner, N. Export of organic carbon from peat soils. *Nature* **2001**, *412*, 785–785.
12. Freeman, C.; Fenner, N.; Ostle, N.J.; Kang, H.; Dowrick, D.J.; Reynolds, B.; Lock, M.A.; Sleep, D.; Hughes, S.; Hudson, J. Export of dissolved organic carbon from peatlands under elevated carbon dioxide levels. *Nature* **2004**, *430*, 195–198.
13. Wilson, H.F.; Saiers, J.E.; Raymond, P.A.; Sobczak, W.V. Hydrologic drivers and seasonality of dissolved organic carbon concentration, nitrogen content, bioavailability, and export in a forested new england stream. *Ecosystems* **2013**, *16*, 604–616.
14. Hongve, D.; Riise, G.; Kristiansen, J.F. Increased colour and organic acid concentrations in norwegian forest lakes and drinking water—A result of increased precipitation? *Aquat. Sci.* **2004**, *66*, 231–238.
15. Anderson, N.; Dietz, R.; Engstrom, D. Land-Use change, not climate, controls organic carbon burial in lakes. *Proc. R. Soc. B Biol. Sci.* **2013**, *280*, doi:10.1098/rspb.2013.1278.
16. Cool, G.; Lebel, A.; Sadiq, R.; Rodriguez, M.J. Impact of catchment geophysical characteristics and climate on the regional variability of dissolved organic carbon (DOC) in surface water. *Sci. Total Environ.* **2014**, *490*, 947–956.
17. McGuire, A.D.; Anderson, L.G.; Christensen, T.R.; Dallimore, S.; Guo, L.D.; Hayes, D.J.; Heimann, M.; Lorenson, T.D.; Macdonald, R.W.; Roulet, N. Sensitivity of the carbon cycle in the arctic to climate change. *Ecol. Monogr.* **2009**, *79*, 523–555.
18. Detwiler, R.P.; Hall, C.A.S. Tropical forests and the global carbon-cycle. *Science* **1988**, *239*, 42–47.
19. Brown, S.; Lugo, A.E. The storage and production of organic-matter in tropical forests and their role in the global carbon-cycle. *Biotropica* **1982**, *14*, 161–187.
20. Spencer, R.G.M.; Hernes, P.J.; Ruf, R.; Baker, A.; Dyda, R.Y.; Stubbins, A.; Six, J. Temporal controls on dissolved organic matter and lignin biogeochemistry in a pristine tropical river, democratic republic of Congo. *J. Geophys. Res. Biogeosci.* **2010**, *115*, doi:10.1029/2009JG001180.
21. Cramer, W.; Bondeau, A.; Schaphoff, S.; Lucht, W.; Smith, B.; Sitch, S. Tropical forests and the global carbon cycle: Impacts of atmospheric carbon dioxide, climate change and rate of deforestation. *Philos. Trans. R. Soc. Lond. Ser. B Biol. Sci.* **2004**, *359*, 331–343.

22. Monteith, D.; Evans, C.; Henrys, P.; Simpson, G.; Malcolm, I. Trends in the hydrochemistry of acid-sensitive surface waters in the UK 1988–2008. *Ecol. Indic.* **2014**, *37*, 287–303.
23. Evans, C.D.; Monteith, D.T.; Reynolds, B.; Clark, J.M. Buffering of recovery from acidification by organic acids. *Sci. Total Environ.* **2008**, *404*, 316–325.
24. Battin, T.J.; Luyssaert, S.; Kaplan, L.A.; Aufdenkampe, A.K.; Richter, A.; Tranvik, L.J. The boundless carbon cycle. *Nat. Geosci.* **2009**, *2*, 598–600.
25. Filella, M. Freshwaters: Which NOM matters? *Environ. Chem. Lett.* **2009**, *7*, 21–35.
26. Thurman, E.M. *Organic Geochemistry of Natural Waters*; Kluwer Academic: Dordrecht, the Netherlands, 1985.
27. McDonald, S.; Bishop, A.G.; Prenzler, P.D.; Robards, K. Analytical chemistry of freshwater humic substances. *Anal. Chim. Acta* **2004**, *527*, 105–124.
28. Leenheer, J.A.; Croue, J.P. Characterizing aquatic dissolved organic matter. *Environ. Sci. Technol.* **2003**, *37*, 18A–26A.
29. Malcolm, R.L. The uniqueness of humic substances in each of soil, stream and marine environments. *Anal. Chim. Acta* **1990**, *232*, 19–30.
30. Drever, J.I. *The Geochemistry of Natural Waters: Surface and Groundwater Environments*; Prentice Hall: Upper Saddle River, NJ, USA, 1997.
31. Evans, C.D.; Monteith, D.T.; Cooper, D.M. Long-Term increases in surface water dissolved organic carbon: Observations, possible causes and environmental impacts. *Environ. Pollut.* **2005**, *137*, 55–71.
32. Stevenson, I.L.; Schnitzer, M. Transmission electron-microscopy of extracted fulvic and humic acids. *Soil Sci.* **1982**, *133*, 179–185.
33. Cory, R.M.; McKnight, D.M. Fluorescence spectroscopy reveals ubiquitous presence of oxidized and reduced quinones in dissolved organic matter. *Environ. Sci. Technol.* **2005**, *39*, 8142–8149.
34. Parlanti, E.; Morin, B.; Vacher, L. Combined 3D-spectrofluorometry, high performance liquid chromatography and capillary electrophoresis for the characterization of dissolved organic matter in natural waters. *Org. Geochem.* **2002**, *33*, 221–236.
35. Abdulla, H.A.N.; Sleighter, R.L.; Hatcher, P.G. Two dimensional correlation analysis of fourier transform ion cyclotron resonance mass spectra of dissolved organic matter: A new graphical analysis of trends. *Anal. Chem.* **2013**, *85*, 3895–3902.
36. Koch, B.P.; Witt, M.R.; Engbrodt, R.; Dittmar, T.; Kattner, G. Molecular formulae of marine and terrigenous dissolved organic matter detected by electrospray ionization fourier transform ion cyclotron resonance mass spectrometry. *Geochim. Cosmochim. Acta* **2005**, *69*, 3299–3308.
37. Gonsior, M.; Peake, B.M.; Cooper, W.T.; Podgorski, D.; D'Andrilli, J.; Cooper, W.J. Photochemically induced changes in dissolved organic matter identified by ultrahigh resolution fourier transform ion cyclotron resonance mass spectrometry. *Environ. Sci. Technol.* **2009**, *43*, 698–703.
38. Brown, T.L.; Rice, J.A. Effect of experimental parameters on the ESI FT-ICR mass spectrum of fulvic acid. *Anal. Chem.* **2000**, *72*, 384–390.
39. Kujawinski, E.B.; Hatcher, P.G.; Freitas, M.A. High-Resolution fourier transform ion cyclotron resonance mass spectrometry of humic and fulvic acids: Improvements and comparisons. *Anal. Chem.* **2002**, *74*, 413–419.

40. Benner, R.; Kaiser, K. Biological and photochemical transformations of amino acids and lignin phenols in riverine dissolved organic matter. *Biogeochemistry* **2011**, *102*, 209–222.
41. Mutabaruka, R.; Hairiah, K.; Cadisch, G. Microbial degradation of hydrolysable and condensed tannin polyphenol-protein complexes in soils from different land-use histories. *Soil Biol. Biochem.* **2007**, *39*, 1479–1492.
42. Fenner, N.; Freeman, C.; Reynolds, B. Hydrological effects on the diversity of phenolic degrading bacteria in a peatland: Implications for carbon cycling. *Soil Biol. Biochem.* **2005**, *37*, 1277–1287.
43. Orth, A.B.; Denny, M.; Tien, M. Overproduction of lignin-degrading enzymes by and isolate of phanerochaete-chrysosporium. *Appl. Environ. Microbiol.* **1991**, *57*, 2591–2596.
44. McKnight, D.M.; Andrews, E.D.; Spaulding, S.A.; Aiken, G.R. Aquatic fulvic-acids in algal-rich antarctic ponds. *Limnol. Oceanogr.* **1994**, *39*, 1972–1979.
45. McKnight, D.M.; Boyer, E.W.; Westerhoff, P.K.; Doran, P.T.; Kulbe, T.; Andersen, D.T. Spectrofluorometric characterization of dissolved organic matter for indication of precursor organic material and aromaticity. *Limnol. Oceanogr.* **2001**, *46*, 38–48.
46. Jaffe, R.; McKnight, D.; Maie, N.; Cory, R.; McDowell, W.H.; Campbell, J.L. Spatial and temporal variations in DOM composition in ecosystems: The importance of long-term monitoring of optical properties. *J. Geophys. Res. Biogeosci.* **2008**, *113*, doi:10.1029/2008JG000683.
47. SanClements, M.D.; Oelsner, G.P.; McKnight, D.M.; Stoddard, J.L.; Nelson, S.J. New insights into the source of decadal increases of dissolved organic matter in acid-sensitive lakes of the northeastern United States. *Environ. Sci. Technol.* **2012**, *46*, 3212–3219.
48. Coble, P.G.; Green, S.A.; Blough, N.V.; Gagosian, R.B. Characterization of dissolved organic-matter in the black-sea by fluorescence spectroscopy. *Nature* **1990**, *348*, 432–435.
49. Coble, P.G. Characterization of marine and terrestrial DOM in seawater using excitation emission matrix spectroscopy. *Mar. Chem.* **1996**, *51*, 325–346.
50. Gone, D.L.; Seidel, J.; Batiot, C.; Bamory, K.; Ligban, R.; Biemi, J. Using fluoresence spectroscopy eem to evaluate the efficiency of organic matter removal during coagulation-flocculation of a tropical surface water (agbo reservoir). *J. Hazard. Mater.* **2009**, *172*, 693–699.
51. Santin, C.; Yamashita, Y.; Otero, X.L.; Alvarez, M.A.; Jaffe, R. Characterizing humic substances from estuarine soils and sediments by excitation-emission matrix spectroscopy and parallel factor analysis. *Biogeochemistry* **2009**, *96*, 131–147.
52. Walker, S.A.; Amon, R.M.W.; Stedmon, C.; Duan, S.; Louchouarn, P. The use of PARAFAC modeling to trace terrestrial dissolved organic matter and fingerprint water masses in coastal Canadian arctic surface waters. *J. Geophys. Res. Biogeosci.* **2009**, *114*, doi:10.1029/2009JG000990.
53. Pagano, T.; Ross, A.D.; Chiarelli, J.; Kenny, J.E. Multidimensional fluorescence studies of the phenolic content of dissolved organic carbon in humic substances. *J. Environ. Monit.* **2012**, *14*, 937–943.
54. Sobek, S.; Tranvik, L.J.; Prairie, Y.T.; Kortelainen, P.; Cole, J.J. Patterns and regulation of dissolved organic carbon: An analysis of 7,500 widely distributed lakes. *Limnol. Oceanogr.* **2007**, *52*, 1208–1219.
55. Williams, C.J.; Yamashita, Y.; Wilson, H.F.; Jaffe, R.; Xenopoulos, M.A. Unraveling the role of land use and microbial activity in shaping dissolved organic matter characteristics in stream ecosystems. *Limnol. Oceanogr.* **2010**, *55*, 1159–1171.

56. Algesten, G.; Sobek, S.; Bergstrom, A.K.; Jonsson, A.; Tranvik, L.J.; Jansson, M. Contribution of sediment respiration to summer CO_2 emission from low productive boreal and subarctic lakes. *Microb. Ecol.* **2005**, *50*, 529–535.
57. Mattsson, T.; Kortelain, P.; Laubel, A.; Evans, D.; Pujo-Pay, M.; Raike, A.; Conan, P. Export of dissolved organic matter in relation to land use along a european climatic gradient. *Sci. Total Environ.* **2009**, *407*, 1967–1976.
58. Stanley, E.H.; Powers, S.M.; Lottig, N.R.; Buffam, I.; Crawford, J.T. Contemporary changes in dissolved organic carbon (DOC) in human-dominated rivers: Is there a role for DOC management? *Freshw. Biol.* **2012**, *57*, 26–42.
59. Jonsson, A.; Meili, M.; Bergstrom, A.K.; Jansson, M. Whole-Lake mineralization of allochthonous and autochthonous organic carbon in a large humic lake (ortrasket, N. sweden). *Limnol. Oceanogr.* **2001**, *46*, 1691–1700.
60. Royer, T.V.; David, M.B. Export of dissolved organic carbon from agricultural streams in Illinois, USA. *Aquat. Sci.* **2005**, *67*, 465–471.
61. Bida, M.R. Quantity and Composition of Stream Dissolved Organic Matter in the Watershed of Conesus Lake, New York. Master's Thesis, Rochester Institute of Technology, Rochester, NY, USA, 15 July 2013.
62. Tipping, E.; Woof, C. Humic substances in acid organic soils—Modeling their release to the soil solution in terms of humic charge. *J. Soil Sci.* **1990**, *41*, 573–586.
63. Pind, A.; Freeman, C.; Lock, M.A. Enzymatic degradation of phenolic materials in peatlands—Measurement of phenol oxidase activity. *Plant Soil* **1994**, *159*, 227–231.
64. Kalbitz, K.; Solinger, S.; Park, J.H.; Michalzik, B.; Matzner, E. Controls on the dynamics of dissolved organic matter in soils: A review. *Soil Sci.* **2000**, *165*, 277–304.
65. Wershaw, R.L. *Evaluation of Conceptual Models of Natural Organic Matter (humus) from a Consideration of the Chemical and Biochemical Processes of Humification*; U.S. Geological Survey: Reston, VA, USA, 2004.
66. Findlay, S.; Sinsabaugh, R.L. *Aquatic Ecosystems: Interactivity of Dissolved Organic Matter*; Academic Press: San Diego, CA, USA, 2003.
67. Aitkenhead-Peterson, J.A.; McDowell, W.H.; Neff, J.C. Sources, production, and regulation of allochthonous dissolved organic matter inputs to surface waters. In *Aquatic Ecosystems: Interactivity of Dissolved Organic Matter*; Findlay, S.E.G., Sinsabaugh, R.L., Eds.; Elsevier Science: San Diego, CA, USA, 2003; pp. 25–70.
68. Gorham, E. Northern peatlands—Role in the carbon-cycle and probable responses to climatic warming. *Ecol. Appl.* **1991**, *1*, 182–195.
69. Fenner, N.; Ostle, N.J.; McNamara, N.; Sparks, T.; Harmens, H.; Reynolds, B.; Freeman, C. Elevated CO_2 effects on peatland plant community carbon dynamics and DOC production. *Ecosystems* **2007**, *10*, 635–647.
70. Holmes, R.M.; McClelland, J.W.; Peterson, B.J.; Tank, S.E.; Bulygina, E.; Eglinton, T.I.; Gordeev, V.V.; Gurtovaya, T.Y.; Raymond, P.A.; Repeta, D.J.; *et al.* Seasonal and annual fluxes of nutrients and organic matter from large rivers to the arctic ocean and surrounding seas. *Estuaries Coasts* **2012**, *35*, 369–382.

71. Kang, H.; Freeman, C.; Jang, I. Global increases in dissolved organic carbon in rivers and their implications. *Korean J. Limnol.* **2010**, *43*, 453–458.
72. Engelhaupt, E.; Bianchi, T.S. Sources and composition of high-molecular-weight dissolved organic carbon in a Southern Louisiana tidal stream (bayou trepagnier). *Limnol. Oceanogr.* **2001**, *46*, 917–926.
73. Skjelkvale, B.L.; Mannio, J.; Wilander, A.; Andersen, T. Recovery from acidification of lakes in Finland, Norway and Sweden 1990–1999. *Hydrol. Earth Syst. Sci.* **2001**, *5*, 327–337.
74. Dillon, P.J.; Skjelkvale, B.L.; Somers, K.M.; Torseth, K. Coherent responses of sulphate concentration in norwegian lakes: Relationships with sulphur deposition and climate indices. *Hydrol. Earth Syst. Sci.* **2003**, *7*, 596–608.
75. Driscoll, C.T.; Driscoll, K.M.; Roy, K.M.; Mitchell, M.J. Chemical response of lakes in the adirondack region of new york to declines in acidic deposition. *Environ. Sci. Technol.* **2003**, *37*, 2036–2042.
76. Hudson, J.J.; Dillon, P.J.; Somers, K.M. Long-Term patterns in dissolved organic carbon in boreal lakes: The role of incident radiation, precipitation, air temperature, southern oscillation and acid deposition. *Hydrol. Earth Syst. Sci.* **2003**, *7*, 390–398.
77. Pastor, J.; Solin, J.; Bridgham, S.D.; Updegraff, K.; Harth, C.; Weishampel, P.; Dewey, B. Global warming and the export of dissolved organic carbon from boreal peatlands. *Oikos* **2003**, *100*, 380–386.
78. Xenopoulos, M.A.; Lodge, D.M.; Frentress, J.; Kreps, T.A.; Bridgham, S.D.; Grossman, E.; Jackson, C.J. Regional comparisons of watershed determinants of dissolved organic carbon in temperate lakes from the upper great lakes region and selected regions globally. *Limnol. Oceanogr.* **2003**, *48*, 2321–2334.
79. Worrall, F.; Burt, T.; Adamson, J. Can climate change explain increases in DOC flux from upland peat catchments? *Sci. Total Environ.* **2004**, *326*, 95–112.
80. Worrall, F.; Harriman, R.; Evans, C.D.; Watts, C.D.; Adamson, J.; Neal, C.; Tipping, E.; Burt, T.; Grieve, I.; Monteith, D.; *et al.* Trends in dissolved organic carbon in UK rivers and lakes. *Biogeochemistry* **2004**, *70*, 369–402.
81. Fahey, T.J.; Siccama, T.G.; Driscoll, C.T.; Likens, G.E.; Campbell, J.; Johnson, C.E.; Battles, J.J.; Aber, J.D.; Cole, J.J.; Fisk, M.C.; *et al.* The biogeochemistry of carbon at hubbard brook. *Biogeochemistry* **2005**, *75*, 109–176.
82. Harrison, J.A.; Caraco, N.; Seitzinger, S.P. Global patterns and sources of dissolved organic matter export to the coastal zone: Results from a spatially explicit, global model. *Glob. Biogeochem. Cycl.* **2005**, *19*, doi:10.1029/2005GB002480.
83. Skjelkvale, B.L.; Borg, H.; Hindar, A.; Wilander, A. Large scale patterns of chemical recovery in lakes in Norway and Sweden: Importance of seasalt episodes and changes in dissolved organic carbon. In Proceedings of the 7th International Conference on Acid Deposition, Prague, Czech Republic, 12–17 June 2005; pp. 1174–1180.
84. Skjelkvale, B.L.; Stoddard, J.L.; Jeffries, D.S.; Torseth, K.; Hogasen, T.; Bowman, J.; Mannio, J.; Monteith, D.T.; Mosello, R.; Rogora, M.; *et al.* Regional scale evidence for improvements in surface water chemistry 1990–2001. *Environ. Pollut.* **2005**, *137*, 165–176.

85. Burns, D.A.; McHale, M.R.; Driscoll, C.T.; Roy, K.M. Response of surface water chemistry to reduced levels of acid precipitation: Comparison of trends in two regions of New York, USA. *Hydrol. Process.* **2006**, *20*, 1611–1627.

86. Gueguen, C.; Guo, L.D.; Wang, D.; Tanaka, N.; Hung, C.C. Chemical characteristics and origin of dissolved organic matter in the Yukon river. *Biogeochemistry* **2006**, *77*, 139–155.

87. Roulet, N.; Moore, T.R. Environmental chemistry—Browning the waters. *Nature* **2006**, *444*, 283–284.

88. Vuorenmaa, J.; Forsius, M.; Mannio, J. Increasing trends of total organic carbon concentrations in small forest lakes in finland from 1987 to 2003. *Sci. Total Environ.* **2006**, *365*, 47–65.

89. Fenner, N.; Freeman, C.; Lock, M.A.; Harmens, H.; Reynolds, B.; Sparks, T. Interactions between elevated CO_2 and warming could amplify DOC exports from peatland catchments. *Environ. Sci. Technol.* **2007**, *41*, 3146–3152.

90. Worrall, F.; Burt, T.P. Flux of dissolved organic carbon from UK rivers. *Glob. Biogeochem. Cycl.* **2007**, *21*, doi:10.1029/2006GB002709.

91. Worrall, F.; Burt, T.P. Trends in DOC concentration in great britain. *J. Hydrol.* **2007**, *346*, 81–92.

92. Worrall, F.; Burt, T.P. The effect of severe drought on the dissolved organic carbon (DOC) concentration and flux from British rivers. *J. Hydrol.* **2008**, *361*, 262–274.

93. Worrall, F.; Guilbert, T.; Besien, T. The flux of carbon from rivers: The case for flux from england and wales. *Biogeochemistry* **2007**, *86*, 63–75.

94. De Wit, H.A.; Mulder, J.; Hindar, A.; Hole, L. Long-Term increase in dissolved organic carbon in streamwaters in Norway is response to reduced acid deposition. *Environ. Sci. Technol.* **2007**, *41*, 7706–7713.

95. Couture, S.; Houle, D.; Gagnon, C. Increases of dissolved organic carbon in temperate and boreal lakes in Quebec, Canada. *Environ. Sci. Pollut. Res.* **2012**, *19*, 361–371.

96. Lawrence, G.B.; Dukett, J.E.; Houck, N.; Snyder, P.; Capone, S. Increases in dissolved organic carbon accelerate loss of toxic al in Adirondack lakes recovering from acidification. *Environ. Sci. Technol.* **2013**, *47*, 7095–7100.

97. Schindler, D.W.; Curtis, P.J.; Bayley, S.E.; Parker, B.R.; Beaty, K.G.; Stainton, M.P. Climate-Induced changes in the dissolved organic carbon budgets of boreal lakes. *Biogeochemistry* **1997**, *36*, 9–28.

98. Hejzlar, J.; Dubrovsky, M.; Buchtele, J.; Ruzicka, M. The apparent and potential effects of climate change on the inferred concentration of dissolved organic matter in a temperate stream (the Malse River, South Bohemia). In Proceedings of the Detecting Environmental Change—Science and Society Conference, London, England, 16–20 July 2001; pp. 143–152.

99. Oulehle, F.; Chuman, T.; Majer, V.; Hruska, J. Chemical recovery of acidified bohemian lakes between 1984 and 2012: The role of acid deposition and bark beetle induced forest disturbance. *Biogeochemistry* **2013**, *116*, 83–101.

100. Frey, K.E.; Smith, L.C. Amplified carbon release from vast west Siberian Peatlands by 2100. *Geophys. Res. Lett.* **2005**, *32*, doi:10.1029/2004GL022025.

101. Prokushkin, A.S.; Kajimoto, T.; Prokushkin, S.G.; McDowell, W.H.; Abaimov, A.P.; Matsuura, Y. Climatic Factors influencing fluxes of dissolved organic carbon from the forest floor in a

continuous-permafrost Siberian Watershed. In Proceedings of the 12th Annual Conference of the International-Boreal-Forest-Research-Association, Fairbanks, AK, USA, 3–6 May 2004; pp. 2130–2140.

102. Gordeev, V.V.; Kravchishina, M.D. River flux of dissolved organic carbon (DOC) and particulate organic carbon (POC) to the Arctic Ocean: What are the consequences of the global changes? In *Influence of Climate Change on the Changing Arctic and Sub-Arctic Conditions*; Nihoul, J.C.J., Kostianoy, A.G., Eds.; Springer: Dordrecht, the Netherlands, 2009; pp. 145–160.
103. Porcal, P.; Koprivnjak, J.F.; Molot, L.A.; Dillon, P.J. Humic substances-part 7: The biogeochemistry of dissolved organic carbon and its interactions with climate change. *Environ. Sci. Pollut. Res.* **2009**, *16*, 714–726.
104. Tranvik, L.J.; Downing, J.A.; Cotner, J.B.; Loiselle, S.A.; Striegl, R.G.; Ballatore, T.J.; Dillon, P.; Finlay, K.; Fortino, K.; Knoll, L.B.; *et al.* Lakes and reservoirs as regulators of carbon cycling and climate. *Limnol. Oceanogr.* **2009**, *54*, 2298–2314.
105. Ritson, J.P.; Graham, N.J.D.; Templeton, M.R.; Clark, J.M.; Gough, R.; Freeman, C. The impact of climate change on the treatability of dissolved organic matter (DOM) in upland water supplies: A UK perspective. *Sci. Total Environ.* **2014**, *473*, 714–730.
106. Stoddard, J.L.; Jeffries, D.S.; Lukewille, A.; Clair, T.A.; Dillon, P.J.; Driscoll, C.T.; Forsius, M.; Johannessen, M.; Kahl, J.S.; Kellogg, J.H.; *et al.* Regional trends in aquatic recovery from acidification in North America and Europe. *Nature* **1999**, *401*, 575–578.
107. Ulrich, K.U.; Paul, L.; Meybohm, A. Response of drinking-water reservoir ecosystems to decreased acidic atmospheric deposition in SE Germany: Trends of chemical reversal. *Environ. Pollut.* **2006**, *141*, 42–53.
108. Rowe, E.C.; Tipping, E.; Posch, M.; Oulehle, F.; Cooper, D.M.; Jones, T.G.; Burden, A.; Hall, J.; Evans, C.D. Predicting nitrogen and acidity effects on long-term dynamics of dissolved organic matter. *Environ. Pollut.* **2014**, *184*, 271–282.
109. Fowler, D.; Smith, R.I.; Muller, J.B.A.; Hayman, G.; Vincent, K.J. Changes in the atmospheric deposition of acidifying compounds in the UK between 1986 and 2001. *Environ. Pollut.* **2005**, *137*, 15–25.
110. Holland, E.A.; Braswell, B.H.; Sulzman, J.; Lamarque, J.F. Nitrogen deposition onto the United States and Western Europe: Synthesis of observations and models. *Ecol. Appl.* **2005**, *15*, 38–57.
111. Pregitzer, K.S.; Zak, D.R.; Burton, A.J.; Ashby, J.A.; MacDonald, N.W. Chronic nitrate additions dramatically increase the export of carbon and nitrogen from northern hardwood ecosystems. *Biogeochemistry* **2004**, *68*, 179–197.
112. Yano, Y.; McDowell, W.H.; Aber, J.D. Biodegradable dissolved organic carbon in forest soil solution and effects of chronic nitrogen deposition. *Soil Biol. Biochem.* **2000**, *32*, 1743–1751.
113. Currie, W.S.; Aber, J.D.; McDowell, W.H.; Boone, R.D.; Magill, A.H. Vertical transport of dissolved organic C and N under long-term N amendments in pine and hardwood forests. *Biogeochemistry* **1996**, *35*, 471–505.
114. Bragazza, L.; Freeman, C.; Jones, T.; Rydin, H.; Limpens, J.; Fenner, N.; Ellis, T.; Gerdol, R.; Hajek, M.; Hajek, T.; *et al.* Atmospheric nitrogen deposition promotes carbon loss from peat bogs. *Proc. Natl. Acad. Sci. USA* **2006**, *103*, 19386–19389.

115. Evans, C.D.; Reynolds, B.; Hinton, C.; Hughes, S.; Norris, D.; Grant, S.; Williams, B. Effects of decreasing acid deposition and climate change on acid extremes in an upland stream. *Hydrol. Earth Syst. Sci.* **2008**, *12*, 337–351.
116. Krug, E.C.; Frink, C.R. Acid-Rain on acid soil—A new perspective. *Science* **1983**, *221*, 520–525.
117. Evans, A.; Zelazny, L.W.; Zipper, C.E. Division s-7—Forest and range soils—Solution parameters influencing dissolved organic-carbon levels in 3 forest soils. *Soil Sci. Soc. Am. J.* **1988**, *52*, 1789–1792.
118. Evans, C.D.; Jones, T.G.; Burden, A.; Ostle, N.; Zielinski, P.; Cooper, M.D.A.; Peacock, M.; Clark, J.M.; Oulehle, F.; Cooper, D.; *et al.* Acidity controls on dissolved organic carbon mobility in organic soils. *Glob. Chang. Biol.* **2012**, *18*, 3317–3331.
119. Worrall, F.; Gibson, H.S.; Burt, T.P. Production *vs.* Solubility in controlling runoff of DOC from peat soils—The use of an event analysis. *J. Hydrol.* **2008**, *358*, 84–95.
120. Solomon, S. *Climate Change 2007: The Physical Science Basis: Part of the Working Group I Contribution to the Fourth Assessment Report of the Intergovernmental Panel on Climate Change*; Cambridge University Press: Cambridge, MA, USA, 2007.
121. Stocker, T.F.; Qin, D.; Plattner, G.-K.; Tignor, M.; Allen, S.K.; Boschung, J.; Nauels, A.; Xia, Y.; Bex, V.; Midgley, P.M. Climate change 2013: The physical science basis. In *Intergovernmental Panel on Climate Change, Working Group I Contribution to the IPCC Fifth Assessment Report (AR5)*; Cambridge University Press: New York, NY, USA, 2013.
122. Houghton, J.T. *Climate Change 2001: The Scientific Basis: Contribution of Working Group I to the Third Assessment Report of the Intergovernmental Panel on Climate Change*; Cambridge University Press: Cambridge, MA, USA, 2001.
123. Harrison, A.F.; Taylor, K.; Scott, A.; Poskitt, J.; Benham, D.; Grace, J.; Chaplow, J.; Rowland, P. Potential effects of climate change on DOC release from three different soil types on the northern pennines UK: Examination using field manipulation experiments. *Glob. Chang. Biol.* **2008**, *14*, 687–702.
124. Pace, M.L.; Cole, J.J. Synchronous variation of dissolved organic carbon and color in lakes. *Limnol. Oceanogr.* **2002**, *47*, 333–342.
125. Tipping, E.; Smith, E.J.; Bryant, C.L.; Adamson, J.K. The organic carbon dynamics of a moorland catchment in NW England. *Biogeochemistry* **2007**, *84*, 171–189.
126. Kominoski, J.S.; Moore, P.A.; Wetzel, R.G.; Tuchman, N.C. Elevated CO_2 alters leaf-litter-derived dissolved organic carbon: Effects on stream periphyton and crayfish feeding preference. *J. North Am. Benthol. Soc.* **2007**, *26*, 663–672.
127. Norby, R.J.; DeLucia, E.H.; Gielen, B.; Calfapietra, C.; Giardina, C.P.; King, J.S.; Ledford, J.; McCarthy, H.R.; Moore, D.J.P.; Ceulemans, R.; *et al.* Forest response to elevated CO_2 is conserved across a broad range of productivity. *Proc. Natl. Acad. Sci. USA* **2005**, *102*, 18052–18056.
128. Freeman, C.; Baxter, R.; Farrar, J.F.; Jones, S.E.; Plum, S.; Ashendon, T.W.; Stirling, C. Could competition between plants and microbes regulate plant nutrition and atmospheric CO_2 concentrations? *Sci. Total Environ.* **1998**, *220*, 181–184.
129. Kane, E.S.; Mazzoleni, L.R.; Kratz, C.J.; Hribljan, J.A.; Johnson, C.P.; Pypker, T.G.; Chimner, R. Peat porewater dissolved organic carbon concentration and lability increase with warming: A field temperature manipulation experiment in a poor-fen. *Biogeochemistry* **2014**, *119*, 161–178.

130. Clark, J.M.; Chapman, P.J.; Adamson, J.K.; Lane, S.N. Influence of drought-induced acidification on the mobility of dissolved organic carbon in peat soils. *Glob. Chang. Biol.* **2005**, *11*, 791–809.
131. Worrall, F.; Burt, T.P.; Adamson, J.K. Linking pulses of atmospheric deposition to DOC release in an upland peat-covered catchment. *Glob. Biogeochem. Cycl.* **2008**, *22*, 15.
132. Wittwer, S.; Robb, W. Carbon dioxide enrichment of greenhouse atmospheres for food crop production. *Econ. Bot.* **1964**, *18*, 34–56.
133. Norby, R.J.; Zak, D.R. Ecological lessons from free-air CO_2 enrichment (FACE) experiments. *Annu. Rev. Ecol. Evol. Syst.* **2011**, *42*, 181–203.
134. Clair, T.A.; Arp, P.; Moore, T.R.; Dalva, M.; Meng, F.R. Gaseous carbon dioxide and methane, as well as dissolved organic carbon losses from a small temperate wetland under a changing climate. In Proceedings of the Advances in Terrestrial Ecosystem: Carbon Inventory Measurements and Monitoring Conference, Raleigh, NC, USA, 3–5 October 2000; pp. S143–S148.
135. Mitsch, W.J.; Gosselink, J.G. *Wetlands*; Van Nostrand Reinhold: New York, NY, USA, 1993.
136. Woodin, S.; Graham, B.; Killick, A.; Skiba, U.; Cresser, M. Nutrient limitation of the long-term response of heather [Calluna-vulgaris (L) Hull] to CO_2 enrichment. *New Phytol.* **1992**, *122*, 635–642.
137. Dawes, M.A.; Hagedorn, F.; Handa, I.T.; Streit, K.; Ekblad, A.; Rixen, C.; Korner, C.; Hattenschwiler, S. An alpine treeline in a carbon dioxide-rich world: Synthesis of a nine-year free-air carbon dioxide enrichment study. *Oecologia* **2013**, *171*, 623–637.
138. Jiang, H.B.; Wang, Y.; Kanost, M.R. Pro-Phenol oxidase activating proteinase from an insect, manduca sexta: A bacteria-inducible protein similar to drosophila Easter. *Proc. Natl. Acad. Sci. USA* **1998**, *95*, 12220–12225.
139. Wetzel, R.G. Gradient-Dominated ecosystems—Sources and regulatory functions of dissolved organic-matter in fresh-water ecosystems. In Proceedings of the Symposium on Dissolved Organic Matter in Lacustrine Ecosystems: Energy Source and System Regulator, Helsinki, Finland, 6–10 August 1990; pp. 181–198.
140. Freeman, C.; Lock, M.A.; Marxsen, J.; Jones, S.E. Inhibitory effects of high-molecular-weight dissolved organic-matter upon metabolic processes in biofilms from contrasting rivers and streams. *Freshw. Biol.* **1990**, *24*, 159–166.
141. Sinsabaugh, R.L.; Antibus, R.K.; Linkins, A.E. An enzymatic approach to the analysis of microbial activity during plant litter decomposition. In Proceedings of the Workshop on Modern Techniques in Soil Ecology, Atlanta, GA, USA, 11–15 September 1989; Elsevier Science Bv: Atlanta, GA, USA; pp. 43–54.
142. Freeman, C.; Ostle, N.; Kang, H. An enzymic "latch" on a global carbon store—A shortage of oxygen locks up carbon in peatlands by restraining a single enzyme. *Nature* **2001**, *409*, 149.
143. Fenner, N.; Freeman, C. Drought-Induced carbon loss in peatlands. *Nat. Geosci.* **2011**, *4*, 895–900.
144. Nistor, C.; Rose, A.; Farre, M.; Stoica, L.; Wollenberger, U.; Ruzgas, T.; Pfeiffer, D.; Barcelo, D.; Gorton, L.; Emneus, J. In-Field monitoring of cleaning efficiency in waste water treatment plants using two phenol-sensitive biosensors. In Proceedings of the 8th Workshop on Biosensors for Environmental Monitoring, Lisbon, Portugal, 11–13 September 2000; pp. 3–17.
145. Tranvik, L.J.; Jansson, M. Climate change—Terrestrial export of organic carbon. *Nature* **2002**, *415*, 861–862.

146. Evans, C.D.; Freeman, C.; Monteith, D.T.; Reynolds, B.; Fenner, N. Climate change—Terrestrial export of organic carbon—Reply. *Nature* **2002**, *415*, 862–862.
147. Freeman, C.; Gresswell, R.; Guasch, H.; Hudson, J.; Lock, M.A.; Reynolds, B.; Sabater, F.; Sabater, S. The role of drought in the impact of climatic-change on the microbiota of peatland streams. *Freshw. Biol.* **1994**, *32*, 223–230.
148. Tang, R.; Clark, J.M.; Bond, T.; Graham, N.; Hughes, D.; Freeman, C. Assessment of potential climate change impacts on peatland dissolved organic carbon release and drinking water treatment from laboratory experiments. *Environ. Pollut.* **2013**, *173*, 270–277.
149. Whitworth, K.L.; Baldwin, D.S.; Kerr, J.L. Drought, floods and water quality: Drivers of a severe hypoxic blackwater event in a major river system (the Southern Murray-Darling Basin, Australia). *J. Hydrol.* **2012**, *450*, 190–198.
150. Hamilton, S.K.; Sippel, S.J.; Calheiros, D.F.; Melack, J.M. An anoxic event and other biogeochemical effects of the pantanal wetland on the Paraguay River. *Limnol. Oceanogr.* **1997**, *42*, 257–272.
151. Fontenot, Q.C.; Rutherford, D.A.; Kelso, W.E. Effects of environmental hypoxia associated with the annual flood pulse on the distribution of larval sunfish and shad in the Atchafalaya River basin, Louisiana. *Trans. Am. Fish. Soc.* **2001**, *130*, 107–116.
152. Marin-Spiotta, E.; Gruley, K.E.; Crawford, J.; Atkinson, E.E.; Miesel, J.R.; Greene, S.; Cardona-Correa, C.; Spencer, R.G.M. Paradigm shifts in soil organic matter research affect interpretations of aquatic carbon cycling: Transcending disciplinary and ecosystem boundaries. *Biogeochemistry* **2014**, *117*, 279–297.
153. Vidon, P.; Wagner, L.E.; Soyeux, E. Changes in the character of DOC in streams during storms in two midwestern watersheds with contrasting land uses. *Biogeochemistry* **2008**, *88*, 257–270.
154. Inamdar, S.; Finger, N.; Singh, S.; Mitchell, M.; Levia, D.; Bais, H.; Scott, D.; McHale, P. Dissolved organic matter (DOM) concentration and quality in a forested mid-Atlantic watershed, USA. *Biogeochemistry* **2012**, *108*, 55–76.
155. Fellman, J.B.; Hood, E.; Edwards, R.T.; D'Amore, D.V. Changes in the concentration, biodegradability, and fluorescent properties of dissolved organic matter during stormflows in coastal temperate watersheds. *J. Geophys. Res. Biogeosci.* **2009**, *114*, doi:10.1029/2008JG000790.
156. Winterdahl, M.; Erlandsson, M.; Futter, M.N.; Weyhenmeyer, G.A.; Bishop, K. Intra-Annual variability of organic carbon concentrations in running waters: Drivers along a climatic gradient. *Glob. Biogeochem. Cycle* **2014**, *28*, 451–464.
157. Sanderman, J.; Lohse, K.A.; Baldock, J.A.; Amundson, R. Linking soils and streams: Sources and chemistry of dissolved organic matter in a small coastal watershed. *Water Resour. Res.* **2009**, *45*, doi:10.1029/2008WR006977.
158. Jencso, K.G.; McGlynn, B.L.; Gooseff, M.N.; Wondzell, S.M.; Bencala, K.E.; Marshall, L.A. Hydrologic connectivity between landscapes and streams: Transferring reach-and plot-scale understanding to the catchment scale. *Water Resour. Res.* **2009**, *45*, doi:10.1029/2008WR007225.
159. Amon, R.M.W.; Rinehart, A.J.; Duan, S.; Louchouarn, P.; Prokushkin, A.; Guggenberger, G.; Bauch, D.; Stedmon, C.; Raymond, P.A.; Holmes, R.M.; *et al.* Dissolved organic matter sources in large arctic rivers. *Geochim. Cosmochim. Acta* **2012**, *94*, 217–237.

160. Walker, S.A.; Amon, R.M.W.; Stedmon, C.A. Variations in high-latitude riverine fluorescent dissolved organic matter: A comparison of large arctic rivers. *J. Geophys. Res. Biogeosci.* **2013**, *118*, 1689–1702.

161. Laudon, H.; Tetzlaff, D.; Soulsby, C.; Carey, S.; Seibert, J.; Buttle, J.; Shanley, J.; McDonnell, J.J.; McGuire, K. Change in winter climate will affect dissolved organic carbon and water fluxes in mid-to-high latitude catchments. *Hydrol. Process.* **2013**, *27*, 700–709.

162. Foley, J.A.; DeFries, R.; Asner, G.P.; Barford, C.; Bonan, G.; Carpenter, S.R.; Chapin, F.S.; Coe, M.T.; Daily, G.C.; Gibbs, H.K.; *et al.* Global consequences of land use. *Science* **2005**, *309*, 570–574.

163. Findlay, S.; Quinn, J.M.; Hickey, C.W.; Burrell, G.; Downes, M. Effects of land use and riparian flowpath on delivery of dissolved organic carbon to streams. *Limnol. Oceanogr.* **2001**, *46*, 345–355.

164. Houghton, R.A. Land-Use change and the carbon-cycle. *Glob. Chang. Biol.* **1995**, *1*, 275–287.

165. Gough, R.; Holliman, P.J.; Willis, N.; Jones, T.G.; Freeman, C. Influence of habitat on the quantity and composition of leachable carbon in the O_2 horizon: Potential implications for potable water treatment. *Lake Reserv. Manag.* **2012**, *28*, 282–292.

166. Yallop, A.R.; Clutterbuck, B.; Thacker, J. Increases in humic dissolved organic carbon export from upland peat catchments: The role of temperature, declining sulphur deposition and changes in land management. *Clim. Res.* **2010**, *45*, 43–56.

167. Bricker, S.B.; Longstaf, B.; Dennison, W.; Jones, A.; Boicourt, K.; Wicks, C.; Woerner, J. Effects of nutrient enrichment in the nation's estuaries: A decade of change. *Harmful Algae* **2008**, *8*, 21–32.

168. Carpenter, S.R.; Caraco, N.F.; Correll, D.L.; Howarth, R.W.; Sharpley, A.N.; Smith, V.H. Nonpoint pollution of surface waters with phosphorus and nitrogen. *Ecol. Appl.* **1998**, *8*, 559–568.

169. Fellman, J.B.; Hood, E.; Spencer, R.G.M. Fluorescence spectroscopy opens new windows into dissolved organic matter dynamics in freshwater ecosystems: A review. *Limnol. Oceanogr.* **2010**, *55*, 2452–2462.

170. Wilson, H.F.; Xenopoulos, M.A. Effects of agricultural land use on the composition of fluvial dissolved organic matter. *Nat. Geosci.* **2009**, *2*, 37–41.

171. Schlesinger, W. *Biogeochemistry: An Analysis of Global Change*; Academic Press: San Diego, CA, USA, 1991.

172. Mierle, G.; Ingram, R. The role of humic substances in the mobilization of mercury from watersheds. *Water Air Soil Pollut.* **1991**, *56*, 349–357.

173. Fabris, R.; Chowa, C.W.K.; Drikas, M.; Eikebrokk, B. Comparison of NOM character in selected Australian and Norwegian drinking waters. *Water Res.* **2008**, *42*, 4188–4196.

174. Chow, A.T.; Guo, F.M.; Gao, S.D.; Breuer, R.S. Trihalomethane reactivity of water- and sodium hydroxide-extractable organic carbon fractions from peat soils. *J. Environ. Qual.* **2006**, *35*, 114–121.

175. Fleck, J.A.; Bossio, D.A.; Fujii, R. Dissolved organic carbon and disinfection by-product precursor release from managed peat soils. *J. Environ. Qual.* **2004**, *33*, 465–475.

176. Diaz, F.J.; Chow, A.T.; O'Geen, A.T.; Dahlgren, R.A.; Wong, P.K. Restored wetlands as a source of disinfection byproduct precursors. *Environ. Sci. Technol.* **2008**, *42*, 5992–5997.
177. Saleh, D.K. *Organic Carbon Trends, Loads, and Yields to the Sacramento-San Joaquin Delta, California, Water Years 1980–2000*, 2nd ed.; Water Resources Investigation Report 03-4070; U.S. Geological Survey: Sacramento, CA, USA, 2007.
178. Ledesma, J.L.; Köhler, S.J.; Futter, M.N. Long-Term dynamics of dissolved organic carbon: Implications for drinking water supply. *Sci. Total Environ.* **2012**, *432*, 1–11.
179. Gough, R.; Holliman, P.J.; Willis, N.; Freeman, C. Dissolved organic carbon and trihalomethane precursor removal at a UK upland water treatment works. *Sci. Total Environ.* **2014**, *468*, 228–239.
180. United States Environmental Protection Agency (US EPA). National primary drinking water regulations: Stage 2 disinfectants and disinfection byproducts rule; National primary and secondary drinking water regulations: Approval of analytical methods for chemical contaminants. *Federal Register.* **2003**, *68*, 49548–49681.
181. United States Environmental Protection Agency (US EPA). National primary drinking water regulations: Stage 2 disinfectants and disinfection byproducts rule. *Federal Register.* **2006**, *71*, 388–493.
182. Haaland, S.; Hongve, D.; Laudon, H.; Riise, G.; Vogt, R.D. Quantifying the drivers of the increasing colored organic matter in boreal surface waters. *Environ. Sci. Technol.* **2010**, *44*, 2975–2980.
183. Eckard, R.S.; Hernes, P.J.; Bergamaschi, B.A.; Stepanauskas, R.; Kendall, C. Landscape scale controls on the vascular plant component of dissolved organic carbon across a freshwater delta. *Geochim. Cosmochim. Acta* **2007**, *71*, 5968–5984.
184. Norwood, D.L.; Johnson, J.D.; Christman, R.F.; Hass, J.R.; Bobenrieth, M.J. Reactions of chlorine with selected aromatic models of aquatic humic material. *Environ. Sci. Technol.* **1980**, *14*, 187–190.
185. Hanna, J.V.; Johnson, W.D.; Quezada, R.A.; Wilson, M.A.; Lu, X.Q. Characterization of aqueous humic substances before and after chlorination. *Environ. Sci. Technol.* **1991**, *25*, 1160–1164.
186. Arnold, W.A.; Bolotin, J.; von Gunten, U.; Hofstetter, T.B. Evaluation of functional groups responsible for chloroform formation during water chlorination using compound specific isotope analysis. *Environ. Sci. Technol.* **2008**, *42*, 7778–7785.
187. Hendricks, D.W. *Water Treatment Unit Processes: Physical and Chemical. Civil and Environmental Engineering*; CRC/Taylor & Francis: Boca Raton, FL, USA, 2006.
188. Archer, A.D.; Singer, P.C. An evaluation of the relationship between SUVA and NOM coagulation using the ICR database. *J. Am. Water Works Assoc.* **2006**, *98*, 110–123.
189. Archer, A.D.; Singer, P.C. Effect of SUVA and enhanced coagulation on removal of TOX precursors. *J. Am. Water Works Assoc.* **2006**, *98*, 97–107.
190. Gang, D.C.; Clevenger, T.E.; Banerji, S.K. Effects of alum coagulation on speciation and distribution of trihalomethanes (THMs) and haloacetic acids (HAAs). *J. Environ. Sci. Health Part A Toxic Hazard. Subst. Environ. Eng.* **2005**, *40*, 521–534.

191. Johnstone, D.W.; Sanchez, N.P.; Miller, C.M. Parallel factor analysis of excitation-emission matrices to assess drinking water disinfection byproduct formation during a peak formation period. *Environ. Eng. Sci.* **2009**, *26*, 1551–1559.
192. Hussain, S.; van Leeuwen, J.; Chow, C.; Beecham, S.; Kamruzzaman, M.; Wang, D.S.; Drikas, M.; Aryal, R. Removal of organic contaminants from river and reservoir waters by three different aluminum-based metal salts: Coagulation adsorption and kinetics studies. *Chem. Eng. J.* **2013**, *225*, 394–405.
193. Reckhow, D.A.; Rees, P.L.S.; Bryan, D. Watershed sources of disinfectant byproduct precursors. *Water Supply*. **2004**, *4*, 61–69.
194. Clark, J.M.; Chapman, P.J.; Heathwaite, A.L.; Adamson, J.K. Suppression of dissolved organic carbon by sulfate induced acidification during simulated droughts. *Environ. Sci. Technol.* **2006**, *40*, 1776–1783.

7

MODIS-Derived Spatiotemporal Changes of Major Lake Surface Areas in Arid Xinjiang, China, 2000–2014

Qingting Li [1], Linlin Lu [1,2,*], Cuizhen Wang [3], Yingkui Li [4], Yue Sui [1] and Huadong Guo [1]

1 Key Laboratory of Digital Earth Science, Institute of Remote Sensing and Digital Earth, Chinese Academy of Sciences, No. 9 Dengzhuang South Road, Haidian District, Beijing 100094, China; E-Mails: liqt@radi.ac.cn (Q.L.); yue11680@163.com (Y.S.); hdguo@radi.ac.cn (H.G.)

2 State Key Laboratory of Remote Sensing Science, Institute of Remote Sensing and Digital Earth, Chinese Academy of Sciences, Beijing 100101, China

3 Department of Geography, University of South Carolina, Columbia, SC 29208, USA; E-Mail: cwang@mailbox.sc.edu

4 Department of Geography, University of Tennessee, Knoxville, TN 37996, USA; E-Mail: yli32@utk.edu

* Author to whom correspondence should be addressed; E-Mail: lull@radi.ac.cn;

Academic Editor: Miklas Scholz

Abstract: Inland water bodies, which are critical freshwater resources for arid and semi-arid areas, are very sensitive to climate change and human disturbance. In this paper, we derived a time series of major lake surface areas across Xinjiang Uygur Autonomous Region (XUAR), China, based on an eight-day MODIS time series in 500 m resolution from 2000 to 2014. A classification approach based on water index and dynamic threshold selection was first developed to accommodate varied spectral features of water pixels at different temporal steps. The overall classification accuracy for a MODIS-derived water body is 97% compared to a water body derived using Landsat imagery. Then, monthly composites of water bodies were derived for the months of April, July, and September to identify seasonal patterns and inter-annual dynamics of 10 major lakes (>100 km^2) in XUAR. Our results indicate that the changing trends of surface area of major lakes varied across the region. The surface areas of the Ebinur and Bosten Lakes showed a significant shrinking trend. The Ulungur-Jili Lake remained relatively stable during the entire period.

For mountain lakes, the Barkol Lake showed a decreasing trend in April and July, but the Sayram Lake showed a significant expanding trend in September. The four plateau lakes exhibited significant expanding trends in all three seasons except for Arkatag Lake in July. The shrinking of major lakes reflects severe anthropogenic impacts due to agricultural and industrial needs, in addition to the impact of climate change. The pattern of lake changes across the XUAR can provide insight into the impact of climate change and human activities on regional water resources in this arid and semi-arid region.

Keywords: water bodies; Xinjiang Uygur Autonomous Region (XUAR); MODIS time series

---

1. Introduction

Inland water bodies are important parts of the hydrosphere, serving as an essential source of freshwater for human consumption, agriculture, industry, and other uses. Due to climate change, uneven distribution of precipitation, and human activities, water resources show tremendous temporal variability worldwide [1]. Lakes and rivers are primary freshwater sources available to the local population and their livestock in arid and semi-arid areas [2,3]. Spatial dynamics and up-to-date information on surface water resources are essential for understanding water resource-related issues in these areas. Temporal water bodies that provide habitats for plant and animal communities in these areas have rarely been included in global datasets, such as the Global Lakes and Wetlands Database (GLWD) [4,5] and Vector Map Level 0 (VMAP0) [6].

With the capability of synoptic view and repeated coverage of the earth's surface, satellite remote sensing is an effective means of extracting water bodies across a variety of spatial and temporal scales. Due to their strong absorption in the near-infrared (NIR) spectrum, optical remote sensing platforms, such as Landsat [7,8], Advanced Spaceborne Thermal Emission and Reflection Radiometer (ASTER) [9,10], and Satellite Pour l'Observation de la Terre (SPOT) [11], have been used to map the area of the water body at various spatial resolutions. However, the high costs, narrow swath, and long revisit intervals of the medium- and high-resolution images limit their applications on monitoring the dynamics of lake systems across large spatial scales. Remote sensing data with high temporal resolution have the advantage of documenting detailed water area variation. Time series data from satellite sensors, such as the SPOT VEGETATION, Moderate Resolution Imaging Spectroradiometer (MODIS), and Advanced Very High Resolution Radiometer (AVHRR), have been applied to seasonal and inter-annual change detection of water bodies over large areas [12–16].

One of the common image processing methods for extracting water extent is based on a threshold of a water detection index. Spectral indexes, such as Normalized Difference Water Index (NDWI) [17], Modified Normalized Difference Water Index (MNDWI) [18], and Normalized Difference Pond Index (NDPI) [11], have been developed for water detection using remote sensing imagery. NDWI uses green band and near-infrared band to distinguish water from vegetation and soil [17]. In order to enhance the ability of water detection, especially for areas with built-up land in the background, the middle infrared band was integrated into MNDWI and NDPI instead of NIR band in NDWI [11,18]. Built-up areas and water bodies show discriminating spectral responses at the MODIS short-wavelength infrared (SWIR) band. A Combined Water Index (CWI) combining SWIR's and NDVI's ability to represent vegetation information was proposed for water body identification using MODIS data [19]. In addition, spectral reflectance of water shows spatiotemporal variability across different scenes and acquisition dates. Thus, delineating water bodies using a standard threshold may become problematic in large-area applications [20]. A strategy for threshold computation of different satellite images is needed.

The Xinjiang Uygur Autonomous Region (XUAR), with a widespread area of 1,660,000 km^2, is the largest autonomous region in China. There are 113 lakes with an area of >1 km^2 in the XUAR. Many lakes are important wetlands for threatened species, five of which are designated as the National Nature Reserves of China [21]. The water levels of inland lakes in XUAR are influenced by the runoff of their inflowing rivers, and are sensitive to climate change and human activities altering the rivers' inflows [22]. It has been reported in past studies that the water extent has been shrinking in past decades, causing severe environmental problems, such as land desertification, salinization, vegetation degradation, water shortage, and biodiversity loss in this arid/semi-arid area [22].

Several studies have examined changes of lakes in the XUAR using satellite imagery, such as Landsat [22–24], SPOT VEGETATION [25], and MODIS [12,26] data. Landsat TM imagery has been used to interpret changes in area of inland lakes in Xinjiang over the period 1975–2007, but only selected lakes in spring and autumn seasons were analyzed [22]. Comparing with nationwide lake surveys undertaken in the 1960s–1980s, Ma *et al.* (2010) reported that 62 lakes vanished in the XUAR from the 1960s to 2000s; one of the completely dried lakes is the Lop Nur Lake with an original lake area of 5500 km^2 [23]. Water extents were found to have decreased significantly due to anthropogenic impacts, such as agricultural water consumption and damming in this region [24]. The Ebinur Lake, the largest salt lake in the XUAR, exhibited a significant inter-annual and inter-seasonal variation based on SPOT VEGETATION data [25]. However, due to the large area of XUAR, previous studies have mainly focused on selected water bodies and annual or seasonal temporal intervals. The temporal fluctuations in surface area of major lakes were monitored with MODIS data in XUAR, but only inter-annual variations were analyzed [12,26]. The temporal and spatial dynamics of lake surface areas across XUAR are rarely documented in detail.

The purpose of this study is to examine the spatiotemporal variation of water bodies from 2000 to 2014 based on MODIS time series data. Detailed objectives include: (1) to develop an automatic approach to extract water bodies from MODIS data; (2) to generate a 15-year water body mask and document the changes of water extent; and (3) to examine the driving factors of the changes in major lakes in the XUAR.

2. Study Area

The XUAR is located in north-western China and encompasses the Altay Mountains, Junngar Basin, Tianshan Mountain, Tarim Basin, and Kunlun Mountains from north to south (Figure 1). It is an arid and semi-arid area with mean annual precipitation ranging between 100 and 200 mm [27]. Vast areas of the XUAR are covered by grassland and desert [28]. Forests are sparsely scattered within high mountains and along rivers. Oasis landscapes characterized by human settlements and agriculture lands are distributed within inland river deltas, alluvial-diluvial plains, and along the edges of diluvial-alluvial fans.

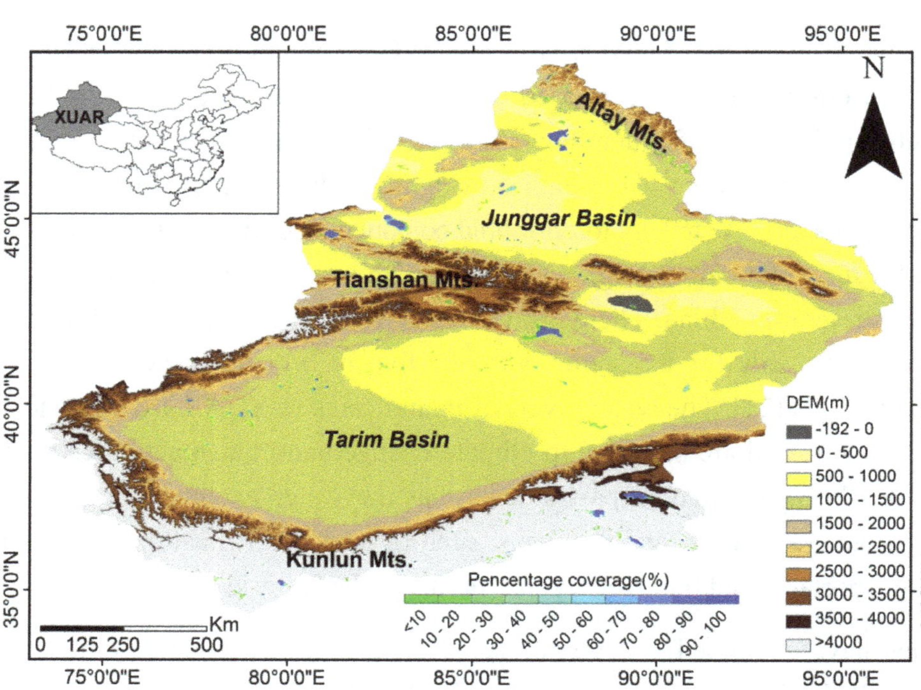

Figure 1. Geographic location and topographic map of study area. The percentage coverage of water bodies was combined for April, July, and September from 2000 to 2014.

According to the topographic characteristics, lakes in the XUAR can be categorized as four types [26]: (1) plateau lakes (>3500 m), with snow and glacier ice melt and surface runoffs as their main charge; (2) mountain lakes (1000–3500 m), with snow and glacier ice melt, underground runoff as their main influx; (3) plain lakes (<1000 m), heavily influenced by human activities; and (4) transition lakes, located at the transition area between mountains and plains.

Among all the lakes in the XUAR, 10 lakes have areas larger than 100 km^2. Their total area accounts for more than 80% of the lake surface area (Table 1) [26].

Table 1. Ten major lakes in the study area (with surface area >100 km^2).

Name	Area (km^2)	Altitude (m)	Mean Water Depth (m)	Region	Type
Ebinur	673.46	194	1.2	Junggar	Plain
Manas	259.81	244	6	Junggar	Plain
Ulungur-Jili	1041.60	478	10.4	Junggar	Plain
Bosten	1004.33	1050	9	Tarim	Transition
Sayram	462.63	2072	46	Junggar	Mountain
Barkol	118.57	1577	0.6	Junggar	Mountain
Ayakkum	200.46	3876	10	Kumukuli	Plateau
Aqqikkol	168.93	4251	8	Kumukuli	Plateau
Arkatag	110.33	4713	8	Kumukuli	Plateau
Aksayquin	88.54	4844	8	Northern Tibet	Plateau

3. Materials and Methods

3.1. Remote Sensing Datasets

Terra MODIS images were selected as the main data source for monitoring water body variation in XUAR. The MODIS Surface Reflectance (MOD09A1) dataset was used to build an image time series from 2000 to 2014 [29]. It provides surface reflectance at bands 1–7 with 500 m spatial resolution and eight-day temporal resolution. The XUAR region is entirely covered by six tiles (h23v04, h23v05, h24v04, h24v05, h25v04, and h25v05). Lakes in XUAR were relatively stable during the spring and autumn seasons. In summer, some were influenced by extensive agricultural irrigation and high evaporation. In winter, some of the water bodies were frozen and may have caused high uncertainty in lake extent extraction. Therefore, we downloaded MODIS data for the months of April, July, and September to analyze the spatiotemporal dynamics of lake surface area. The Shuttle Radar Topography Mission (SRTM) digital elevation model with a spatial resolution of 90 m was used to correct the water body extraction affected by shadow and snow in mountainous regions [14,30].

Landsat data of 30 m resolution was used to assess the accuracy of remote sensing products with 250 m [31] and 1 km resolution [32]. In our study, Landsat TM, ETM+, and OLI images were used to validate the results of water extraction from MODIS data. A total of 71 Landsat image scenes (including Landsat 5 TM, Landsat 7 ETM+, and Landsat8 OLI_TIRS) were processed for the validation (Table 2).

Table 2. Landsat data used in our study.

Lake	Ulungur	Manas	Bosten	Ayakekumu
Sensors	Landsat5 TM Landsat7 ETM+ Landsat8 OLI_TIRS	Landsat5 TM Landsat7 ETM+ Landsat8 OLI_TIRS	Landsat5 TM Landsat7 ETM+ Landsat8 OLI_TIRS	Landsat5 TM Landsat7 ETM+ Landsat8 OLI_TIRS
Path/Row	143/27, 144/27	144/28	143/31	140/34
Date	7 July 2000	16 March 2000	25 March 2000	4 September 2000
	24 September 2000	6 July 2000	31 July 2000	29 July 2001
	13 April 2001	24 September 2000	17 September 2000	1 October 2001
	27 July 2002	19 March 2001	1 August2006	11 April 2002
	14 September 2002	10 April 2003	10 September 2006	1 August2002
	10 April 2003	31 July 2006	14 April 2007	4 October 2002
	24 July 2006	17 September 2006	27 July 2007	14 April 2003
	17 September 2006	4 September 2007	13 September 2007	19 April 2005
	6 April 2007	23 July 2009	2 May 2008	27 July 2006
	23 July 2009	9 September 2009	16 July 2009	13 September 2006
	18 September 2009	10 May 2011	3 April 2009	6 July 2007
	10 May 2011	13 July 2011	6 July 2011	25 April 2010
	13 July 2011	1 October 2011	24 September 2011	30 July 2010
	15 September 2011	13 April 2013	22 April 2013	31 August2010
	13 April 2013	2 July 2013	27 July 2013	8 May 2012
	11 July 2013	4 September 2013	29 September 2013	7 August2013
	29 September 2013	2 May 2014	31 August2014	24 September 2013
	25 April 2014	21 July 2014		6 May 2014

The MODIS 500-m land-cover product (MCD12Q1) was used to identify primary land covers in the study area [33]. The MCD12Q1 is produced using an ensemble supervised classification algorithm with MODIS band 1–7 surface reflectance, an enhanced vegetation index, and land surface temperature as the main input. Post-processing refinements with ancillary datasets were also conducted. The MCD12Q1 product of 2012 covering the study area was downloaded. With the International Geosphere-Biosphere Program (IGBP) classification scheme, the land cover types were mapped and used to assess the impacts of human activities on lake changes in this study.

3.2. MODIS-Based Water Body Extraction

In XUAR, large areas of deserts and bare rocks have high spectral responses in short-wavelength infrared (SWIR) band [34]. Water usually has low reflectance along the spectrum. In this study we use CWI to detect water bodies from MODIS data. The computation formula of CWI is as follows [19]:

$$CWI = (NDVI + SWIR + \mathrm{A}) \times \mathrm{C} \quad (1)$$

$$NDVI = \frac{\mathrm{b2} - \mathrm{b1}}{\mathrm{b2} + \mathrm{b1}} \quad (2)$$

$$SWIR = \frac{\mathrm{b7}}{\overline{\mathrm{b7}}}, \quad (3)$$

where b1, b2, and b7 represent the reflectance of band 1 (Red band, 620–670 nm), band 2 (NIR band, 871–876 nm), and band 7 (SWIR band, 2105–2155 nm) of the MOD09 data, respectively. A and C are correction factors to adjust the data ranges of *CWI* values. They are empirically determined by comparing *CWI* values between water pixels and background in the study area. We set A as 0.4 and C as 100 in our study [19]. Figure 2 illustrates the procedure for extracting water bodies based on MODIS data.

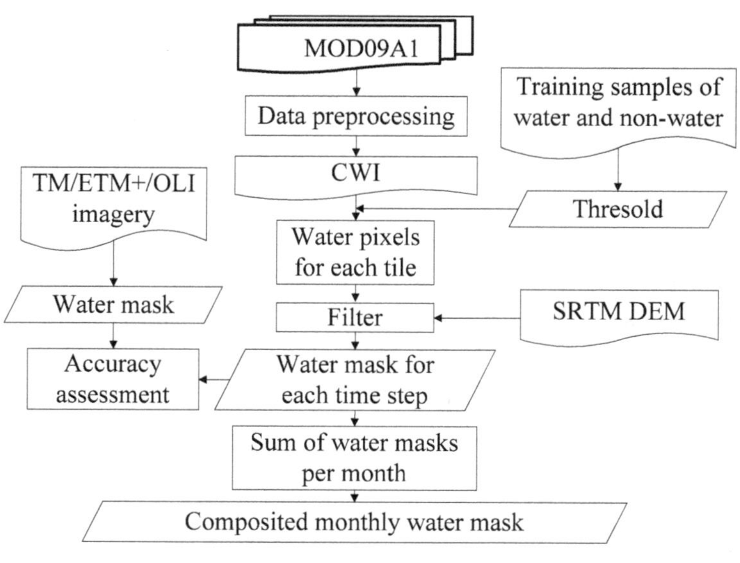

Figure 2. Workflow for water body extraction in our study.

The MODIS tiled data in Sinusoidal projection were mosaicked, re-projected using the nearest neighbor re-sampling method, and saved as GeoTIFF format using the MODIS Reprojection Tool. The

DEM data is resampled to the pixel size of MODIS to refine the water detection results in the following steps.

The quality assessment information of MODIS data was used to exclude pixels labeled as cloudy or snow/ice. Then CWI was calculated for each MODIS tile. The atmosphere condition, water depth, and chlorophyll content all have influence on the spectral features of water on remote sensing images. A single threshold value derived for one image might not be suitable for another. Since there is no standard threshold for the whole study period, an optimized threshold must be identified for each scene or each month. In this study, we set a different threshold for each time step and extract water pixels. For the threshold selection, the training datasets that as pixels were covered by water for all time steps were collected manually from MODIS data in July. The statistics of CWI values were calculated based on training samples. We choose two standard deviations of the mean CWI value as the threshold value and classify pixels within it as water and *vice versa*.

Shadows in mountainous areas can lead to confusion with water bodies. To eliminate the snow and shadow effect, a slope map was used to refine the water extent. All pixels with a slope >1° were removed from the classification result because water bodies usually have flat surfaces. In this way, misclassification of shadow in mountainous areas can be corrected. In addition, to reduce the noise caused by small and temporary water bodies in mountainous areas, we removed water bodies smaller than 4 km^2 in the detection results. After these steps, a binary mask with water and non-water pixels for each time step was derived.

A binary water mask was calculated for each eight-day interval. For each month, we summed up the binary masks at four time steps. For all pixels, we obtained the number of times they were classified as water. Only pixels classified as water three out of four times in a month were marked as water [14]. In this way, composited monthly water masks were generated for April, July, and September, 2000–2014. The seasonality of each lake was calculated using the maximum/minimum ratio of area extent of one representative year for each lake [35]. Finally, combining the monthly masks for April, July, and September from 2000 to 2014, a percentage coverage layer was derived. In this layer, pixel values show the percentage of times a pixel was classified as water from 2000 to 2014. A 100% percentage means the pixel was identified as water at all the 45 monthly masks from 2000 to 2014.

3.3. Accuracy Assessment

Water extents classified from Landsat images served as our validation sources in this study. An integrated water body mapping method combining the NDVI, NDWI, NIR, and slope layers was applied [36]. The commonly used threshold method was used to calculate threshold values and segment water bodies [36]. After that, the water bodies detected from Landsat images were resampled to MODIS pixel size to perform a pixel-to-pixel comparison. Confusion matrices were calculated to represent the accuracy of the classification results [37]. Three measurements, namely user's accuracy, producer's accuracy, and overall accuracy were calculated to assess the accuracies of MODIS detection. The user's accuracy is defined as the number of correctly classified water pixels divided by the total number of classified water pixels in the MODIS detection results. The producer's accuracy is defined as the total number of correctly classified water pixels divided by the total number of water pixels in the Landsat detection results. Overall accuracy is defined as the sum of all correctly classified water/non-water pixels divided by the total number of validation samples.

We selected four water bodies to conduct accuracy assessment for MODIS water detection results. For each lake, Landsat data acquired in April, July, and September or the nearest month were collected for each year. The MODIS results on the nearest neighboring date were selected and compared with the Landsat interpreted results to ensure the images were consistent in acquisition time.

It is worth noting that classification error exists in the water extent mapped from Landsat data and may influence the accuracy assessment results. For the 30-m resolution Landsat images, mixed pixels of lake fringes and small water bodies contain significant spectral response from backgrounds such as grasslands and croplands. In addition, the turbidity of water can increase its response at the near-infrared band and lead to a lower NDWI value [38]. A more accurate validation exercise could be conducted based on comparison with manually collected validation samples, which is very time-consuming and was not conducted in this study.

3.4. Data Sources of Climate and Human Activities

We also analyzed possible drivers of lake area variations in XUAR. Specifically, annual mean temperature and annual precipitation were used as indicators of regional climate. Cropland and built-up areas were used as indicators of human activities. The monthly air temperatures and precipitations for XUAR were obtained from 55 meteorological stations (National Meteorological Information Center of China Meteorological Administration; http://cdc.nmic.cn/home.do). Monthly values were averaged or summed (for temperature and precipitation, respectively) to acquire annual values. For each variable, annual time series graphs were plotted for the 15-year study period, 2000–2014. The linear regression method and t-test are used to estimate the changing trend and test statistical significance of the changing trends of climate data. If the P value derived from the statistical analysis is less than the significance level, a significant changing trend is observed. Areas of irrigated croplands and built-up areas from 2000 to 2013 were collected from the XUAR Statistical Year Book [39].

4. Results and Discussion

4.1. Intra- and Inter-Annual Dynamics of Water Bodies

The water body mapping results of the Ebinur Lake using NDWI, MNDWI, and CWI from MODIS data were compared in Figure 3. The spectral response of water is very similar to bare lands around the lake for NDWI (Figure 3b). For MNDWI, the spectral signatures of dried-up lake and snow at the lower right of the image are identical with water pixels (Figure 3d). We applied a threshold value of zero to extract water pixels from the NDWI and MNDWI images [18]. The mapping result from NDWI shows an obvious underestimation of the lake surface area (Figure 3c). An overestimation of lake surface area is observed from the MNDWI results (Figure 3e). Comparing with water bodies extracted from NDWI and MNDWI visually, the water pixels identified from CWI have higher accuracy (Figure 3g).

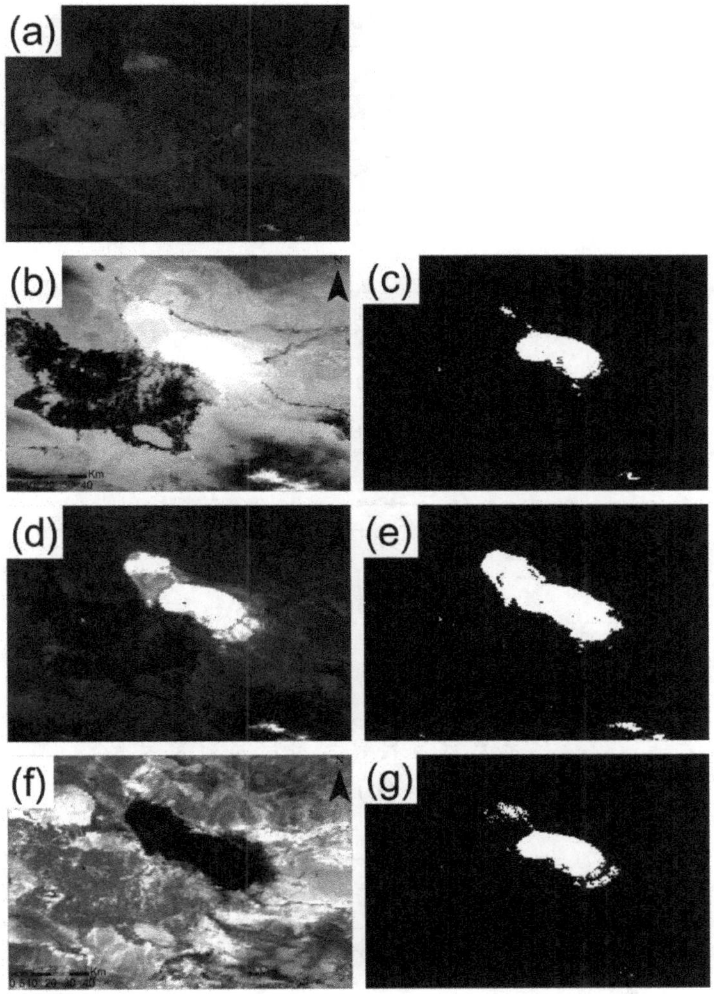

Figure 3. Water body detection results from MODIS data. (**a**) Landsat image; (**b**) NDWI [17]; (**c**) extracted water bodies from NDWI; (**d**) MNDWI [18]; (**e**) extracted water bodies from MNDWI; (**f**) CWI; and (**g**) extracted water bodies from CWI. Water bodies are shown in white in (**c**), (**e**), and (**g**).

In our study, the water body detection results from 2000 to 2014 in April, July, and September were combined and illustrated in Figure 1. In addition to the permanent water bodies that were detected at each time step, many temporal pools that were rarely mapped in global datasets were revealed. For lakes located in high elevation areas, they were frozen and cannot be detected in winter and spring months. For lakes located in arid and semi-arid areas, some of them dried up in the summer months.

We calculated the total area of 10 large lakes and analyzed their intra- and inter-annual dynamics in XUAR. Their total surface area varied from 4217 km^2 (September 2010) to 5014.75 km^2 (April 2003), with high seasonal variability. The largest surface area is usually in April and the smallest value in September. A decreasing trend of total lake surface area can also be observed for all three months from 2000 to 2014, even though the trend is insignificant with a 0.1 significance level (Figure 4). Significant shrinking trends were found for April (34.65 km^2/year), July (22.91 km^2/year), and September (31.02 km^2/year) from 2000 to 2010 with a 0.1 significance level.

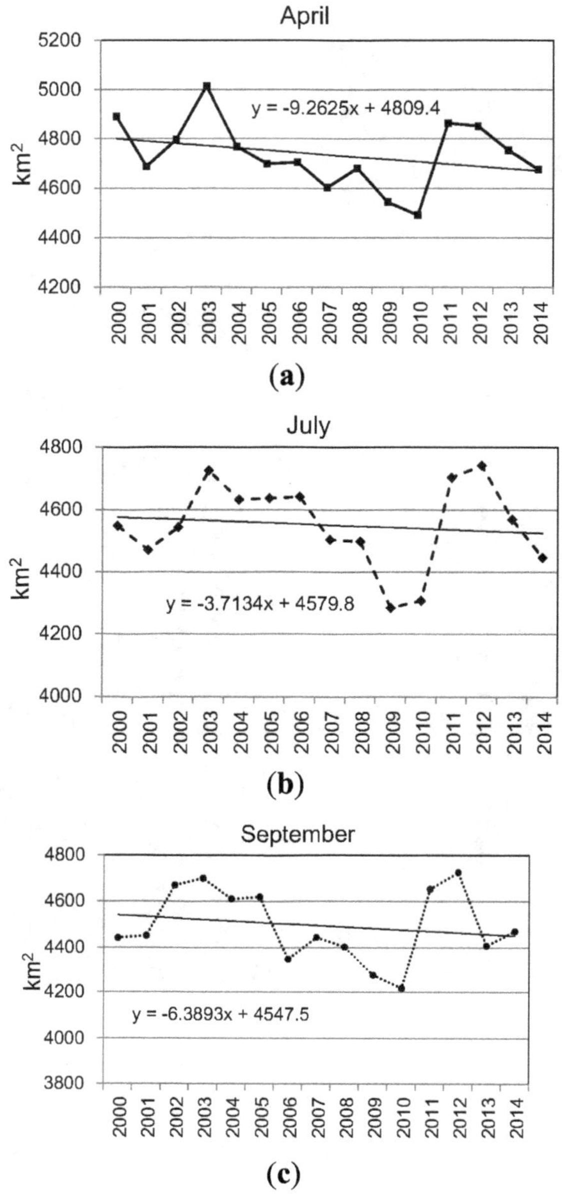

Figure 4. Inter-annual variation of the total area of major lakes for the months of (**a**) April; (**b**) July; and (**c**) September in XUAR from 2000 to 2014.

4.2. Temporal Variations of Major Water Bodies

The inter-annual variation in the surface area of each major lake was illustrated for April, July, and September (Figure 5). The changing trend in the surface area of each lake in July was calculated (Figure 6). The statistics of inter-annual variations of surface area for each lake were listed in Table 3. These results revealed that the lakes along the Tianshan Mountains (central Xinjiang) are shrinking, while the lakes in the northern and southern parts are expanding. This may indicate that human activities cause the lake to shrink, since most human settlements and agricultural lands are distributed along the Tianshan Mountains (Figure 6). Lakes where there is less human influence exhibited expanding trends, mainly due to changes in climate variables.

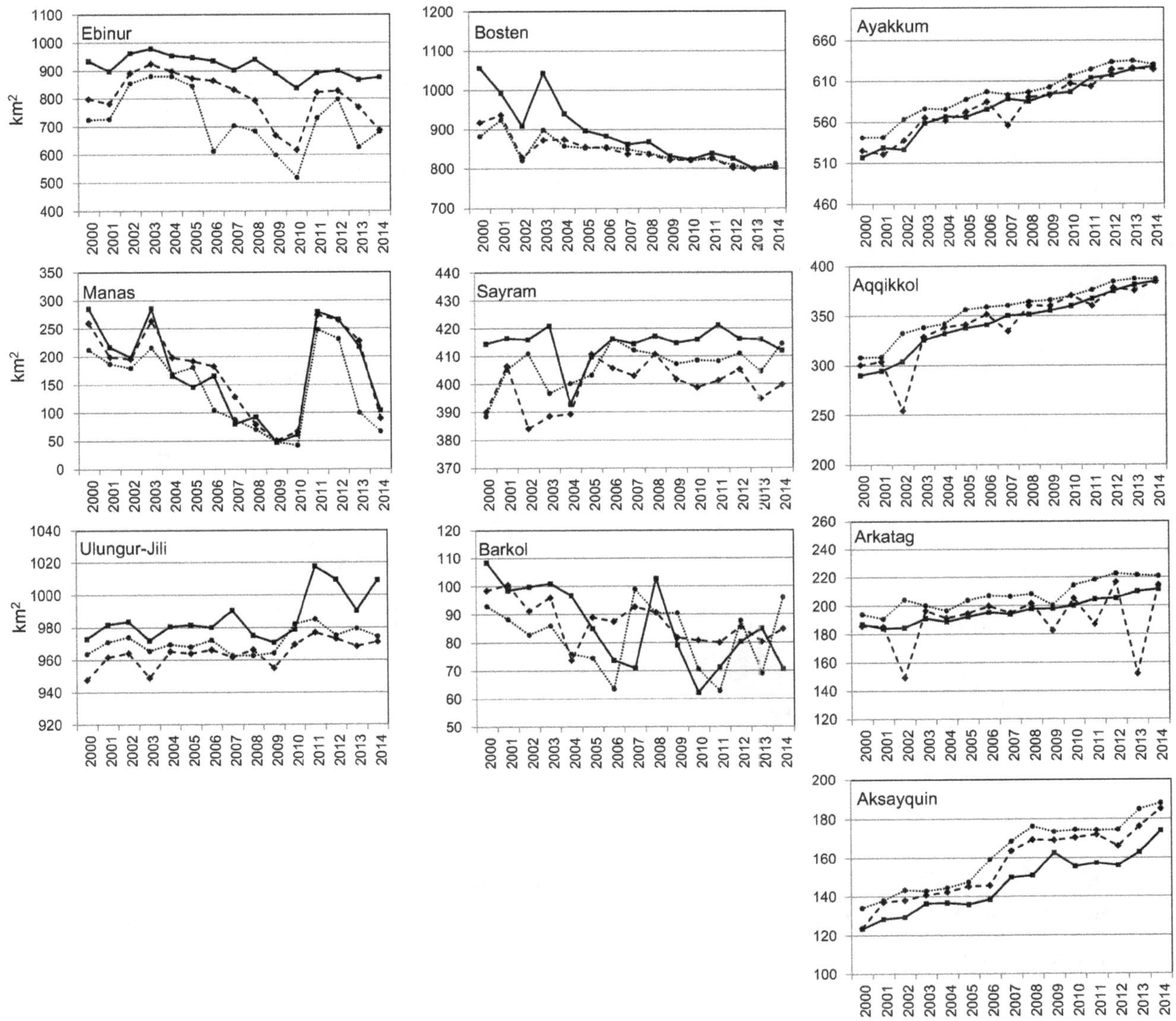

Figure 5. Inter-annual variation in the surface area of major lakes for the months of April, July, and September in XUAR from 2000 to 2014. Solid, dashed, and dotted lines represent April, July, and September, respectively.

Table 3. Changing trend in the surface area of major lakes in XUAR from 2000 to 2014.

Name	April	July	September	Seasonality (max/min ratio)
Ebinur	−6.09	−10.36	−11.74	1.29
Manas	−5.53	−5.08	−6.59	1.52
Ulungur-Jili	2.08	1.25	0.81	1.03
Bosten	−16.71	−7.78	−6.36	1.06
Sayram	0.22	0.61	0.90	1.03
Barkol	−2.32	−1.07	−0.43	1.38
Ayakkum	7.80	7.48	6.72	1.05
Aqqikkol	6.55	6.82	5.52	1.06
Arkatag	1.9	1.13	2.08	1.04
Aksayquin	3.21	3.88	3.92	1.15

Note: The values in bold indicate significant trends (p value <0.1).

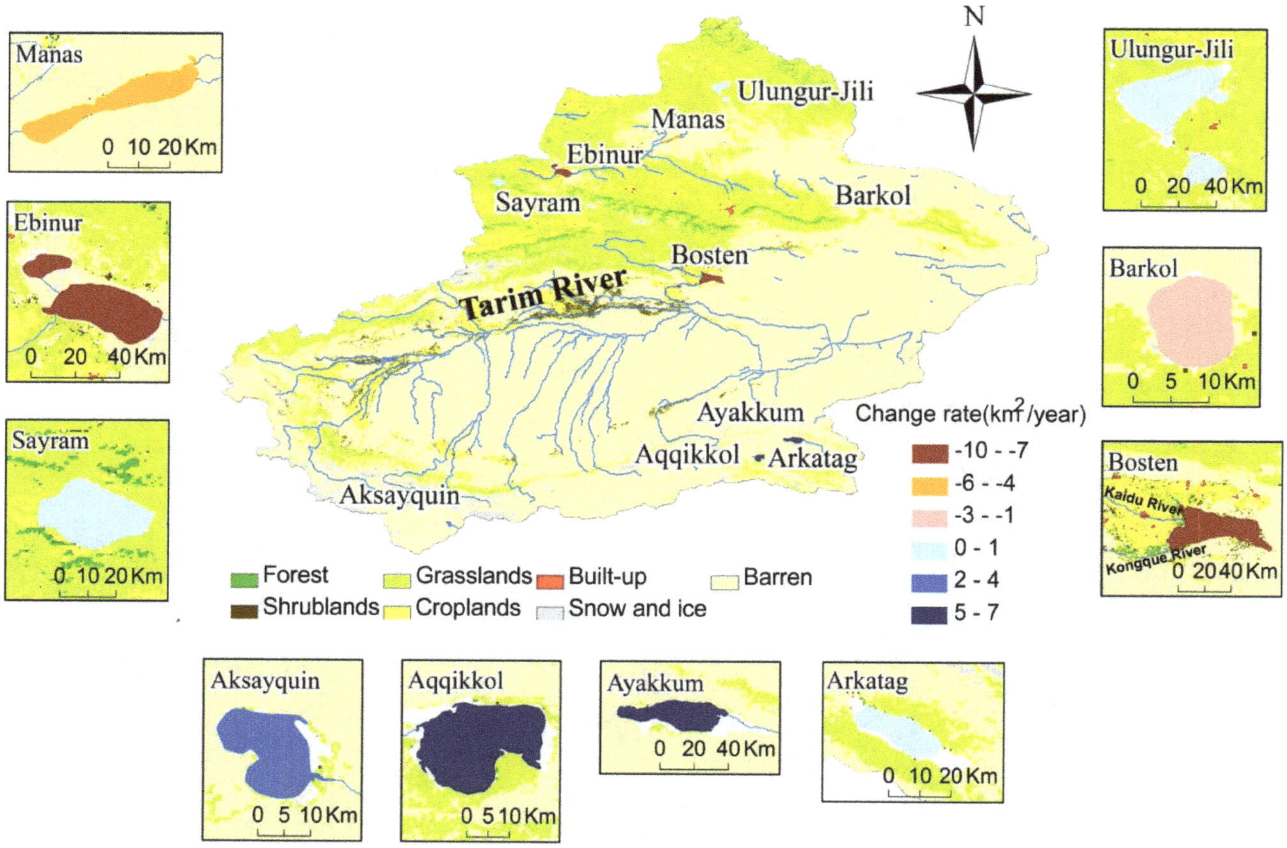

Figure 6. Changing trends and rates in the surface area of major lakes in XUAR in July from 2001 to 2014. Spatial distribution of land cover/use types was extracted from MODIS land cover type product in 2012 [33].

The highest seasonality is at Manas Lake, Barkol Lake, and Ebinur Lake. The lowest seasonality exists at Ulungur-Jili Lake, Sayram Lake, and Arkatag Lake. The seasonal dynamics of Ebinur Lake, Manas Lake, Barkol Lake, and Aksayquin Lake, which have the highest seasonality, are illustrated in Figure 7. For each pixel, we summed the number of times it was classified as water for April, July, and September over the entire study period. The monthly composite was classified into four intervals to indicate areas that stayed stable over the entire time series, and areas that were covered by water for several years during the 2000–2014 period. Different seasonal behaviors were observed for the four lakes. Ebinur Lake showed the largest area in April and the smallest in September. Manas Lake was largest in July and smallest in September.

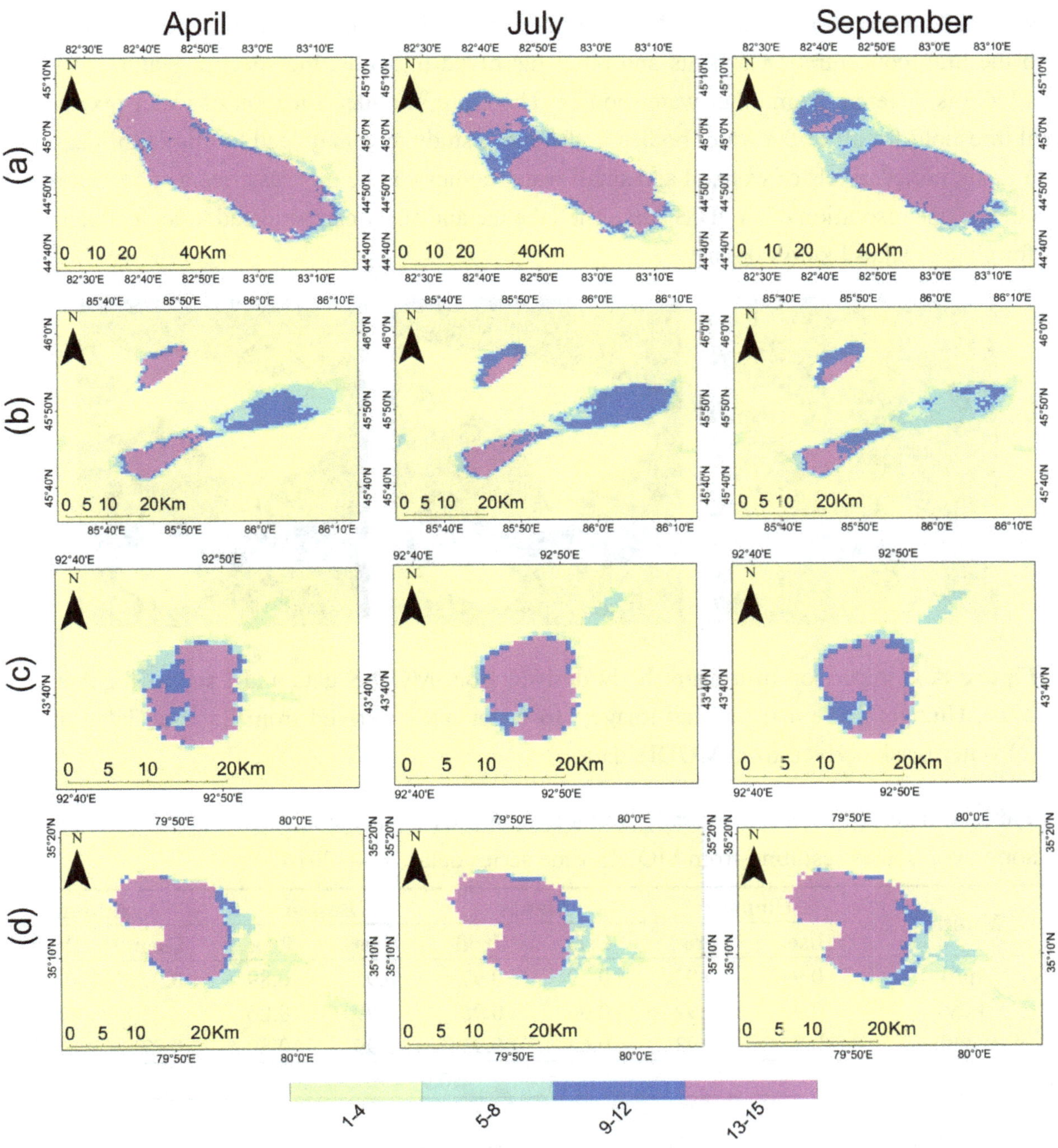

Figure 7. Typical seasonal water bodies of XUAR and their temporal dynamics for the months of April, July, and September from 2000 to 2014. (**a**) Ebinur; (**b**) Manas; (**c**) Barkol and (**d**) Aksayquin Lakes. The numbers at each pixel indicate the number of times it was detected as a water body during the 2000–2014 period.

4.3. Accuracy Assessment

Figure 8 illustrates the comparison of a monthly water mask derived from MODIS data with a water mask derived from Landsat images of the same month. The small water bodies that were detected from Landsat data cannot be extracted from MODIS imagery at 500 m resolution. This leads to a partial disagreement of detection results between Landsat and MODIS data. Based on our assessment, an overall accuracy of 0.97 is obtained (Table 4), which is adequate for the dynamic analysis of lake surface area. The detection results for Manas Lake showed lower user accuracy than the other lakes. Due to the high concentration of salts and other dissolved minerals, the spectral feature of water in Manas Lake is different from other water bodies. Therefore, the identification of water pixels using a general threshold for all of the water bodies in the entire study area may lead to misclassifications. For Bosten Lake, mixed pixels of wetland and small water bodies can be misclassified as non-water bodies due to the coarse resolution of MODIS data. The lake ice and snow can also lead to lower accuracies in April for plateau lakes like Ayakkum.

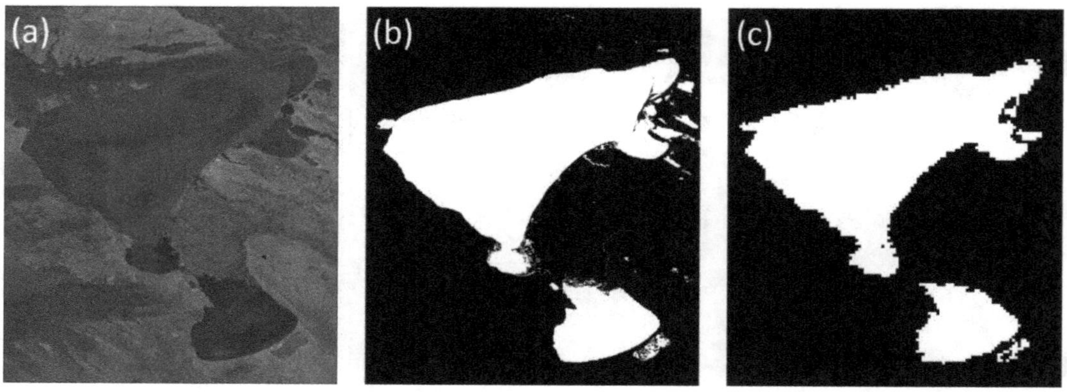

Figure 8. Comparison of a water body derived from MODIS data and Landsat for April 2014, Ulungur Lake. (**a**) Landsat image; (**b**) water mask derived from Landsat data; and (**c**) water mask derived from MODIS data.

Table 4. The error matrix, overall, producer's, and user's accuracies of water bodies and non-water bodies resulting from MODIS time series data, 2000–2014.

Month	Ulungur		Manas		Bosten		Ayakkum	
	User	Prod	User	Prod	User	Prod	User	Prod
April	0.92	0.93	0.72	0.97	0.95	0.89	0.97	0.95
July	0.96	0.92	0.66	0.95	0.99	0.86	0.95	0.94
September	0.97	0.93	0.63	0.93	0.99	0.89	0.97	0.97
Overall Accuracy	0.97							

4.4. *Effects of Regional Climate and Human Activities on Lake Changes*

The linear trend of temperature and precipitation are not statistically significant at the 0.1 significance level for 2000–2014. In our study period, an increase in precipitation occurred in 2010, which corresponds well to the extending lake area in 2011 and 2012 (Figure 9). The area of cropland increased from 3.39 million hectares in 2000 to 5.21 million hectares in 2013. Built-up areas increased from 473 km^2 in 2000 to 1065 km^2 in 2013. As we can see in Figure 5, the oases comprised of agricultural lands and built-up areas are distributed in river plains near major lakes. The water demands of oases can influence aerial changes in lakes that have river runoff as their main inflows, such as Ebinur, Bosten, and Manas Lakes. For a regional comparison, the changing trends of temperature and precipitation at meteorological stations near each major lake were analyzed (Table 5). Their relationships with lake area changes were also analyzed. Due to the lack of observed climate data, the analysis was not performed for the four plateau lakes.

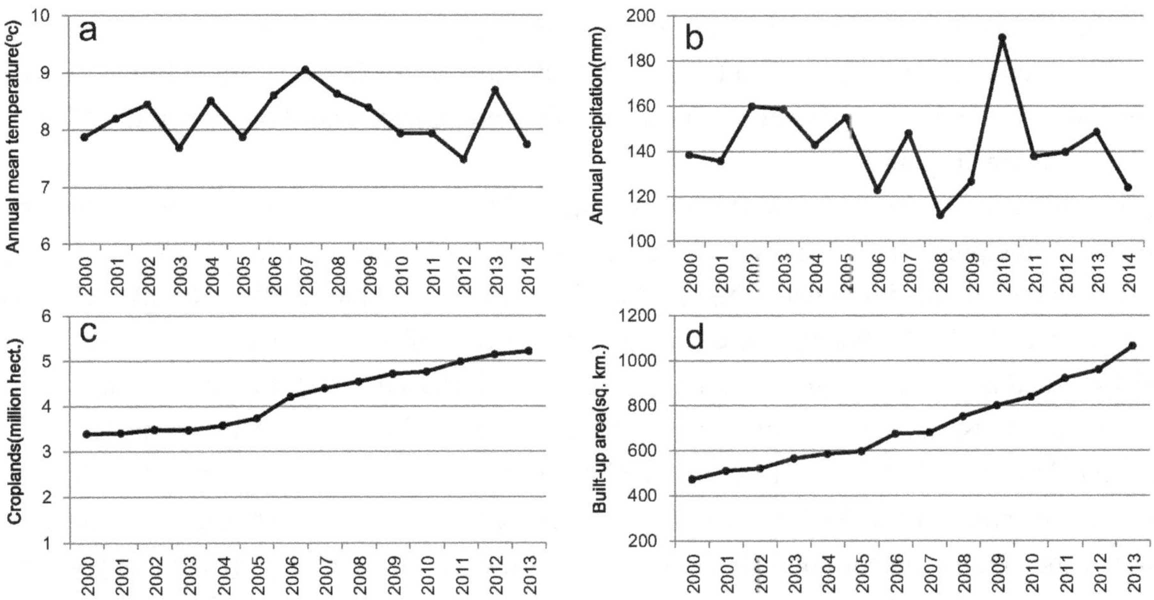

Figure 9. Variations in (**a**) annual mean temperature; (**b**) annual precipitation; (**c**) area of croplands; and (**d**) built-up areas in XUAR over the study period.

Table 5. Changing trends in temperature and precipitation at meteorological stations near the major lakes from 2000 to 2014.

Climate Variable	Ebinur	Manas	Ulungur-Jili	Bosten	Sayram	Barkol
Annual Mean Temperature (°C)	−0.20	−0.06	0.31	−0.26	−0.27	0.51
Annual Precipitation (mm)	−0.28	−1.70	−2.28	−2.72	−3.21	2.82

Among the major lakes, three plain lake or lake systems, Ebinur, Manas, and Ulungur-Jili, were analyzed. Located in low elevation plain areas, their main supply is river runoff, which can be influenced tremendously by human activities. Ebinur is the largest saline lake in XUAR. With a maximum depth of 3.5 m and a mean depth of 1.2 m, it is a closed lake without outlet. Its hydrologic input is mainly from the Bo and Jing rivers, which originate from precipitation in mountainous areas. In their analysis of multi-temporal VGT-S10 data from 1 April 1998 to 31 December 2005, Ma *et al.* (2007) revealed that this lake had a peak area of 903 km^2 in 2003 and subsequently decreased to an area of 847 km^2 in 2004 [23]. Our study showed a decreasing trend of surface area from 2000 to 2014 in all three seasons, with a short increase in 2002–2004 (Figure 5). The significant decreasing trend of the Ebinur Lake is 6.09 km^2/year for April, 10.36 km^2/year for July, and 11.74 km^2/year for September, respectively. A slight decreasing trend of precipitation and temperature was revealed (Table 5). According to Zhang *et al.* (2015), human factors, such as population growth and agricultural development, led to the increase in desertification area in the Ebinur Lake region between 1990 and 2010 [40].

Manas Lake is the terminal lake of Manas River. This lake is about 55 km long and 15–20 km wide, with an area of about 550 km^2 and an average depth 6 m [41]. Our results show a decreasing trend of Manas Lake from 2000 to 2014. The smallest lake size occurred in 2009 and 2010 (Figure 5). A decreasing trend of precipitation and temperature was observed at the meteorological station near Manas Lake (Table 5). Manas River Valley is a representative area for oasis exploitation in Xinjiang as a primary agricultural production region and core region of economic development of the northern slope of the Tianshan [41]. The expansion of oasis in Manas River Valley was characterized by the spread of settlements and agricultural lands. In addition to reclaiming agricultural lands, many hydrological constructions, including reservoirs, wells, and canals, were built along the Manas River for irrigation of the croplands.

Ulungur-Jili lake system is a closed inland lake, with its water supplied by the Ulungur River, groundwater, precipitation, and snow melting. As the second largest lake in XUAR, the Ulungur-Jili lake system can be divided into two sections, the Ulungur Lake and the smaller Jili Lake, connected by a narrow channel. In 2011, Ulungur Lake had an average depth of 10.4 m and a surface area of 859 km^2, and Jili Lake had an average depth of 8.8 m and an area of 169 km^2 [42]. Based on an analysis of the lake area with Landsat images acquired in August, a stable or slightly increasing trend of lake surface area is revealed from 2000 to 2011 [43]. A slight extending trend of Ulungur-Jili Lake was also observed in our study from 2000 to 2014. Increasing temperature and decreasing precipitation were observed at the nearby meteorological stations. Since snow and ice melting off the Altay Mountain is the main runoff supply of Irtysh River, snow and glacier changes caused by the increasing temperature contributed to the variation of lake area of the Ulungur-Jili lake system.

Bosten Lake, as a transition lake located between the mountains and plains, is the largest inland freshwater lake in China. As an open catchment lake with outlets, Bosten Lake lies at the end of Kaidu River and the beginning of Kongque River. The lake inflow mainly comes from Kaidu River, which contributes about 95% of the total water inflow [44,45]. The main outflow is Kongque River and evaporation [46]. With a large area in 2001 and 2003 (Figure 5), our study revealed an obvious decreasing trend in the surface area of Bosten Lake from 2000 to 2014. The nearby meteorological station observed the decreasing trend of precipitation and temperature. According to the monthly mean lake level data, the lake level decreased dramatically from 2002 to 2010 [46]. The sharply decreasing lake level during 2003 to 2010 was reported to be caused by the emergency project of transferring water to Tarim River and increasing lake outflow in addition to the reduced precipitation [46].

Sayram Lake is the largest and highest alpine lake in XUAR. The lake is located in a mountain basin in the western part of the Tian Shan with an average water depth of 46.1 m. It has a frozen period extending for six months from October to May [47]. Though a decreasing trend of precipitation and temperature was observed at the nearby meteorological station, a slight increasing trend was observed from 2000 to 2014 in our study. Located in a natural environment, Sayram Lake was less impacted by anthropogenic disturbances and showed a stable surface area from 1975 to 2007 [22].

Barkol Lake is a closed saline lake with an elevation of about 1580 m. Located in the Barkol Basin, Barkol Lake is bordered by the Barkol Range, the eastern Tianshan to the south, and the Moqinwula Range to the north. The average water depth is only 0.6 m, with a maximum water depth of about 1 m [48]. In our study, a shrinking trend of water surface area was observed from 2000 to 2014, while an increasing trend of precipitation was observed (Table 5). The primary cause for lake size decrease may be attributed to human interference. Due to the mirabilite production and reducing precipitation near the lake, the lake area shrunk from 233 km^2 in the 1950s to 60 km^2 in 2011 [49]. The severe mineralization of lake water led to large area degradation of the surrounding wetlands [49].

Ayakkum, Aqqikkol, and Arkatag Lakes are located in the Kumkol Basin between the Altyn Mountains and the Kunlun Mountains. Aksayquin Lake is located in the western Kunlun Mountains of the northern Tibetan Plateau. The main supply of the four lakes is river runoff from the melting of glaciers and snow. In our study, significant expanding trends were observed for these four lakes from 2000 to 2014 except Arkatag Lake in July. Based on Landsat image analysis over the entire Tibetan Plateau, Aqqikkol had undergone surface extent increases in excess of 10% from 2000 to 2011 [8]. The expansion of lake areas and increasing trend of lake level on northern Tibetan Plateau was observed and documented by several studies [8,50,51]. Due to the increasing temperature, the water recharge increased a lot from accelerated melting of glaciers and perennial snow cover [52], and permafrost degradation [53,54]. However, for areas where surface water resources are generated mainly in the mountain glaciers of XUAR, the increasing flows caused by melting of mountain glaciers cannot be sustained in the long term if the glaciers disappear due to increasing temperature [55].

5. Conclusions

This study presented a time series analysis of lake water surfaces across XUAR using MODIS data from 2000 to 2014. A classification approach based on water index calculation and dynamic threshold selection was developed. Compositing water detection results of each time step, water masks were derived for the months of April, July, and September. The major lakes with an area of >100 km^2 were categorized into four classes based on their topographic locations. The seasonal and inter-annual surface area variation of the 10 major lakes was revealed in detail.

For plain lakes, the surface area of Ebinur Lake showed a significant shrinking trend and Manas lake an insignificant shrinking trend. They are both influenced by the expanding oasis and increasing water consumption. The Ulungur-Jili Lake had a stable area throughout the entire time period. The decreasing area of Bosten Lake may have been caused by the construction of hydrological projects in addition to the reduced precipitation. For mountain lakes, overexploitation has caused the shrinking of the Barkol Lake and the degradation of surrounding wetlands. As the largest and highest alpine lake, Sayram Lake showed a significant expanding trend in September. The four plateau lakes exhibited significant expanding trends for all three seasons except Arkatag Lake in July.

The lake dynamics revealed by MODIS time series are useful for ecological assessment of XUAR. Further studies are needed to use satellite imagery with different spatiotemporal resolutions, such as AVHRR and Landsat data since the 1970–80s, to capture the long-term dynamics of lakes in XUAR. It is also important to integrate the analysis of satellite data and climatic datasets for better understanding of the impact of climate change on water bodies in this arid and semi-arid region.

Acknowledgments

This research was supported by the National Natural Science Foundation of China under Grant No. 41471369, the Open Research Fund of State Key Laboratory of Remote Sensing Science under Grant No. OFSLRSS201413, and the Major International Cooperation and Exchange Project "Comparative study on global environmental change using remote sensing technology" under Grant No. 4112011400. The authors appreciate the financial assistance provided by the China Scholarship Council.

Author Contributions

Yingkui Li and Cuizhen Wang designed the research. Qingting Li and Linlin Lu conducted the water body detection and dynamic analysis. Yue Sui collected the MODIS data and performed quality assessment. All authors interpreted and reviewed the data. Qingting Li, Linlin Lu, Cuizhen Wang, and Yingkui Li wrote the final manuscript.

References

1. Oki, T.; Shinjiro, K. Global hydrological cycles and world water resources. *Science* **2006**, *313*, 1068–1072.
2. Soti, V.; Tran, A.; Bailly, J.S.; Puech, C.; Seen, D.L.; Bégué, A. Assessing optical earth observation systems for mapping and monitoring temporary ponds in arid areas. *Int. J. Appl. Earth Obs.* **2009**, *11*, 344–351.
3. Haas, E.M.; Bartholomé, E.; Lambin, E.F.; Vanacker, V. Remotely sensed surface water extent as an indicator of short-term changes in ecohydrological processes in sub-Saharan Western Africa. *Remote Sens. Environ.* **2011**, *115*, 3436–3445.
4. Haas, E.M.; Bartholomé, E.; Combal, B. Time series analysis of optical remote sensing data for the mapping of temporary surface water bodies in sub-Saharan western Africa. *J. Hydrol.* **2009**, *370*, 52–63.
5. Lehner, B.; Doll, P. Development and validation of a global database of lakes, reservoirs and wetlands. *J. Hydrol.* **2004**, *296*, 1–22.
6. Vector Map Level 0 (VMap0). Available online: http://earth-info.nga.mil/publications/vmap0.html (accessed on 16 February 2015).
7. Li, W.; Du, Z.; Ling, F.; Zhou, D.; Wang, H.; Gui, Y.; Sun, B.; Zhang, X. A comparison of land surface water mapping using the normalized difference water index from TM, ETM+ and ALI. *Remote Sens.* **2013**, *5*, 5530–5549.
8. Song, C.; Huang, B.; Ke, L. Modeling and analysis of lake water storage changes on the Tibetan Plateau using multi-mission satellite data. *Remote Sens. Environ.* **2013**, *135*, 25–35.
9. Fujita, K.; Sakai, A.; Nuimura, T.; Ymaguchi, S.; Sharma, R.R. Recent changes in Imja Glacial Lake and its damming moraine in the Nepal Himalaya revealed by in situ surveys and multi-temporal ASTER imagery. *Environ. Res. Lett.* **2009**, *4*, doi:10.1088/1748-9326/4/4/045205.
10. Lira, J. Segmentation and morphology of open water bodies from multispectral images. *Int. J. Remote Sens.* **2006**, *27*, 4015–4038.
11. Lacaux, J.P.; Tourre, Y.M.; Vignolles, C.; Ndione, J.A.; Lafaye, M. Classification of ponds from high-spatial resolution remote sensing: Application to rift valley fever epidemics in Senegal. *Remote Sens. Environ.* **2007**, *106*, 66–74.
12. Chipman, J.W.; Lillesand, T.M. Satellite-based assessment of the dynamics of new lakes in southern Egypt. *Int. J. Remote Sens.* **2007**, *28*, 4365–4379.
13. Sun, F.; Zhao, Y.; Gong, P.; Ma, R.; Dai, Y. Monitoring dynamic changes of global land cover types: Fluctuations of major lakes in China every 8 days during 2000–2010. *Chin. Sci. Bull.* **2014**, *59*, 171–189.

14. Klein, I.; Dietz, A.J.; Gessner, U.; Galayeva, A.; Myrzakhmetov, A.; Kuenzer, C. Evaluation of seasonal water body extents in Central Asia over the past 27 years derived from medium-resolution remote sensing data. *Int. J. Appl. Earth Obs.* **2014**, *26*, 335–349.
15. Deus, D.; Gloaguen, R. Remote sensing analysis of lake dynamics in semi-arid regions: Implication for water resource management. Lake Manyara, East African Rift, Northern Tanzania. *Water* **2013**, *5*, 698–727.
16. Wang, J.; Sheng, Y.; Tong, T.S.D. Monitoring decadal lake dynamics across the Yangtze Basin downstream of Three Gorges Dam. *Remote Sens. Environ.* **2014**, *152*, 251–269.
17. Mcfeeters, S.K. The use of normalized difference water index (NDWI) in the delineation of open water features. *Int. J. Remote Sens.* **1996**, *17*, 1425–1432.
18. Xu, H. Modification of normalized difference water index (NDWI) to enhance open water features in remotely sensed imagery. *Int. J. Remote Sens.* **2006**, *27*, 3025–3033.
19. Mo, W.; Sun, H.; Zhong, S.; Huang, Y.; He, L. Research on the CIWI model and its application. *Remote Sens. Inf.* **2007**, *5*, 16–21.
20. Song, C.; Huang, B.; Ke, L.; Richards, K.S. Remote sensing of alpine lake water environment changes on the Tibetan Plateau and surroundings: A review. *ISPRS J. Photogramm. Remote Sens.* **2014**, *92*, 26–37.
21. Chinese Academy of Forestry. The List of National Nature Reserve in China. 2013. Available online: http://www.papc.cn/ (accessed on 16 August 2015).
22. Bai, J.; Chen, X.; Li, J.; Yang, L.; Fang, H. Changes in the area of inland lakes in arid regions of central Asia during the past 30 years. *Environ. Monit. Assess.* **2011**, 178, 247–256.
23. Ma, M. Change in area of Ebinur Lake during the 1998–2005 period. *Int. J. Remote Sens.* **2007**, *28*, 5523–5533.
24. Ma, R.; Duan, H.; Hu, C.; Feng, X.; Li, A.; Ju, W. A half-century of changes in China's lakes: Global warming or human influence? *Geophys. Res. Lett.* **2010**, *37*, doi:10.1029/2010GL045514.
25. Yang, X.K; Lu, X.X. Drastic change in China's lakes and reservoirs over the past decades. *Sci. Rep.* **2014**, *4*, doi:10.1038/srep06041.
26. Bai, R.; He, L.; Wu, J. Analysis on recent change of water area of the main lakes in Xinjiang based on MODIS data. *Arid Zone Res.* **2012**, *29*, 561–566.
27. Wang, T.; Yan, C.Z.; Song, X.; Xie, J.L. Monitoring recent trends in the area of Aeolian desertified land using Landsat images in China's Xinjiang region. *ISPRS J. Photogramm. Remote Sens.* **2012**, *68*, 184–190.
28. Lu, L.; Kuenzer, C.; Wang, C.; Guo, H.; Li, Q. Evaluation of three MODIS-derived vegetation index time series for dryland vegetation dynamics monitoring. *Remote Sens.* **2015**, *7*, 7597–7614.
29. Vermote, E.F.; Kotchenova, S.Y.; Ray, J.P. MODIS surface reflectance user's guide. Available online: http://modis-sr.ltdri.org/guide/MOD09_UserGuide_v1_3.pdf (accessed on 16 August 2015).
30. Huang, S.; Li, J.; Xu, M. Water surface variations monitoring and flood hazard analysis in Dongting Lake area using long-term Terra/MODIS data time series. *Nat. Hazards* **2012**, *62*, 93–100.
31. Caroll, M.L.; Townshend, J.R.; DiMiceli, C.W.; Noojipady, P.; Sohlberg, R.A. A new global raster water mask at 250 m resolution. *Int. J. Digit. Earth* **2009**, *2*, 291–308.

32. Huesler, F.; Jonas, T.; Wunderle, S.; Albrecht, S. Validation of a modified snow cover retrieval algorithm from historical 1-km AVHRR data over the European Alps. *Remote Sens. Environ.* **2012**, *121*, 497–515.
33. Friedl, M.A.; Sulla-Menashe, D.; Tan, B.; Schneider, A.; Ramankutty, N.; Sibley, A.; Huang, X. MODIS Collection 5 global land cover: Algorithm refinements and characterization of new datasets. *Remote Sens. Environ.* **2010**, *114*, 168–182.
34. Lu, L.; Kuenzer, C.; Guo, H.; Li, Q.; Long, T.; Li, X. A novel land cover classification map based on a MODIS time-series in Xinjiang, China. *Remote Sens.* **2014**, *6*, 3387–3408.
35. Feng, L.; Hu, C.; Chen, X.; Cai, X.; Tian, L.; Gan, W. Assessment of inundation changes of Poyang Lake using MODIS observations between 2000 and 2010. *Remote Sens. Environ.* **2012**, *121*, 80–92.
36. Lu, S.; Wu, B.; Yan, N.; Wang, H. Water body mapping method with HJ-1A/B satellite imagery. *Int. J. Appl. Earth Obs.* **2011**, *13*, 428–434.
37. Congalton, R.G.; Green, K. *Assessing the Accuracy of Remotely Sensed Data: Principles and Practices*, 2nd ed.; CRC Press: Boca Raton, FL, USA, 2009.
38. Frazier, P.S.; Page, K.J. Water body detection and delineation with Landsat TM data. *Photogramm. Remote Sens.* **2000**, *66*, 1461–1467.
39. Xinjiang Uygur Autonomous Region Bureau of Statistics. *Xinjiang Uygur Autonomous Region Statistical Yearbook*; China Statistics Press: Beijing, China, 2001–2014.
40. Zhang, F.; Tiyip, T.; Johnson, V.C.; Kung, H.; Ding, J.; Zhou, M.; Fan, Y.; Kelimu, A.; Nurmuhammat, I. Evaluation of land desertification from 1990 to 2010 and its causes in Ebinur Lake region, Xinjiang China. *Environ. Earth Sci.* **2015**, *73*, 5731–5745.
41. Cheng, W.; Zhou, C.; Liu, H.; Zhang, Y.; Jiang, Y.; Zhang, Y.; Yao, Y. The oasis expansion and eco-environment change over the last 50 years in Manas River Valley, Xinjiang. *Sci. China Ser. D* **2006**, *49*, 163–175.
42. Wu, J.; Zeng, H.; Ma, L.; Bai, R. Recent changes of selected lake water resources in arid Xinjiang, northwestern China. *Quat. Sci.* **2012**, *32*, 142–150.
43. Maitiniyazi, A.; Kasimu, A. Change trend of surface water resources in Altai, Xinjiang China. *J. Desert Res.* **2014**, *34*, 1393–1401.
44. Gao, H.; Yao, Y. Quantitative effect of human activities on water level change of Bosten Lake in recent 50 years. *Sci. Geogr. Sin.* **2005**, *25*, 305–309.
45. Liu, L.; Zhao, J.; Zhang, J.; Peng, W.; Fan, J.; Zhang, T. Water balance of Lake Bosten using annual water-budget method for the past 50 years. *Arid Land Geogr.* **2013**, *36*, 33–40.
46. Guo, M.; Wu, W.; Zhou, X.; Chen, Y.; Li, J. Investigation of the dramatic changes in lake level of the Bosten Lake in northwestern China. *Theor. Appl. Climatol.* **2015**, *119*, 341–351.
47. Ma, D.; Zhang, L.; Wang, Q.; Zeng, Q.; Jiang, F.; Wang, Y.; Hu, R. Influence of warm-wet climate on Sailimu Lake. *J. Glaciol. Geocryol.* **2003**, *25*, 219–223.
48. Xue, J.; Zhong, W. Holocene climate variation denoted by Barkol lake sediments in northeastern Xinjiang and its possible linkage to the high and low latitude climates. *Sci. China Earth Sci.* **2011**, *54*, 603–614.
49. Zheng, S.; Luo, L. Variation of water quality in Balikun Lake in the last 18 years and water resource protection strategies. *Environ. Sci. Tech.* **2011**, *34*, 85–88.

50. Zhang, G.; Xie, H.; Kang, S.; Yi, D.; Ackley, S.F. Monitoring lake level changes on the Tibetan Plateau using ICESat altimetry data (2003–2009). *Remote Sens. Environ.* **2011**, *115*, 1733–1742.
51. Liao, J.; Shen, G.; Li, Y. Lake variations in response to climate change in the Tibetan Plateau in the past 40 years. *Int. J. Digit. Earth* **2013**, *6*, 534–549.
52. Yao, T.; Pu, J.; Lu, A.; Wang, Y.; Yu, W. Recent glacial retreat and its impact on hydrological processes on the Tibetan Plateau, China, and surrounding regions. *Arct. Antarct. Alp. Res.* **2007**, *39*, 642–650.
53. Cheng, G.; Wu, T. Responses of permafrost to climate change and their environmental significance, Qinghai-Tibet Plateau. *J. Geophys. Res.* **2007**, *112*, doi:10.1029/2006JF000631.
54. Li, Y.; Liao, J.; Guo, H.; Liu, Z.; Shen, G. Patterns and potential drivers of dramatic changes in Tibetan lakes, 1972–2010. *PLoS ONE* **2014**, *9*, doi:10.1016/j.jaridenv.2009.04.022.
55. Lioubimtseva, E.; Henebry, G.M. Climate and environmental change in arid Central Asia: Impacts, vulnerability, and adaptations. *J. Arid Environ.* **2009**, *73*, 963–977.

8

Water Resources Response to Changes in Temperature, Rainfall and CO_2 Concentration

Ricardo Arias [1], M. Luz Rodríguez-Blanco [1,*], M. Mercedes Taboada-Castro [1], Joao Pedro Nunes [2], Jan Jacob Keizer [2] and M. Teresa Taboada-Castro [1]

[1] Facultad de Ciencias and Centro de Investigaciones Científicas Avanzadas (CICA), University of A Coruña, Campus de A Coruña, A Coruña 15071, Spain; E-Mails: ricardo.arias@udc.es, (R.A.); mtaboada@udc.es (M.M.T.-C.); teresa.taboada@udc.es (M.T.T.-C.)

[2] Centre for Environmental and Marine Studies (CESAM) and Department Environment & Planning, University of Aveiro, Aveiro 3810-193, Portugal; E-Mails: jpcn@ua.pt (J.P.N.); jjkeizer@ua.pt (J.J.K.)

* Author to whom correspondence should be addressed; E-Mail: mrodriguezbl@udc.es;

External Editor: Julia Piantadosi

Abstract: Assessment of the diverse responses of water resources to climate change and high concentrations of CO_2 is crucial for the appropriate management of natural ecosystems. Despite numerous studies on the impact of climate change on different regions, it is still necessary to evaluate the impact of these changes at the local scale. In this study, the Soil and Water Assessment Tool (SWAT) model was used to evaluate the potential impact of changes in temperature, rainfall and CO_2 concentration on water resources in a rural catchment in NW Spain for the periods 2031–2060 and 2069–2098, using 1981–2010 as a reference period. For the simulations we used compiled regional climate models of the ENSEMBLES project for future climate input data and two CO_2 concentration scenarios (550 and 660 ppm). The results showed that changes in the concentration of CO_2 and climate had a significant effect on water resources. Overall, the results suggest a decrease in streamflow of 16% for the period 2031–2060 (intermediate future) and 35% by the end of the 21st century as a consequence of decreasing rainfall (2031–2060: −6%; 2069–2098: −15%) and increasing temperature (2031–2060: 1.1 °C; 2069–2098: 2.2 °C).

Keywords: SWAT; climate change; water resources; NW Spain

1. Introduction

The study of the effects of climate change on the quantity and quality of water resources has attracted a great deal of attention in recent years, particularly at a regional and global scale [1]. There is a general consensus that the Earth will be subject to warming, leading to changes in global climate patterns [1]. Different responses to global warming are expected from different regions of the world. For Europe, predictions on the evolution of temperature and rainfall, based on models of varying resolution and uncertainty, warn of the possibility of increased aridity in the coming decades. This effect is particularly evident for the southern regions where future climate change is projected to worsen conditions in a region already vulnerable to climate variability as well as lower water availability [1]. In general, the models (e.g., ENSEMBLES) predict a rise in mean temperature and a reduction in rainfall for Spain, with an increase in the number of extreme years in terms of maximum temperature, flood and droughts [2,3], although there are divergences depending on the model and the greenhouse gas emission scenario used. Moreover, projections of the future evolution of rainfall are more speculative than those for temperature, especially for smaller regions, but rainfall patterns are expected to alter in intensity and amount [1,2]. Changes in the temporal and spatial distribution of rainfall can also increase inter-annual rainfall variability as well as the risk of heavy rainfall events and droughts. These changes in temperature and rainfall are expected to impact on the hydrological cycle and alter the different processes occurring at catchment scale, including changes in surface runoff, evapotranspiration rates, nutrient enrichment, erosion and sediment transport [4,5], with concomitant effects on human activities and welfare. Human welfare would be impacted by changes in water supplies and demand; changes in opportunities for non-consumptive uses of the environment for recreation and tourism; changes in loss of property and lives from extreme climate phenomena; and changes in human health [1]. An assessment of the vulnerability of water resources to climate change allows anticipating potential negative impacts, and thus planning and establishing preventive actions with time.

Various approaches to assess the effects of climate change on water resources have been used [5–8]. In general, studies have reported that an increase in CO_2, as long as temperature and rainfall remain constant, will cause increases in water yield due to the marked decrease of the stomatal conductance of plants, thus decreasing evapotranspiration [8,9]. On the contrary, others have shown that higher temperatures lead to increased evaporation rates, reductions in streamflow and more frequent droughts [4,5,10]. All these studies indicate that catchment processes may be very sensitive to changes in temperature, rainfall and higher concentrations of CO_2 in the atmosphere. In Europe, most of the investigations on climate change impact on water resources have predicted a general reduction in the annual streamflow (more intense in the dry season) in southern Europe due to lower rainfall amount and higher temperatures [4].

The Iberian Peninsula is located in an area particularly vulnerable to climate change, where climate projections indicate an increase in arid conditions [1]. Several researches have evaluated the effects of climate change on river regimes in the Iberian Peninsula [10–12], but few studies have been conducted in North-Western Spain [13]. In this area, especially in Galicia, most rivers have little or no flow data available for such analyses. Records of river discharges are relatively recent in the region where water resources have been little considered, since Galicia has been considered to have these in abundance and water availability is not a limiting factor for economic and social development. This is undoubtedly the main reason why the research carried out in the Iberian Peninsula in this field has been focused primarily on Mediterranean, semi-arid environments where the scarcity of water gives rise to serious problems [14]. However, the recent floods in autumn 2006 and winter 2013–14 or the extended drought of summer 2007 occurred in Galicia caused significant ecological, economic and social impacts, highlighting the vulnerability of aquatic ecosystems and the resources that depend on them, e.g., aquaculture. This underlines the need to improve the knowledge of river dynamics as a basis for developing watershed management models in the region in order to prevent the risks associated with these natural phenomena, which seem to have increased in intensity recently [15].

In view of the above, this study has attempted to provide an initial estimate of the effects of potential changes in temperature, rainfall and CO_2 concentration on water resources in the Corbeira catchment, a minimally disturbed area located in Galicia (NW Spain). The analysis was performed using the Soil and Water Assessment Tool (SWAT) hydrological model. The SWAT model is widely used for different purposes (modelling runoff, sediment transport, nutrient, pesticides cycle, *etc.*) around the world and has been applied to many different size catchments and under considerably different conditions, usually with satisfactory results [16]. Among its multiple applications, SWAT has been extensively used to evaluate climate change effects on hydrological processes at catchment scale [9,11,16,17] because of its capability to incorporate future climate predictions from climatic models as inputs to the model, and to account for the effects of increased CO_2 on plant development and evapotranspiration. The interest of the study area is due to its location upstream from the Cecebre reservoir, which is an ecosystem of great ecological interest, being a EU Natura 2000 site, as well as the only source of water supply for the city of A Coruña and the surrounding municipalities (450,000 inhabitants). This fact gives special relevance to the findings of this study, from which a trend can be extracted for policy design purposes for the Cecebre reservoir.

2. Materials and Methods

2.1. Description of the Study Area

The study site for this research is the Corbeira catchment, a headwater catchment of the Mero River Basin, the most important water source for the city of A Coruña, NW Spain (Figure 1a). This catchment has an area of 16 km^2. The altitude ranges from 60 to 474 m (Figure 1b). The slopes are generally steep, with a mean value of 19%. The most common soils (Figure 1c) are Umbrisol and Cambisols [18] with silt and silt-loam texture settled on basic schists of the "Órdenes Complex" [19].

The dominant land uses in this catchment are forestry (65%) and agriculture (30%) (Figure 1d). The forest area mainly consists of commercial eucalyptus plantations, whereas the agriculture area is a patchwork of croplands (4% of total area), growing maize and winter cereal mostly interspersed with grassland (26% of total area). Impervious built-up areas and roads cover about 5% of the whole catchment area and are mainly distributed in the agriculture zone.

The climate is temperate humid, with a mean annual temperature of 13 °C and approximately 1050 mm mean annual rainfall (1983–2009), of which more than 67% falls from October to March.

Figure 1. (**a**) Location of the study area; (**b**) Digital elevation map; (**c**) Soil types map; (**d**) Land use map.

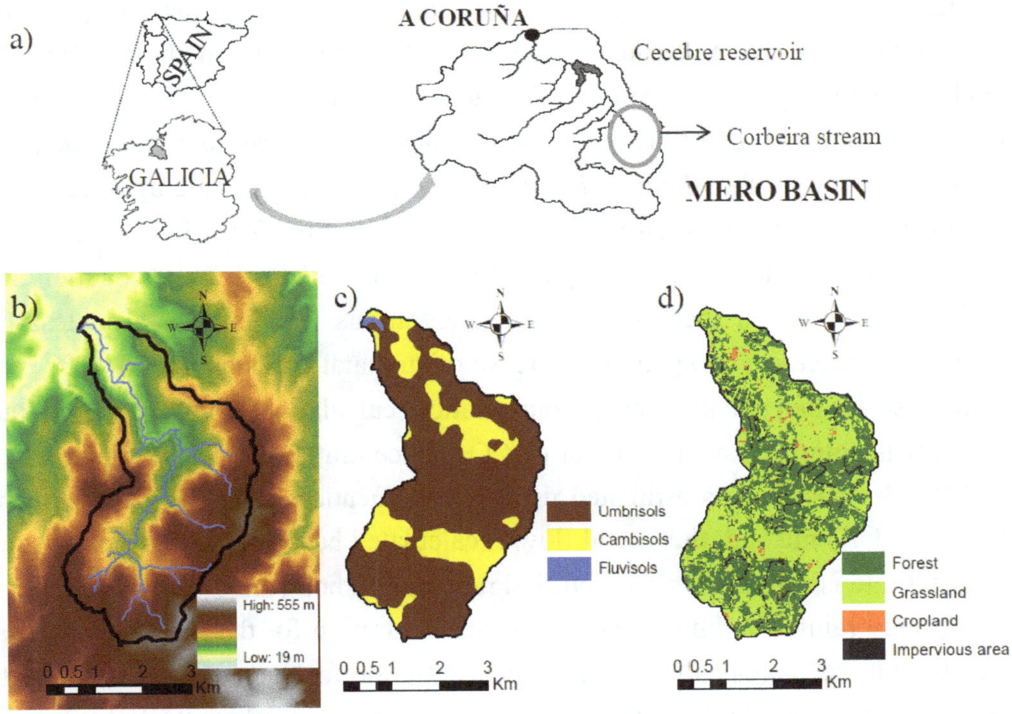

2.2. Data and Model Setup

2.2.1. The SWAT Hydrological Model

The SWAT model (Soil and Water Assessment Tool) is a process-based and spatially semi-distributed model developed by Agricultural Research Service of the United States Department of Agriculture (USDA) to assess the impact of agricultural management practices on water, sediment and chemical yields in large complex catchments [20], but it is also able to predict water, sediment and nutrient fluxes under different climate change scenarios [8,13,16]. SWAT is a basin-scale, continuous time model operating on a daily step, but it is not designed to simulate detailed, single-event flood routing [21]. The model was selected because it is a dynamic simulation model able to simulate streamflow response to climate change, it is in the public domain, and the generation of input files is eased by GIS-based tools. Although it is frequently applied to medium and large catchments with reasonably good results, it has also been calibrated for small forest catchments with good results [7,22], *i.e.*, with semi-natural land use. So, the SWAT model can be applied for hydrologic simulation in small catchments under different climatic conditions.

In SWAT, the watershed is divided into sub-basins, which are further separated into hydrological response units (HRUs), *i.e.*, areas with a specific combination of soil type, land use and management.

Most of the calculations are done at HRU scale and the results integrated at sub-basin scale. SWAT model simulations are divided into two parts: (i) the land phase and (ii) the routing through the river network. The land phase controls the amount of water, sediment and nutrients reaching the main channel, while the second phase defines the movement of water and other elements through the channel to the catchment outlet. The underlying theory detailing transport processes included in the model is available in the SWAT documentation [21].

The simulation of the hydrological cycle is based on the water balance, which is carried out taking into account precipitation, evapotranspiration, surface, lateral and base flow and deep aquifer recharge. The evapotranspiration can be calculated using one of these three methods: Penman-Monteith, Hargreaves and Priestley-Taylor. The Pemman-Monteith method was selected in this case because it uses more physical parameters (daily maximum and minimum temperature, wind speed, humidity and solar radiation). In addition, the Penman-Monteith option in SWAT incorporates the effects of increased CO_2 concentration on plant growth and evapotranspiration. The effect of CO_2 concentration change on plant stomatal conductance is computed by SWAT model using the equation developed by Easterling *et al.* [23], in which increased CO_2 concentrations lead to decreased leaf conductance (doubled CO_2 concentration leads to general decrease of stomatal conductance by 40%) which in turn results in a decrease in the potential evapotranspiration calculation. The change in radiation use efficiency of plants is simulated as a function of CO_2 concentration using the method developed by Stockle *et al.* [24]. Surface flow is estimated using a modification of the Soil Conservation Service Curve Number (SCC CN) method; the lateral flow is calculated based on the kinematic storage model, and the peak runoff rate is estimated by a modified rational method (the peak runoff rate is a function of the fraction of daily rainfall falling in the time of concentration for the sub-basin, the daily surface runoff volume, the sub-basin area and the time of concentration for the sub-basin). The water reaching the river network is then routed to the downstream sub-basin of the catchment. Water is routed through the channel using either the variable storage routing method or the Muskingum river routing method, both of which are variations of the kinematic wave model. In this research, the Muskingum method was used.

2.2.2. Model Inputs

The model required an extensive dataset of meteorological data, topography, soil types and land use and management practices. The meteorological data were acquired from the Galicia Meteorological Service (MeteoGalicia), selecting the closest station to the study area (coordinates: 560019 UTMX-29T ED-50, 4788103 UTMY-29T ED-50). The data included daily rainfall, maximum and minimum temperatures, relative humidity, solar radiation and wind speed. The SWAT model includes the weather generator (WXGEN) model to generate climatic data or to fill in gaps in the measured records. To implement this weather generator, SWAT requires long-term monthly statistical information (e.g., mean and standard deviation) for rainfall, maximum and minimum temperature, dew point temperature, solar radiation, and wind speed. Due to the reduced length of the data series of wind speed, relative humidity and solar radiation, for which consistent 30-year time series are not available, SWAT weather generator used in this study was generated on the basis of the climatic data from a meteorological station (1387E) belonging to the Spanish Meteorological Agency (AEMET) located at about 20 km from the study area whose data keep good correspondence with climatic data acquired from the station of MeteoGalicia (R^2 of 0.82, 0.94, 0.91, 0.72, 0.76 and 0.83 for rainfall, maximum temperature, minimum temperature, wind velocity, solar radiation and relative humidity, respectively). The digital elevation map (DEM) was created from the digital level curves (5 m) provided by the Territorial Information System of Galicia and used to provide elevation details for the SWAT and to delineate catchment boundaries. HRUs were delineated from (i) soil maps (1: 50,000) published by the Environment Department of the Xunta de Galicia, based on the FAO classification [18]; (ii) land use map was obtained from the digital processing (using ER Mapper software) of satellite images Landsat (resolution of 25 m) and aerial photographs (flight from summer 2004, with a resolution of 1 m), provided by the Territorial Information System of Galicia (Xunta of Galicia), and their subsequent field validation, which allowed distinguishing 4 classes of land uses (cropland, grassland, forest and impervious areas); and (iii) three classes of slopes (0%–13%; 13%–25%; >25%). Thresholds of 3% for land use and 20% for soil type were used to limit the number of HRUs in the catchment. This resulted in 12 HRUs. Only one sub-basin was defined in the Corbeira catchment.

The input data of physical soil properties were obtained from experimental works conducted in the catchment, whereas hydrological characterization of soils was built from literature data [25,26] or by estimating the parameters from data of texture and organic matter using pedo-transference functions [27,28].

The characteristics of the grasslands and croplands (maize) were taken from the SWAT database (SWAT plant codes used to represent grasslands and maize land covers were *meadow bromegrass* and *corn*, respectively), while eucalyptus characteristics, not included in the SWAT database, were derived from literature [11,29]. Information on agricultural management, such as dates for planting (maize: 1–10 May) and harvesting (grassland: May, August, November; maize: 20–30 September), was compiled from notes recorded during field research after interviewing farmers. Irrigation is rarely carried out in the catchment; hence, it was not modelled.

2.2.3. Calibration, Validation and Evaluation of the Model

The SWAT model was calibrated using daily stream discharge data measured at the Corbeira catchment outlet. Streamflow was separated into two components (baseflow and direct runoff) using a digital filter [30]. The model was set up from March 2001 to October 2010; the first 3 years were treated as the warm-up period for the model. The period from October 2005 to September 2008 was used for calibration and the period from October 2008 to September 2010 for validation.

The most sensitive model parameters were chosen in the calibration procedure based on preliminary sensitivity analysis using the Latin Hypercube One-factor-At-a-Time approach provided in the SWAT sensitivity analysis interface and SWAT model documentation [21]. The performance of the model presented in Table 1 shows that the simulations generated good results in comparison with observed streamflow data according to the evaluation criteria set by Motovilov *et al.* [31] and Moriasi *et al.* [32], who consider that model performance is satisfactory when the regression coefficient (R^2) is higher than 0.75 and Nash-Sutcliffe efficiency (NSE) is higher than 0.5, and if percentage of bias (PBIAS) is between −25% and +25%. Furthermore, the statistical values for the validation period slightly exceeded those of the calibration period, indicating a low over-parameterization. These differences can be due to the different hydrological conditions of the years used in the study, since hydrological models generally reproduce better normal than extreme hydrological conditions and, in our case, the calibration period comprises more extreme conditions than the validation period.

Table 1. Calibration and validation statistics for daily streamflow. R^2: regression coefficient; PBIAS: percentage of bias; NSE: Nash-Sutcliffe efficiency.

Parameter	Calibration	Validation
R^2	0.80	0.84
PBIAS	−1.8	−3.3
NSE	0.80	0.83
Observed mean and range ($m^3\ s^{-1}$)	0.18 (0.02–1.42)	0.24 (0.02–1.20)

The model successfully reproduces the measured streamflow and its trend over time (Figure 2). However, there are some cases in which the model does not agree well with measured streamflow, since it underestimates some peak flows during high-flow periods, e.g., in late March 2006 and in middle December 2006, both coinciding with a flood period. This could be attributed to the inability to simulate the soil moisture conditions during heavy rainfall periods. Curve Number only defines three antecedent moisture conditions: I-dry (wilting point), II-mean moisture, and III-wet (field capacity), and for each of them it assumes a unique relationship between rainfall and runoff, despite one same condition comprises different soil moisture contents. However, in this catchment, the relationship between rainfall and runoff increases as the antecedent soil moisture content rises, and consequently, the hydrological behaviour differs depending on the amount of water stored in the soil [33,34]. It could also be because the method used for simulation of runoff in SWAT (Curve Number) does not account for saturation near-stream zones and is not sensitive to rainfall intensity, so given a same amount of rainfall, SWAT computes the same amount of runoff regardless of intensity and duration of rainfall. The model also overestimates some peak discharge during some rainfall events (e.g., October 2006). This effect has been attributed to different causes, such as the short warm-up period of the model, the underestimation of evapotranspiration and overestimation of soil water content [35].

Figure 2. Rainfall, observed and predicted daily streamflow during the calibration (October 2005–Septmber 2008) and validation period (October 2008–September 2010).

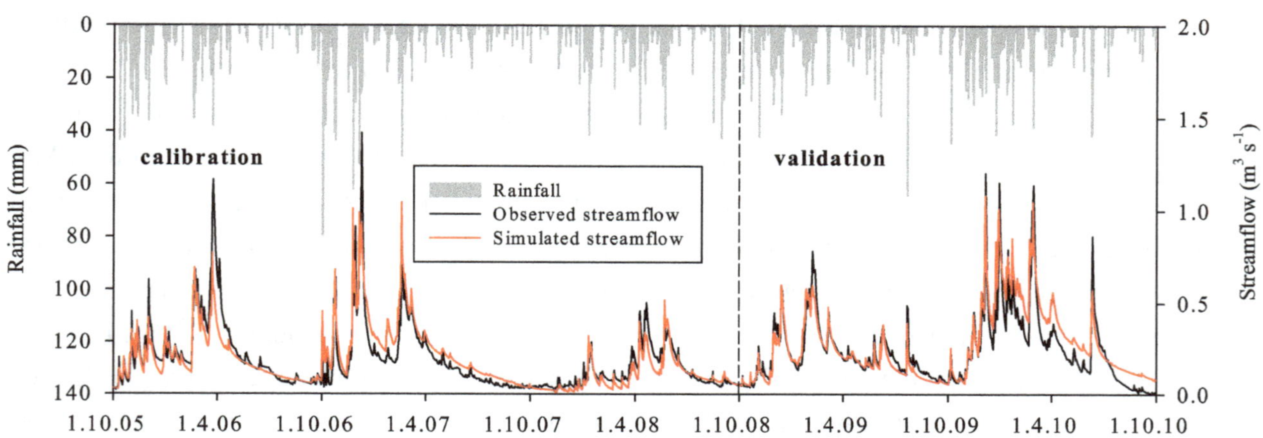

The above results indicate that SWAT is a suitable model to estimate streamflow in the study area; therefore, the calibrated model was used to assess the response of streamflow to future climate change. Several studies have highlighted the degree of uncertainty associated with the evaluation of climate change impacts on hydrology, pointing out that model calibration with present data may result in a bias when applying the model in the future [36,37]. The fact that the model adequately estimates the streamflow either during wet (July 2006) or dry years (August 2007) is a good indicator of its suitability for evaluating the impact of climate change; therefore, it can be used to assess the effects of climate change on streamflow with a reasonable degree of confidence. Anyway, it is appropriate to interpret the effects of climate change in terms of trends, not specific situations.

2.3. Scenarios of Temperature, Rainfall and CO_2 Concentration Changes

In this study, the analysis of climate change impact on water resources has been focused on predicting the potential effects that changes in temperature, rainfall and CO_2 concentration will cause on streamflow. For this purpose, two simulation sets were performed. The first evaluates the effect caused in streamflow by changes in each of the variables, *i.e.*, temperature, rainfall or CO_2 concentration. The second analyses the response of streamflow to simultaneous changes in the three variables.

At present, various methods are used for generating future climate scenarios. These methods include downscaling, change factor method, *etc.* A common procedure to downscale monthly temperature and precipitation of global climate model (GCM) projections to daily time series is stochastic downscaling using a weather generator [38]. Stochastic weather generators can be modified to generate future daily values of climate variables by adjusting historical weather patterns based on predicted future alterations from GCMs or RCMs [39]. In this study, the weather generator included in SWAT [21] was used to create 30-year climate series with changes in input variables. WXGEN uses a first-order Markov chain model to define the day as dry or wet. The daily rainfall amount is estimated based on a skewed or exponential distribution. Daily maximum and minimum temperatures, solar radiation and relative humidity are then generated based on the presence or absence of rain for the day [40]. WXGEN stochastic weather generator is widely used for climate change studies [11,41,42].

To accomplish the analysis of climate change impact on water resources the following steps were performed:

- Once calibrated the SWAT to represent the control conditions (control scenario), the model was run using climate series produced by the weather generator for reference period 1981-2010. Then, the degree of correspondence between simulated streamflow using the stochastic weather generator and simulated streamflow using observed meteorological data were verified, all with the purpose of checking the performance of the SWAT model using a stochastic weather generator to estimate streamflow in the study area. The results of statistical indicators (r^2 = 0.86 and NSE = 0.76), interpreted according to the criteria proposed by Motovilov *et al.* [31] and Moriasi *et al.* [32], suggest that model performance is satisfactory, indicating that the SWAT weather generator model can be used with a reasonable degree of confidence to analyze climate change scenarios.

- Subsequently, the different climate scenarios were created from the information provided by regional models of the ENSEMBLES project, using the change factors to modify the values of the parameters of the weather generator in a monthly-specific manner.

- Finally, the results for the control scenario were compared with the results of the climate change scenarios in order to quantify the changes on water resources.

The data for the reference period (1981–2001) were obtained from the 1387E meteorological station (AEMET). The different scenarios selected for this study are based on the information provided by the regional models of the ENSEMBLES project [2] for this station (Table 2), for the periods 2031–2060 (intermediate future) and 2069–2098 (distant future). Specifically, the selected scenarios for temperature and rainfall represent the mean and maximum forecasts of the ENSEMBLES project models (socio-economic A1B scenario) for these variables in the study area. These change factors were used to modify the rainfall and temperature parameters of the weather generator. Other climate variables, such as wind speed, solar radiation, relative humidity and dew point were assumed to be constant throughout future simulation periods. Climate modifications are given as a percentage change in rainfall (rainfall is multiplied by a given factor). Change factors for rainfall were used to alter the frequency and intensity of rainfall. This modification was performed by adjusting the probability of a rainy day followed by another rainy day in the month, and the probability of a rainy day followed by a dry day. These probabilities were obtained by multiplying the baseline probabilities by fifty percent of change factor for rainfall [adjusted probability = baseline probability + (baseline probability × 1/2 of change factor for rainfall)]. Temperature modifications were applied by adding the prescribed change to the weather generator temperature parameters derived from baseline data.

Table 2. Summary of characteristics of the selected RCMs of the ENSEMBLES project.

Name	Institute	GCM	RCM	Time Period
C4IRCA3	C4I (1)	HadCM3Q16	RCA3	1951–2099
CNRM/RM5.1	CNRM (2)	ARPEGE RM5.1	Aladin	1950–2100
DMI/ARPEGE	DMI (3)	ARPEGE	HIRHAM	1951–2100
DMI/BCM DMI	DMI (3)	BCM	DMI-HIRHAM5	1961–2098
DMI/ECHAM5-r3	DMI (3)	ECHAM5-r3	DMI-HIRHAM5	1951–2099
ETHZ/CLM	ETHZ (4)	HadCM3Q0	CLM	1951–2099
ICTP/RegCM3	ICTP (5)	ECHAM5-r3	RegCM3	1951–2100
KNMI/RACMO2	KNMI (6)	ECHAM5-r3	RACMO	1950–2100
MPIM/REMO	MPI (7)	ECHAM5-r3	REMO	1951–2100
SMHI/BCM	SMHI (8)	BCM	RCA	1961–2100
SMHI/ECHAM5-r3	SMHI (8)	ECHAM5-r3	RCA	1951–2100
SMHI/HadCM3Q3	SMHI (8)	HadCM3Q3	RCA	1951–2100

Notes: GCM: global climate models; RCM: regional climate models; (1) Rossby Centre, Swedish Meteorological and Hydrological Institute; (2) National Center of Meteorological. Research, France; (3) Danish Meteorological Institute; (4) Swiss Federal Institute of Technology Zürich; (5) Abdus Salam International Centre for Theoretical Physics, Italy; (6) Royal Netherlands Meteorological Institute; (7) Max Planck Institute for Meteorology, Germany; (8) Swedish Meteorological and Hydrological Institute.

Projected changes in temperature and rainfall in the study area are presented in Figure 3 for the two-time periods (2031–2060 and 2069–2098). All projections show an increase in annual mean temperature and a decrease in rainfall for the two periods, although there is a wide variability among projections, indicating highly uncertain results. The projected temperature changes vary from

0.4–1.8 °C in 2031–2060 to 1.6–3.9 °C in 2069–2098, depending on the climate models. With regard to rainfall, the projections indicate a decrease between 2%–14% in 2031–2060 and 6%–27% in 2069–2098. The mean values of all the projections used predict that temperature will rise by 1.1 °C during 2031–2060 and by 2.2 °C in 2069–2098, while rainfall will decrease by 6% and 27% at the mid and end of the 21st century. CO_2 scenarios change by an increase of 1.5 and 2 times the current CO_2 concentration (330 ppm). It is thought that the selected CO_2 concentrations (550 and 660 ppm) give a reasonable representation of future CO_2 conditions for the middle and end of the 21st century under the A1B scenario [3]. Table 3 shows all climate change scenarios used in the SWAT simulations. All climate change scenarios were run for a 30-year period. Land use/land cover was assumed to remain unchanged throughout the simulation.

Figure 3. Climate change scenarios for annual mean temperature and rainfall for: (**a**) 2031–2060 and (**b**) 2069–2098 (ENSEMBLES project).

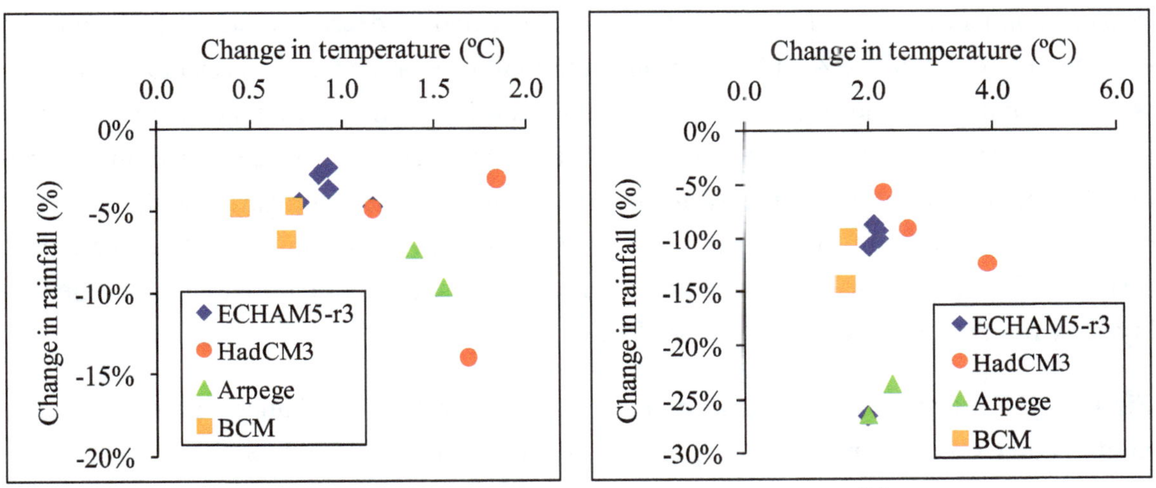

(**a**) (**b**)

Table 3. Climate change scenarios used for SWAT simulations.

Scenario	Temperature (°C)	Rainfall (%)	CO_2 Concentration (ppm)
1	1.1 (mean 2031–2060)	0	330
2	1.7 (maximum 2031–2060)	0	330
3	2.2 (mean 2069–2098)	0	330
4	3.9 (maximum 2069–2098)	0	330
5	0	−6 (mean 2031–2060)	330
6	0	−14 (maximum 2031–2060)	330
7	0	−15 (mean 2069–2098)	330
8	0	−27 (maximum 2069–2098)	330
9	0	0	550
10	0	0	660
11	1.1	−6	550
12	1.7	−14	550
13	2.2	−15	660
14	3.9	−27	660
15	1.1	−6	330
16	2.2	−15	330

Note: 0: means no change in the variable.

T-tests were performed to assess if the streamflow estimated from the climate change scenarios and the reference scenario are statistically different from each other. The significance of statistical test was set at $p < 0.01$.

3. Results and Discussion

3.1. Vulnerability of Streamflow to Change in Temperature, Rainfall or CO_2 Concentration

Figure 4 shows the responses of evapotranspiration and streamflow to climate parameter changes. Streamflow will be significantly altered as a result of changes in temperature, rainfall or CO_2 concentration. Streamflow significantly ($p < 0.01$) decreased with the increase in temperature (a larger amount of water is lost through evapotranspiration) and lower rainfall. In both cases, the impact was more pronounced in the period 2069–2098, which showed strong deviation of the climate variables compared to the current conditions (Table 2).

Increasing temperature by 1.1 and 2.2 °C (scenarios 1 and 3) decreased streamflow rates by 13% and 29%, respectively; while 6% and 15% drops in rainfall (scenarios 5 and 7) resulted in a streamflow decrease of 9% and 25%. These results suggest that streamflow in the Corbeira catchment will be more sensitive to the average increase in temperature than to the average decrease in rainfall, highlighting the role of evapotranspiration in the water cycle. However, when compared with the worst case scenarios (scenario 4: T^a + 3.9 °C, scenario 8: P − 27%), streamflow is more sensitive to a reduction in rainfall (Figure 4), which is in accordance with earlier findings in the literature that underline the major role played by rainfall on streamflow changes [5].

Figure 4. Response of evapotranspiration and streamflow to changes in temperature, rainfall and CO_2 concentration based on the scenarios defined in Table 3.

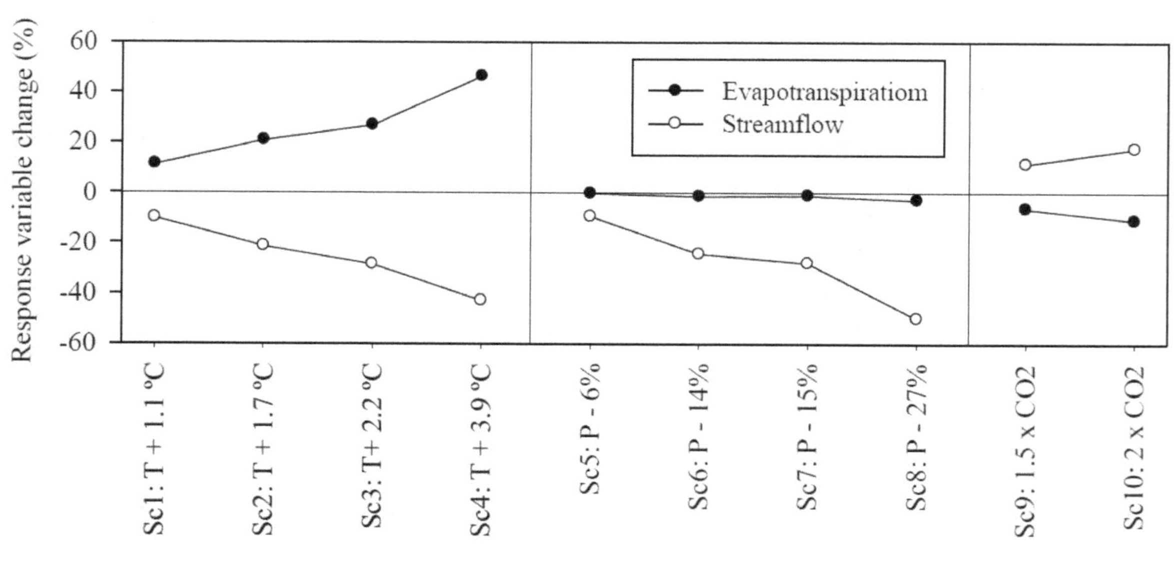

Increase in temperature led to increase in biomass production in some cases and decrease in others, depending on the crop type [6]. This behaviour depends on the temperature reached regarding the optimum, minimum and maximum temperatures associated with plant growth, because temperature is one of the most important factors governing plant growth. For the Corbeira catchment, forest biomass production (forest: 65% of the study area) increased with increasing temperature; however, in grassland and crops, biomass production decreased with temperature increases (Figure 5). This could explain the increase in evapotranspiration (2031–2060: 14%, 21%; 2069–2098: 27%, 46%) and consequently the decrease of streamflow with temperature rise (Figure 4). At present, the main limitations for eucalyptus cultivation in Galicia, the main forest specie in the Corbeira catchment, are low temperatures and frost. The temperature rise and consequent decreased risk of frost increase forest productivity, especially in spring. In summer, the biomass growth rate is lower than in other seasons, which may be associated with water or nutrient limitations. However, the model, under these scenarios, estimated only a slight increase in the number of days with water stress.

Figure 5. Response of vegetation biomass to changes in temperature, rainfall and CO_2 concentration based on the scenarios defined in Table 3.

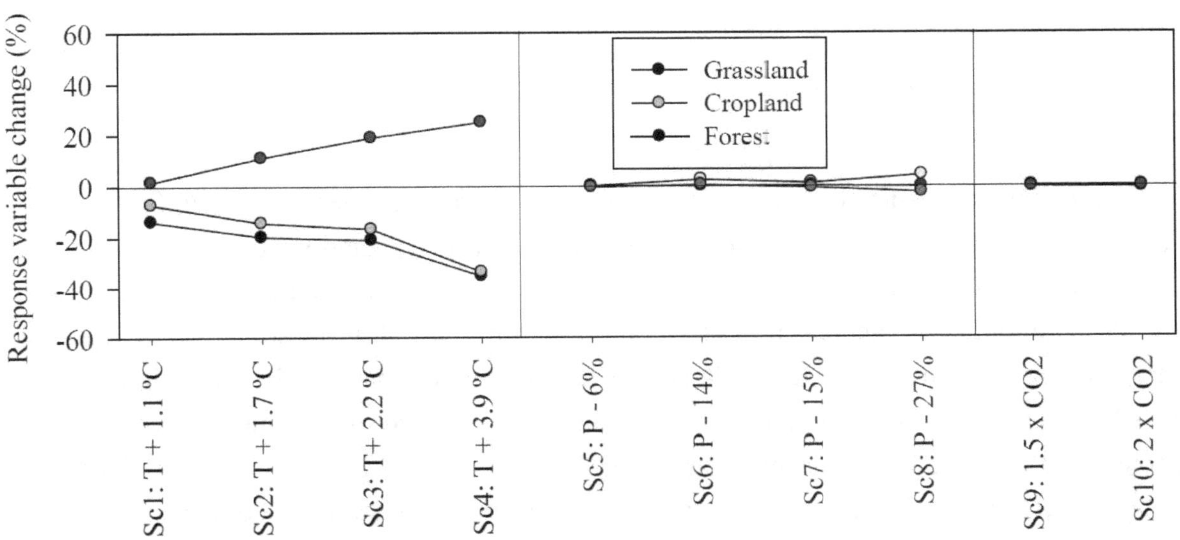

Flow components are affected differently by changes in temperature and rainfall (Figure 6). Thus, a temperature increase had greater impact on subsurface flow (groundwater + lateral flow) because of soil water loss by evapotranspiration, while rainfall decrease had a greater impact on surface runoff. This explains a higher sensitivity of soil water content to temperature changes (Figure 6). These results differ from those of Nunes *et al.* [11] for the Guadiana Basin (southwest Iberian Peninsula) where subsurface flow was mostly affected by reduced rainfall due to the extremely shallow soils in the basin. The soils of the Corbeira catchment, however, are deep and favour the diversion of soil water to evapotranspiration, hence temperature increases mainly affect the subsurface flow.

The streamflow decrease is more significant than that of rainfall (Figures 4 and 6), showing it is not a linear process. It is estimated that for every 1% decrease in rainfall, streamflow falls by approximately 1.5%. These results are close to those obtained by Pruski and Nearing [43], who analysed the effect of rainfall changes on agricultural slopes in different regions of the United States, using the WEPP model. These authors predicted a fall in runoff of 1.97% for every 1% decrease in rainfall. Similarly, Nunes *et al.* [11] reported a fall in runoff of 1.9% and 2.1% for the Guadiana and Tejo basins, respectively.

Figure 6. Response of different flow components and soil water content to changes in temperature, rainfall and CO_2 concentration based on the scenarios defined in Table 3.

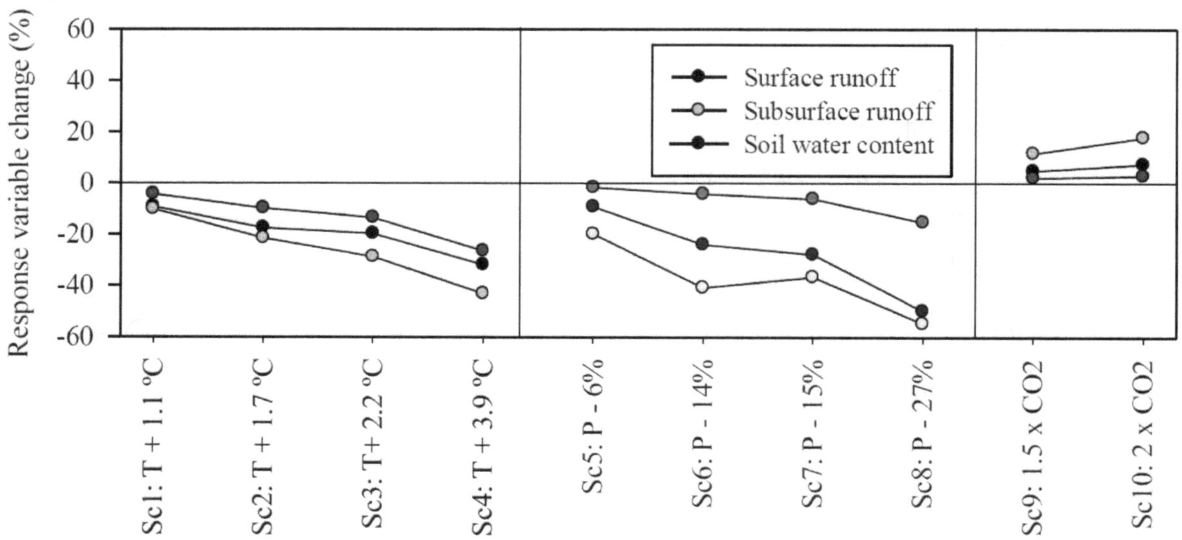

Increasing CO_2 concentration in the atmosphere led to increases in streamflow (10%–15%, Figure 4). The rise in CO_2 concentration could increase vegetation biomass production and evapotranspiration [44]. However, in this study no effect on the vegetation biomass was observed (Figure 5), although a decrease was found in the evapotranspiration (2 × CO_2: −11% evapotranspiration, Figure 4), resulting in higher soil water content and, in turn, higher streamflow (Figure 4). This reduction in the evapotranspiration could be related to stomatal closure of plant leaves in response to increasing CO_2 concentration, as in SWAT a doubled CO_2 concentration leads to a 40% reduction in leaf conductance for all plant species. According to the stomatal conductance optimization hypothesis, the plant stomas are simultaneously maximizing the carbon gain rate while minimizing the rate of water loss [45], *i.e.*, as adaptation mechanism, plants tend to reduce stomatal conductance and suppress transpiration under a high concentration of CO_2. This would lead to greater water-use efficiency by plants (ratios of CO_2 molecules fixed by the plant in relation to the number of water molecules lost by transpiration) allowing a larger amount of water to be available for runoff and recharge.

Moreover, higher CO_2 concentrations can enhance the photosynthesis rate and consequently the vegetation biomass, although it was not observed in this study (Figure 5). This effect, known as the CO_2 fertilization effect, leads to a higher leaf area index (LAI) in the vegetation, which can reduce the radiation reaching the soil surface, thereby reducing soil evaporation, and increasing the streamflow. However, Bunce [46] in a review work concludes that an increase in CO_2 concentrations rarely leads to higher LAI, unless ventilation is artificial, such as it occurs in chambers and greenhouses. In addition, the author indicates that LAI increases above 3–4 m^2 m^{-2} have a minimal effect on evapotranspiration as a result of shade and higher canopy humidity. This conclusion is based on studies of crops in which nutrients are not a limiting factor and, therefore, an even lower response can be expected in natural ecosystems, because nutrients frequently limit plant productivity, thus their responsiveness to CO_2 concentrations.

3.2. *Vulnerability of Streamflow to Simultaneous Changes in Climate Parameters*

Figure 7 shows the response of evapotranspiration and streamflow to combined simultaneous changes in temperature, rainfall and CO_2 concentration. In comparison with the results obtained when changing climate parameters singly (temperature or rainfall or CO_2 concentration), coupled climate parameter changes had a synergistic effect on streamflow, causing an increase in the vulnerability to change. For scenarios 11 and 13 (mean values), a streamflow decrease of 16% and 35% for the periods 2031–2060 and 2069–2098, respectively is forecasted. For scenarios 12 and 14 (maximum values, unlikely) decreases of 46% and 51% for the same horizons are predicted. Although these results are indicative and should be taken as trend indicators, they show the high vulnerability of the Corbeira stream to climate change, even though climatic variations are relatively low (Table 3). This is consistent with most of the studies on the impact of climate change carried out in the Iberian Peninsula, which have predicted a decline in water resources [7,11,12]. The expected decrease in streamflow in the Corbeira catchment is higher than that estimated for other catchments in NW Spain [12,13], reflecting a greater vulnerability of small catchments to changes in climatic variables, as noted by Beguería *et al.* [47] in other regions of Spain. However, these results should be interpreted with caution. Although the trend seems to be clear, the change percent will vary according to the climate change scenarios considered and the catchment characteristics.

Figure 7. Response of evapotranspiration and streamflow to simultaneous changes in temperature, rainfall and CO_2 concentration based on the scenarios defined in Table 3.

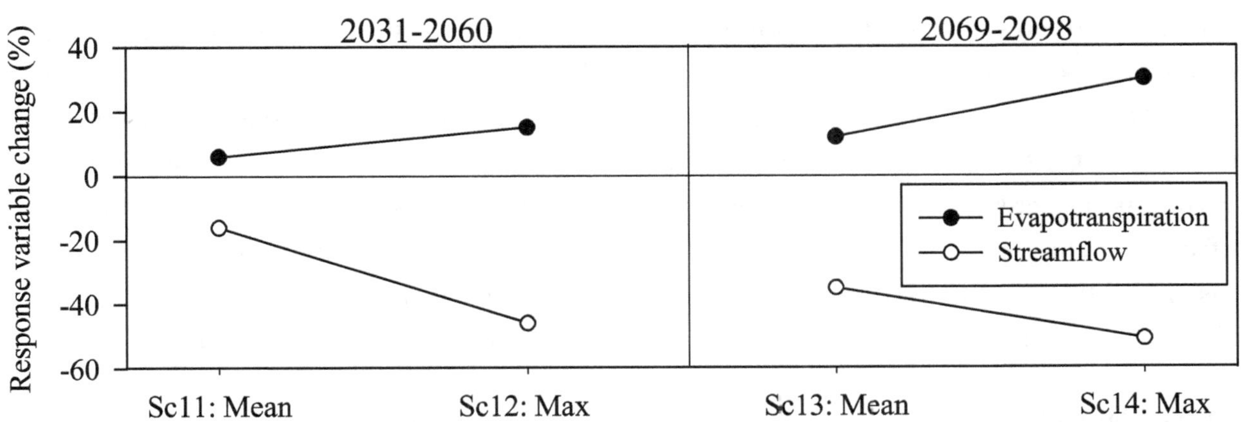

It is generally recognized that the positive effects exerted by an increase in CO_2 concentration on water-plant relationships would be offset by a greater evaporative demand at higher temperatures. Numerous studies indicate that changes in temperature and rainfall alter and, in many cases, limit the effects of CO_2 on plants [11,13]. For example, high temperatures during the flowering period could mitigate the effects of high CO_2 concentration, since they could limit the number, size and grain quality [48]. In order to confirm the effect of a CO_2 increase on streamflow, the model was run using two new scenarios, Sc15, Sc16, (Table 3) with the same rainfall and temperature changes as scenarios 11 and 13, respectively; but with no changes in CO_2 concentration, and both results (with and without increased CO_2) were compared. Not taking into account the effects of CO_2 on plants, there was a decrease in streamflow of 24% (2031–2060, Sc15) and 46% (2069–2098, Sc16) compared to 16%

(2031–2060, Sc11) and 35% (2069–2098, Sc13) respectively when enhanced plant photosynthetic water use efficiency (greater stomatal efficiency of plants in response to increased CO_2 concentration) was considered, as for evapotranspiration calculation SWAT takes into account variation of radiation-use efficiency, plant growth, and plant transpiration due to changes in atmospheric CO_2 concentrations. Given the importance exerted by an increase of CO_2 concentration on water resources, this parameter should be considered in any assessment of climate change impact. However, it should be noted that the effects of CO_2 on streamflow might be overestimated, because the SWAT does not assume that leaf area increases with CO_2 concentrations.

4. Conclusions

This work demonstrated the high vulnerability of streamflow to changes in temperature and rainfall in the Corbeira catchment. Furthermore, it was found that an increase in the concentration of CO_2 in the atmosphere could slightly attenuate the effects of climatic variables on water resources. Similarly, both medium and long-term effects of climate change on streamflow can be significant if the forecast temperature and rainfall changes included in this study are met. Overall, the decrease in rainfall was accompanied by a large increase in the evapotranspiration. The combination of these two trends is likely to result in decreased availability of water for crops and natural vegetation. A moderate decrease in streamflow of 16% and 35% is expected for the periods 2031–2060 and 2069–2098, respectively.

In general, it may indicate that this study provides an example of the possible effects of climate change on water resources in the NW Spain, so it can be used as a starting point to improve the understanding of how climate change will impact water resources in this area and provide some data to decision makers. Hydrology and distribution of land uses in the Corbeira catchment are similar to those of the upper and middle Mero River Basin. Therefore, if climate change scenarios adopted in this work occur in the future, significant changes may also occur in the Mero River, affecting the Cecebre reservoir. With increasingly limited water resources and water consumption increasing annually, all the facts point to a situation of greater water unsustainability and therefore to greater environmental unsustainability. This implies that measures, able to solve this situation, should be taken in order to avoid the consequences of a decrease of water resources against an increased demand.

Acknowledgments

This investigation was carried out within the projects REN2003-08143, funded by the Spanish Ministry of Education and Science, and PGIDIT05RAG10303PR and 10MDS103031, financed by Xunta of Galicia. The second author was awarded a Postdoctoral research contract (Xunta of Galicia). We appreciate the detailed review and suggestions of two anonymous reviewers, which improved the manuscript.

Author Contributions

All authors contributed to the design and development of this manuscript under the supervision of M. Teresa Taboada Castro.

References

1. Intergovernmental Panel on Climate Change (IPCC). *Impacts, Adaptation, and Vulnerability*. Part A: Global and Sectoral Aspects. Contribution of Working Group II to the Fifth Assessment Report of the Intergovernmental Panel on Climate Change; Field, C.B., Barros, V.R., Dokken, D.J., Mach, K.J., Mastrandrea, M.D., Bilir, T.E., Chatterjee, M., Ebi, K.L., Estrada, Y.O., *et al*, Eds.; Cambridge University Press: Cambridge, UK; New York, NY, USA, 2014.
2. Van der Linden, P.; Mitchell, J.F.B. *ENSEMBLES: Climate Change and its Impacts: Summary of Research and Results from the ENSEMBLES Project*; Met Office Hadley Centre: Exeter, UK, 2009; p. 160.
3. Houghton, J.T.; Ding, Y.; Griggs, D.J.; Noguer, M.; van der Linden, P.J.; Dai, X.; Maskell, K.; Johnson, C.A. Climate Change: The Scientific Basis. In *Contribution of Working Group I to the Third Assessment Report of the Intergovernmental Panel on Climate Change*; Cambridge University Press: Cambridge, United Kingdom and New York, NY, USA, 2001; p. 881.
4. Lehner, B.; Doll, P.; Alcamo, J.; Henrichs, T.; Kaspar, F. Estimating the impact of global change on flood and drought risks in Europe: A continental integrated analysis. *Clim. Chang.* **2006**, *75*, 273–299.
5. Kundzewics, Z.; Mata, L.; Arnell, N.; Döll, P.; Jimenez, B.; Miller, K.; Oki, T.; Sen, Z.; Shiklomanov, I. The implications of projected climate change for freshwater resources and their management. *Hydrolog. Sci. J.* **2008**, *53*, 3–10.
6. Pruski, F.F.; Nearing, M.A. Climate-induced changes in erosion during the 21st century for eight U.S. locations. *Water Resour. Res.* **2002**, *38*, 34-1–34-12.
7. Zabaleta, A.; Meaurio, M.; Ruiz, E.; Antigüedad, I. Simulation climate change impact on runoff and sediment yield in a small watershed in the Basque Country, northern Spain. *J. Environ. Qual.* **2013**, *43*, 235–245.
8. Chaplot, V. Water and soil resources response to rising levels of atmospheric CO_2 concentration and to changes in precipitation and air temperature. *J. Hydrol.* **2007**, *337*, 159–171.
9. Butcher, J.P.; Johnson, T.E.; Nover, D.; Sarkar, S. Incorporating the effects of increased atmospheric CO_2 in watershed model projections of climate change impacts. *J. Hydrol.* **2014**, *513*, 322–334.
10. López-Moreno, J.I.; Goyette, S.; Beniston, M. Climate change prediction over complex areas: Spatial variability of uncertainties and expected changes over the Pyrenees from a set of regional climate models. *Int. J. Climatol.* **2008**, *28*, 1535–1550.
11. Nunes, J.P.; Seixas, J.; Pacheco, N.R. Vulnerability of water resources, vegetation productivity and soil erosion to climate change in Mediterranean watersheds. *Hydrol. Process.* **2008**, *22*, 3115–3134.
12. Estrela, T.; Pérez-Martin, M.A.; Vargas, E. Impacts of climate change on water resources in Spain. *Hydrolog. Sci. J.* **2012**, *57*, 1154–1167.
13. Raposo, J.R.; Dafonte, J.; Molinero, J. Assessing the impact of future climate change on groundwater recharge in Galicia-Costa, Spain. *Hydrogeol. J.* **2013**, *21*, 459–479.
14. García Ruiz, J.M.; López-Moreno, J.I.; Vicente-Serrano, S.M.; Lasanta-Martínez, T.; Beguería, S. Mediterranean water resources in a global change scenario. *Earth Sci. Rev.* **2011**, *105*, 121–139.

15. Taboada, J.J. Riesgos asociados a fenómenos meteorológicos extremos. In *Riesgos Naturales en Galicia. El Encuentro Entre Naturaleza y Sociedad*; Fra-Paleo, U., Ed.; Universidad de Santiago de Compostela, Servicio de Publicaciones e Intercambio Científico: Santiago de Compostela, Spain, 2010; pp. 25–45.
16. Gassman, P.W.; Reyes, M.R.; Green, C.H.; Arnold, J.G. The Soil and Water Assessment Tool: Historical development, applications, and future research directions. *Trans. ASABE* **2007**, *50*, 1211–1250.
17. Wu, Y.; Liu, S.; Gallant, A.L. Predicting impacts of increased CO_2 and climate change on the water cycle and water quality in the semiarid James River Basin of the Midwestern USA. *Sci. Total Environ.* **2012**, *430*, 150–160.
18. Food and Agriculture Organization (FAO). *World Reference Base for Soil Resources*; World Soil Resources Reports 106; FAO: Rome, Italy, 2014.
19. Instituto Tecnológico Geominero de España (IGME). *Mapa Geológico de España, 1:50000. Hoja 45. Betanzos*; Servicio de Publicaciones del Ministerio de Industria y Energía: Madrid, Spain, 1981.
20. Arnold, J.G.; Srinivasan, R.; Muttiah, R.S.; Williams, J.R. Large area hydrologic modeling and assessment part I: Model development. *J. Am. Water Res. Assoc.* **1998**, *34*, 73–89.
21. Neitsch, S.L.; Arnold, J.G.; Srinivasan, R.; Williams, J.R. *Soil and Water Assessment Tool User's Manual*; Texas Water Resources Institute: Collegue Station, TX, USA, 2002; p. 506.
22. Green, C.; van Griensven, A. Autocalibration in hydrologic modeling: Using SWAT 2005 in small-scale watersheds. *Environ. Model. Softw.* **2008**, *23*, 422–434.
23. Easterling, W.E.; Rosenberg, N.J.; McKenney, M.S.; Jones, C.A.; Dyke, P.T.; Williams, J.R. Preparing the erosion productivity impact calculator (EPIC) model to simulate crop response to climate change and the direct effects of CO_2. *Agric. For. Meteorol.* **1992**, *59*, 17–34.
24. Stockle, C.O.; Williams, J.R.; Rosenberg, N.J.; Jones, C.A. A method for estimating the direct and climatic effects of rising atmospheric carbon dioxide on growth and yield of crops: Part 1—Modification of the EPIC model for climate change analysis. *Agric. Syst.* **1992**, *38*, 225–238.
25. Martínez-Cortizas, A.; Castillo-Rodríguez, F.; Pérez-Alberti, A. Factores que intervienen en la precipitación y el balance de agua en Galicia. *Bol. Asoc. Geógr. Esp.* **1994**, *18*, 79–96.
26. Taboada-Castro, M.M.; Lado-Liñares, M.; Diéguez, A.; Paz, A. Evolución temporal de la infiltración superficial a escala de parcela. In *Avances Sobre el Estudio de la Erosión Hídrica*; Paz, A., Taboada, M.T., Eds.; Universidade da Coruña: A Coruña, Spain, 1999; pp. 101–127.
27. Saxton, K.E.; Rawls, W.J.; Romberger, J.S.; Papendick, R.L. Estimating generalized soil-water characteristics from texture. *Soil Sci. Soc. Am. J.* **1986**, *50*, 1031–1036.
28. Ferrer Julià, M.; Estrela, M.T.; Sánchez, J.A.; García, M. Constructing a saturated hydraulic conductivity map of Spain using pedotransfer functions and spatial prediction. *Geoderma* **2004**, *123*, 257–277.
29. Rodríguez-Suárez, J.A.; Soto, B.; Iglesias, M.L.; Díaz-Fierros, F. Application of the 3PG forest growth model to a Eucalyptus globulus plantation in Northwest Spain. *Eur. J. For. Res.* **2010**, *129*, 573–583.
30. Arnold, J.G.; Allen, P.M.; Muttiah, R.; Bernhardt, G. Automated base flow separation and recession analysis techniques. *Ground Water* **1995**, *33*, 1010–1018.

31. Motovilov, Y.; Gottschalk, G.L.; Engeland, K.; Rodhe, A. Validation of distributed hydrological model against spatial observations. *Agri. For. Meteorol.* **1999**, *98*, 257–277.
32. Moriasi, D.N.; Arnold, J.G.; van Liew, M.W.; Bingner, R.L.; Harmel, R.D.; Veith, T.L. Model evaluation guidelines for systematic quantification of accuracy in watershed simulations. *Transl. ASABE* **2007**, *50*, 885–900.
33. Rodríguez-Blanco, M.L.; Taboada-Castro, M.M.; Taboada-Castro, M.T. Rainfall runoff response and event-based runoff coefficients in a humid area (northwest Spain). *Hydrolog. Sci. J.* **2012**, *57*, 445–459.
34. Palleiro, L.; Rodríguez-Blanco, M.L.; Taboada-Castro, M.M.; Taboada-Castro, M.T. Hydroclimatic response of a humid agroforestry catchment at different times scales. *Hydrol. Process* **2014**, *28*, 1677–1688.
35. Benaman, J.; Shoemaker, C.A.; Haith, D.A. Calibration and validation of soil and water assessment tool on an agricultural watershed in upstate New York. *J. Hydrol.* **2005**, *10*, 363–374.
36. Kirchner, J.W. Getting the right answers for the right reasons: Linking measurements, analyses, and models to advance the science of hydrology. *Water Resour. Res.* **2006**, *42*, doi:10.1029/2005WR004362.
37. Zhang, X.; Xu, Y.P.; Fu, G. Uncertainties in SWAT extreme flow simulation under climate change. *J. Hydrol.* **2014**, *515*, 205–222.
38. Winkler, J.A.; Guentchev, G.S.; Perdinan; Tan, P.N.; Zhong, S.; Liszewska, M.; Abraham, Z.; Niedzwiedz, T.; Ustrnul, Z. Climate scenario development and applications for local/regional climate change impact assessments: An overview for the non−climate scientist. Part I: Scenario development using downscaling methods. *Geogr. Compass* **2011**, *5*, 275–300.
39. Fowler, H.J.S.; Blenkinsop, S.; Tebaldi, C. Linking climate change modelling to impacts studies: Recent advances in downscaling techniques for hydrological modelling. *Int. J. Climatol.* **2007**, *27*, 1547–1578.
40. Sharpley, A.N.; Williams, J.R. *EPIC-Erosion Productivity Impact Calculator, I. Model Documentation*; U.S. Department of Agriculture, Agricultural Research Service: Washington, DC, USA, 1990; p. 235.
41. Ficklin, D.L.; Luo, Y.; Luedeling, E.; Zhang, M. Climate change sensitivity assessment of a highly agricultural watershed using SWAT model. *J. Hydrol.* **2009**, *374*, 16–29.
42. Zhang, H.; Huang, G.H.; Wang, D.; Zhang, X. Uncertainty assessment of climate change impacts on the hydrology of small prairie wetlands. *J. Hydrol.* **2011**, *396*, 94–103.
43. Pruski, F.F.; Nearing, M.A. Runoff and soil-loss responses to changes in precipitation: A computer simulation study. *J. Soil Water Conserv.* **2002**, *57*, 7–16.
44. Rosenzweig, C.; Hillel, D. *Climate Change and the Global Harvest. Potential Impacts of the Greenhouse Effect on Agriculture*; Oxford University Press: New York, NY, USA, 1998; pp. 324.
45. Katul, G.; Manzoni, S.; Palmroth, S.; Oren, R. A stomatal optimization theory to describe the effects of atmospheric CO_2 on leaf photosynthesis and transpiration. *Ann. Bot.* **2009**, *105*, 431–442.
46. Bunce, J.A. Carbon dioxide effects on stomatal responses to the environment and water use by crops under field conditions. *Oecologia* **2004**, *140*, 1–10.

47. Beguería, S.; López-Moreno, J.I.; Lorente, A.; Seeger, M.; García-Ruiz, J.M. Assessing the effect of climate change and land-use changes on streamflow in the central Spanish Pyrenees. *Ambio* **2003**, *32*, 283–286.
48. Hamilton, E.W.; Heckathorn, S.A.; Joshi, P.; Wang, D.; Barua, D. Interactive effects of elevated CO_2 and growth temperature on the tolerance of photosynthesis to acute heat stress in C3 and C4 species. *J. Integr. Plant Biol.* **2008**, *50*, 1375–1387.

9

Impact of Climate Change on the Irrigation Water Requirement in Northern Taiwan

Jyun-Long Lee † **and Wen-Cheng Huang** †,*

Department of Harbor and River Engineering, National Taiwan Ocean University, Keelung 20224, Taiwan; E-Mail: m98520064@ntou.edu.tw

† These authors contributed equally to this work.

* Author to whom correspondence should be addressed; E-Mail: b0137@mail.ntou.edu.tw;

External Editor: Simon Beecham

Abstract: The requirement for irrigation water would be affected by the variation of meteorological effects under the conditions of climate change, and irrigation water will always be the major portion of the water consumption in Taiwan. This study tries to assess the impact on irrigation water by climate change in Taoyuan in northern Taiwan. Projected rainfall and temperature during 2046–2065 are adopted from five downscaled general circulation models. The future evapotranspiration is derived from the Hamon method and corrected with the quadrant transformation method. Based on the projections and a water balance model in paddy fields, the future crop water requirement, effective rainfall and the demand for water for irrigation can be calculated. A comparison between the present (2004–2011) and the future (2046–2065) clearly shows that climate change would lead both rainfall and the temperature to rise; this would cause effective rainfall and crop water requirement to increase during cropping seasons in the future. Overall, growing effective rainfall neutralizes increasing crop water requirement, the difference of average irrigation water requirement between the present and future is insignificant (<2.5%). However, based on a five year return period, the future irrigation requirement is 7.1% more than the present in the first cropping season, but it is insignificantly less (2.1%) than the present in the second cropping season.

Keywords: climate change, evapotranspiration, irrigation water requirement

---

1. Introduction

The fourth assessment report of the Intergovernmental Panel on Climate Change indicates that the observations of global average temperature during 1995–2006 have increased, and heavy rainfall events have become much frequent. This report also predicts the global average surface temperature during 2080–2099 may rise between 1.1 °C and 6.4 °C more than the period during 1980–1999, and cause crop productivity to increase [1]. It clearly shows the affection of climate change.

General circulation models (GCMs) are the most advanced tools available to simulate the response of the global climate system to increasing greenhouse gas concentrations. With the models, an assessment of the future climate would be possible [2,3], and the problem of uncertainty may be mitigated by considering multiple models [4,5]. According to the results of the cited researches, the rainfall distribution would be different and temperature would rise under climate change. It would make challenges for water resources management.

Taiwan is a small island in the north-west Pacific. Analyzing the historical meteorological data in Taiwan over the past hundred years, the annual rainfall increased in the northern regions, decreased in central and southern regions, and exhibited no clear tendency in the eastern regions [6]; moreover, the surface temperature rose 0.8–1.6 °C in each region [7]. There is 22.7% of the area that has been cultivated in Taiwan in recent years. The annual water consumption in Taiwan is 17,064 million m^3, of which is 11,088 million m^3 consumed by irrigation. The proportion of irrigation water is about 65% of the total consumption; in other words, the irrigation requirement is the main demand factor.

Under climate change, the variation of rainfall and temperature would also impact the irrigation water demand. There are many methods to determine the irrigation water requirement, for example: the Erosion Productivity Impact Calculator [8,9], the Global Irrigation Model [10,11], the CROPWAT model [12,13], and the Stochastic Crop Water Production Functions [14]. The basis of these models is to capture the characteristics of crop water consumption in different periods. Therefore, in the given growth characteristic of crops, rainfall and temperature distribution, and geology of a region, according to the water balance model, the irrigation water requirement would be determined by simulation.

Although the average annual rainfall is about 2500 mm in Taiwan, high rainfall intensity along with a steep slope of river makes water resource storage difficult. Since the supply of irrigation water is one of the most significant tasks for water management, an impact evaluation on irrigation water under climate change in Taiwan is essential. That is the purpose of this study.

2. Materials and Method

First of all, daily meteorological data such as rainfall and temperature either from observation or projection are needed for estimating the evapotranspiration of crops. Second, the effective rainfall and irrigation water on paddy fields could be estimated by simulation method based on the water balance. In addition, data concerning the crop coefficient, percolation rate, conveyance loss rate, and farming area are collected. In this study, the present and future are represented by the periods 2004–2011 and 2046–2065, respectively. The flowchart is shown in Figure 1.

Figure 1. Flow chart of this study.

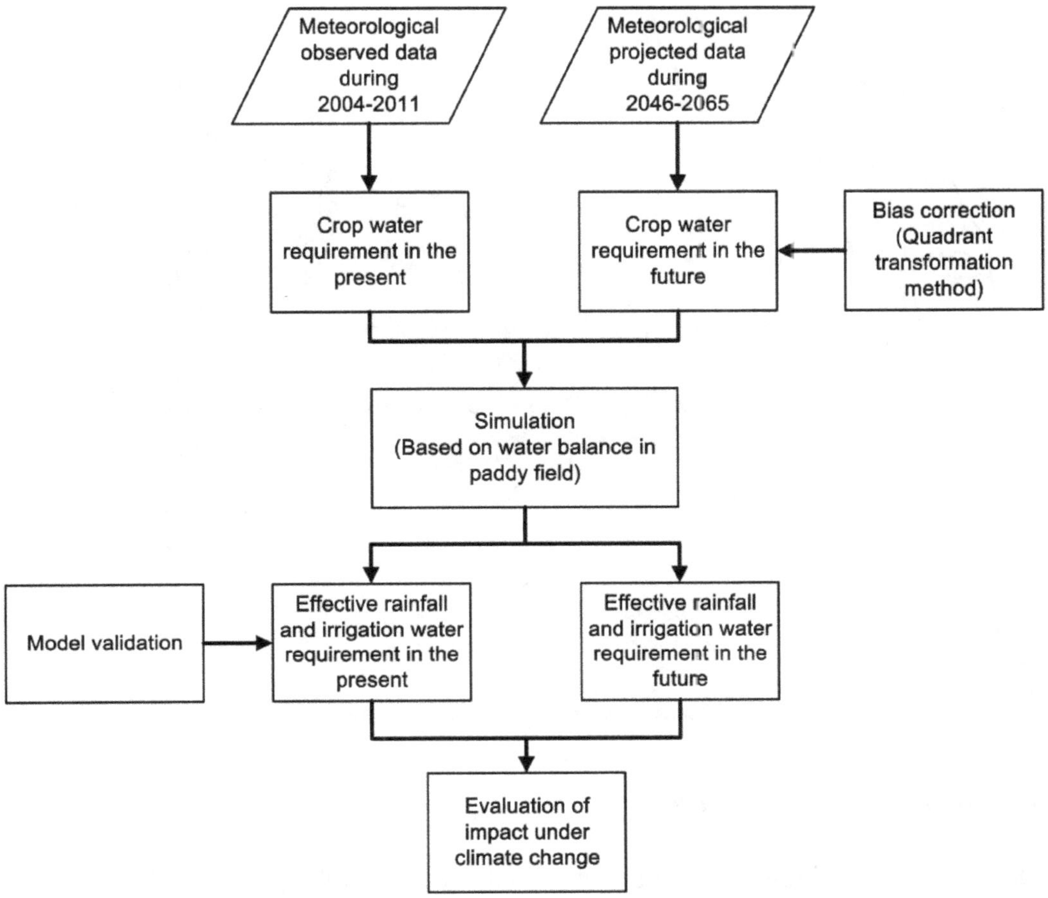

2.1. Study Area

In this study, the irrigation district in northern Taiwan, governed by the Taoyuan Irrigation Association (TIA), is chosen as the study area (see Figure 2). In addition, Shihmen reservoir in the upper reach supplies TIA's irrigation water. The meteorological station adopted here is located in the middle of the irrigation district and operated by the Agricultural Engineering Research Center. The meteorological data include air temperature, dewpoint temperature, solar radiation, sunshine duration, wind speed and rainfall. On average, the air temperature is 22.49 °C, solar radiation equals 10.07 MJ/m^2·day, sunshine duration lasts for 7.12 h, wind speed is 2.25 m/s and the annual rainfall is 1876 mm.

According to the irrigation plan, the area available for farming in TIA is 24,233 ha. There are four types of soil: clayey loam, sandy loam, sand clay loam and light clay. The percentages of each soil are about 41%, 22%, 18% and 19% respectively. In the area, average percolation rate on paddy fields is 8.14 mm/day, and average water conveyance loss is 12.6% [15].

Paddy is the main crop in Taiwan, the proportion of paddy fields to total farming area of TIA is about 95%. The subtropical climate makes two harvests of paddy rice in a year in Taiwan possible. This study assumes the first and second cropping season start between 1 March to 28 June (from the

7th to 18th day) and 1 August to 28 November (from the 22th to 33th day), respectively. That is, a 120 day period for paddy growth is required in each cropping season.

Figure 2. Location of study area and meteorological station.

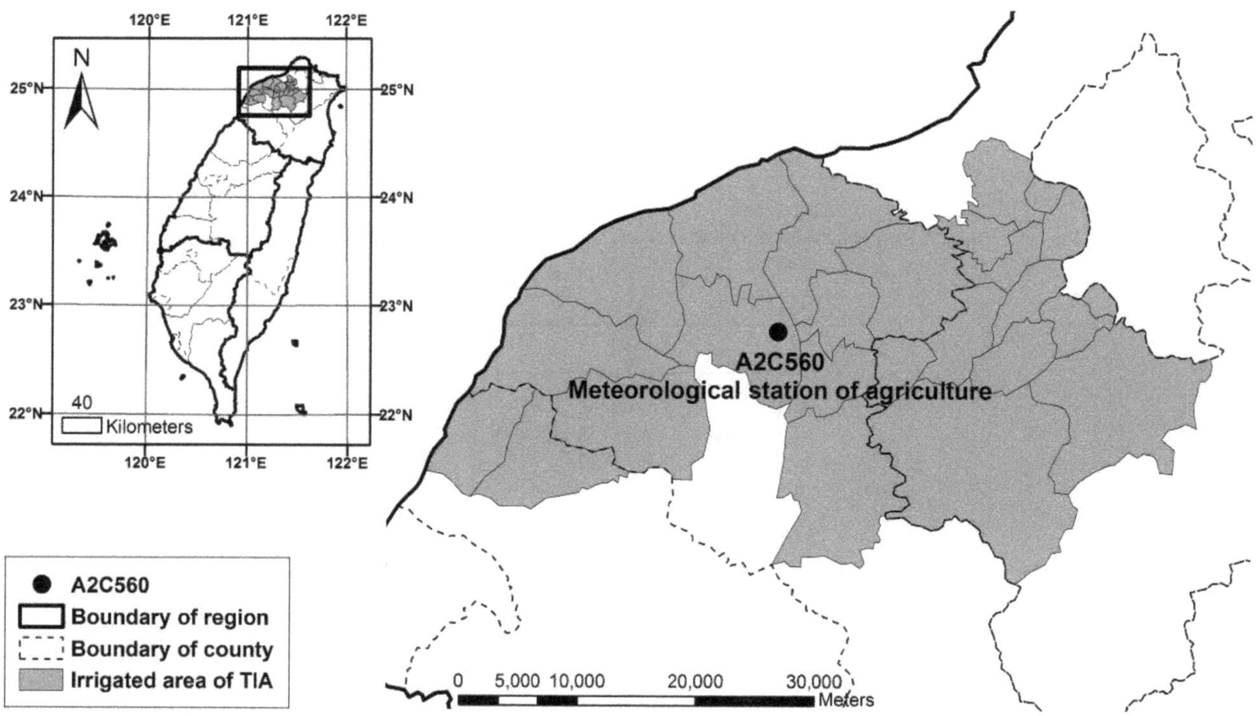

2.2. Projected Rainfall and Temperature

The projected rainfall and temperature under climate change in the period of 2046–2065 came from five GCMs: CGCm3 from the Canadian Center for Climate Modeling and Analysis (CCCma), Cm3 from the Center National de Recherches Meteorologiques (CNRM), Mk3.0 from Australia's Commonwealth Scientific and Industrial Research Organization (CSIRO), Cm2.0 from the Geophysical Fluid Dynamics Laboratory (GFDL) and FGOALS-g1.0 from the State Key Laboratory of Numerical Modeling for Atmospheric Sciences and Geophysical Fluid Dynamics (LASG), which are based on SRES A1B scenarios. The A1B scenario describes a future world of very rapid economic growth, global population that peaks in mid-century and declines thereafter, and the rapid introduction of new and more efficient technologies. Moreover, the A1B scenario is distinguished by its technological emphasis: a balance between fossil and other energy sources [16].

Because of coarse resolution from GCM projection, statistical downscaling of GCM scenario-run outputs to local climate stations were needed and applied. All of the data have been downscaled by the Global Change Research Center of National Taiwan University. Briefly, the process of the downscaling technique would be done in three stages [17]: first, the GCM outputs near Taiwan were adjusted with respect to the NCEP reanalysis data [18] during the training period by linking the normalized probability distribution functions of the mean climate parameters; second, a transfer function (*i.e.*, a multiple-variant linear regression) was established to link NCEP reanalysis variants with local climatic observations during the training period; third, the projected temperature and precipitation data at each station during the verification period were adjusted with respect to the local observation data by the procedure in the first stage. The linkage established was then extended to adjust outputs for the years of projections. If more details about the downscaling technique are needed, please refer to [17,19].

2.3. Paddy Water Requirement

The paddy evapotranspiration in this study is assumed under standard conditions, which means the paddy is grown in large fields with disease-free and well-fertilized conditions. The crop water requirement equals crop evapotranspiration under standard conditions, and it is expressed as [20]:

$$ET_c = K_c \times ET_o \tag{1}$$

where ET_c is crop evapotranspiration under standard conditions (mm/day); K_c is crop coefficient (dimensionless); ET_o is reference evapotranspiration (mm/day).

Notice that K_c varies during the cropping season and depends also on the type of crops. The K_c value that is commonly used in Taiwan at different growth stages for the first and second cropping seasons is shown in Figure 3 [21].

Figure 3. Crop coefficient (K_c) of paddy at each day during cropping season.

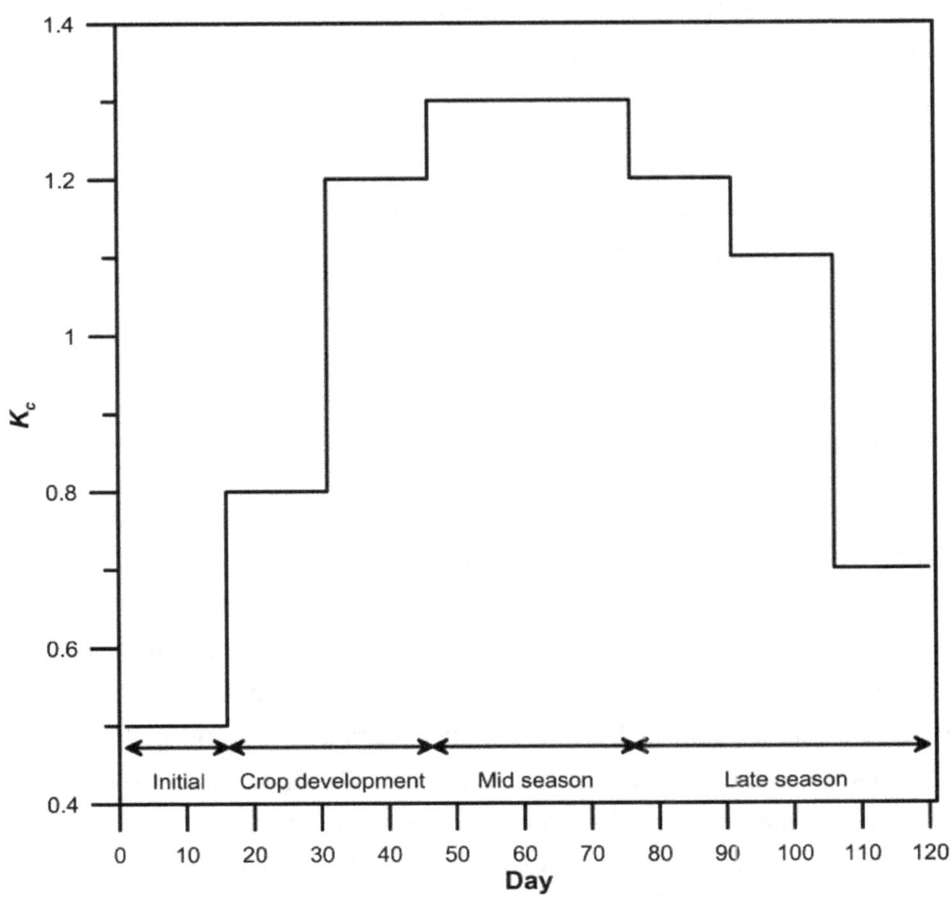

2.3.1. FAO Penman-Monteith Equation

The FAO Penman-Monteith (PM) equation is adopted here for estimating the reference evapotranspiration. The use of the PM equation is recommended as a standard for reference evapotranspiration to provide more consistent values with actual crop water use data worldwide [20]. The equation is expressed as:

$$ET_o = \frac{0.408\Delta(R_n - G) + \gamma\frac{900}{T+273}u_2(e_s - e_a)}{\Delta + \gamma(1+0.34u_2)} \tag{2}$$

where Δ is slope vapor pressure curve (kPa/°C); R_n is net radiation at the crop surface (MJ/m^2·day); G is soil heat flux density (MJ/m^2·day); γ is psychrometric constant (kPa/°C); T is mean daily air temperature at 2 m height (°C); e_s is saturation vapor pressure (kPa); e_a is actual vapor pressure (kPa); u_2 is wind speed at 2 m height (m/s). A detailed explanation of this equation can be found in the literature [20].

2.3.2. Hamon Method

Since only projected temperature and rainfall are available from the output of GCMs in this study, the PM equation could not be used for estimating evapotranspiration in the future. In order to solve this problem, the Hamon method, a temperature-based equation, is adopted. Hamon considered temperature and vapor pressure are the important factors that affect evapotranspiration [22]. The modified Hamon equation is expressed as [23]:

$$PE = 29.8N\frac{e_s}{T+273.2} \tag{3}$$

where *PE* is potential evapotranspiration by Hamon (mm/day); *N* is sunshine duration (h).

Daily sunshine duration *N* can be calculated through the sunset hour angle (ω_s) in theoretically [20]:

$$N = \frac{24}{\pi}\omega_s \tag{4}$$

2.3.3. Bias Correction

As mentioned above, the PM equation and Hamon method are adopted for estimating the evapotranspiration of the present and projected respectively. The basic assumption of these method are different; hence, there would be a mismatch between ET_o and *PE*. In addition, the evapotranspiration of *PE* may need correction. A conversion method, the quadrant transformation method [5,24], is applied here for bias correction. The concept of quadrant transformation method is shown in Figure 4. Here the difference between the 1st and 4th quadrants is estimated either by ET_o or by *PE*, while the difference between the 3rd and 4th quadrants is due to climate change. The duration curve for the corrected *PE* of the future (2046–2065) in the 2nd quadrant is built by the other three quadrants' conversion.

The procedure for correction would be expressed by following steps: (1) constructing the daily duration curves, a cumulative frequency curve that show the percent of time specified rainfall were equaled or exceeded during a given period [25], for the 1st, 3rd and 4th quadrant by the evapotranspiration of PM in the present, Hamon in the future and Hamon in the present, respectively; (2) confirming the corresponding percentile of evapotranspiration at the specific day for the duration curve of the 3rd quadrant; (3) finding the corresponding percentile for the 4th quadrant by the evapotranspiration for the 3rd quadrant; (4) using the percentile by above to find a new corresponding evapotranspiration for the 1st quadrant, and which would be the corrected daily value.

By repeating steps (2) to (4) we would obtain the duration curve for the corrected data in 2nd quadrant. Considering the seasonal variation, this correction method is based on monthly duration curves of evapotranspiration.

Figure 4. Bias correction of evapotranspiration through quadrant transformation.

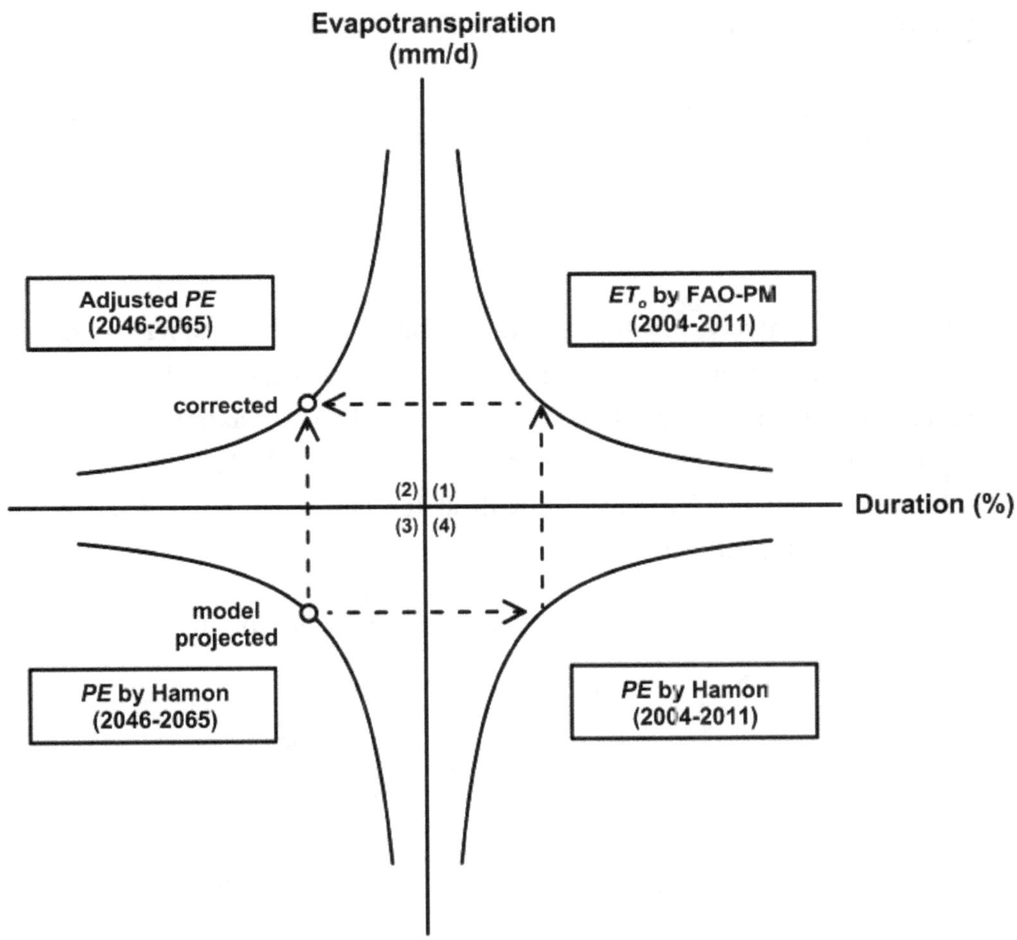

2.4. Calculation of Irrigation Water Requirement

Since the rainfall, paddy water requirement and geology are understood, the irrigation water requirement can be calculated with a water balance model of the paddy fields. Figure 5 shows the factors affecting water balance in the paddy fields. Irrigation should supply the deficiency of water that paddy growth needs. Considering different soil types on paddy fields and water conveyance loss of irrigation canals, the continuity equation in paddy fields is expressed as:

$$S_{t+1} = S_t + P_t - ET_{ct} - \sum_j f_j A_j^{\circ} + \frac{IR_t}{\sum_i A_i^*(1+CL_i)} \tag{5}$$

where S_t, P_t, ET_{ct}, IR_t, respectively, indicate the daily water storage, rainfall, crop evapotranspiration, and irrigation water requirement in the paddy fields at time t (mm/day); f_j is the percolation rate for jth soil type (mm/day), A_j° is the percentage of jth soil type area (%); A_i^* gives the percentage of total farming area controlled by the ith canal (%), and CL_i shows the average water conveyance loss for the area controlled by ith canal (%).

The irrigation water requirement must be externally supplied to fill the deficit for paddy growth when rainfall and storage do not satisfy the water consumption in paddy fields; in contrast, the irrigation water requirement will be 0 while the consumption has been satisfied. Therefore, it could be rewritten as:

$$IR_t = \sum_i A_i^*(1+CL^i)\cdot(ET_{ct} + \sum_j f^j A_j^{\circ} + S_t^{\min} - S_t - P_t) \qquad if \;\; S_t + P_t - ET_{ct} - \sum_j f^j A_j^{\circ} < S_t^{\min} \tag{6}$$

The rainfall stored in paddy fields for growth is called effective rainfall (P_t^*), and it used in the paddy fields at time t equals:

$$P_t^* = \min\left\{ S_t^{\max} - S_t + ET_{ct} + \sum_j f_j A_j^{\circ}, P_t \right\} \tag{7}$$

where, S_t^{max} and S_t^{min}, respectively, are the maximum and minimum ponding storage in the paddy fields during different growth stages (mm). Here, we assume the water is abundant for a continued irrigation to keep the fields in an appropriate state of water depth. The maximum and minimum storage during different growth stages for paddy are shown in Figure 6 [26,27]. Please note that emptying out the storage would be recommended for root growth at the specific time. Therefore, the minimum ponding storage in some days would be 0.

Figure 5. Concept of water balance in paddy field.

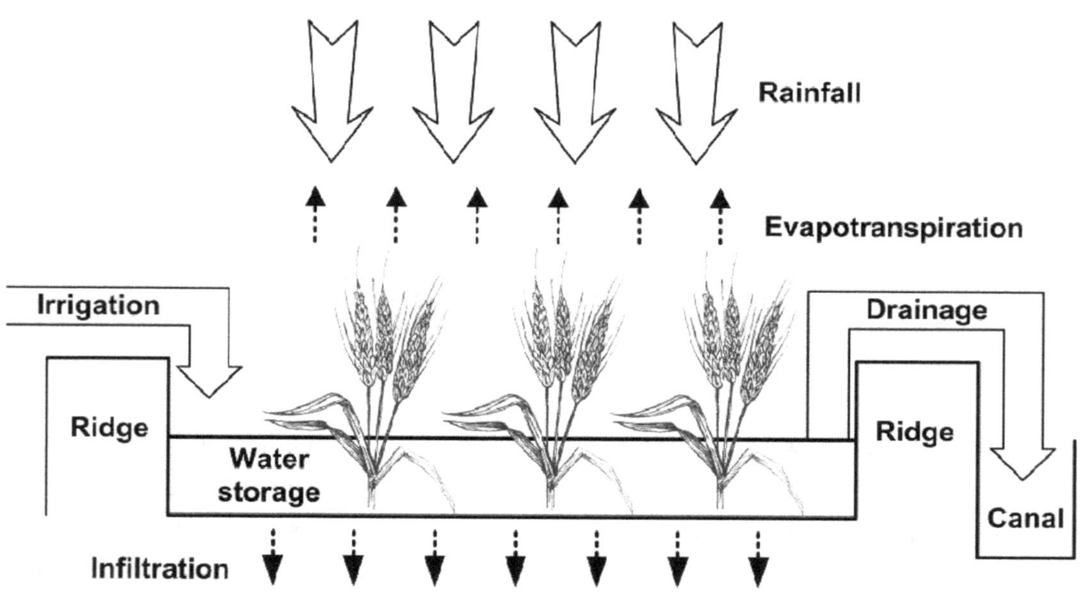

Figure 6. Suggested water depth in paddy field during cropping season.

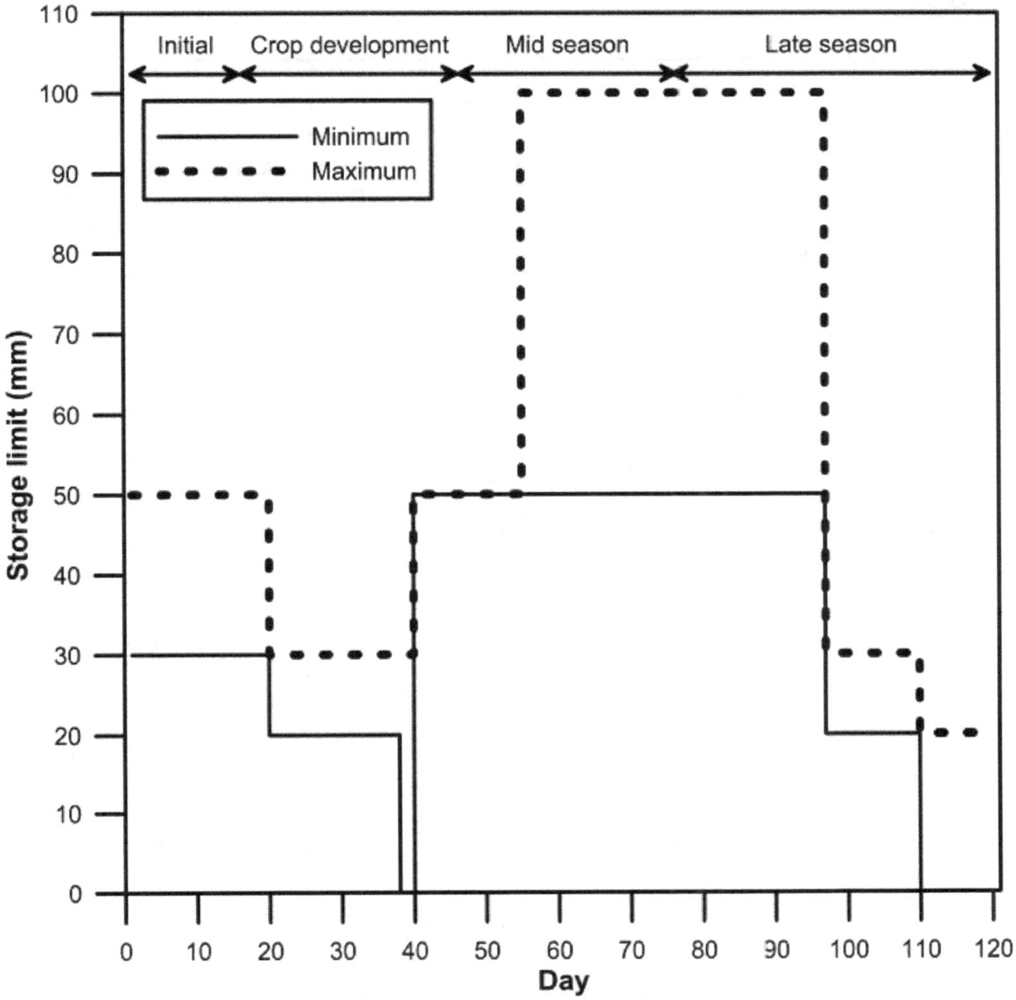

3. Result and Discussion

For the A1B emissions scenario in the period of 2046–2065, five GCMs produce different patterns of change on temperature and rainfall. In this study, the mean value of models is used to show the result for the following. It represents the average of the results by five GCMs, instead of the result of average temperature and rainfall by these GCMs.

3.1. Result of Bias Correction

As seen in Figure 7, ET_o values in the 1st quadrant are inconsistent with *PE* values in the 4th quadrant. *PE* assessed by the Hamon method seems underestimated as evapotranspiration exceeds 2.9 mm/day, and vice versa. Apparently, the biases between the 1st/4th quadrants in the present (2004–2011) and the 2nd/3rd quadrants in the future (2046–2065) are similar. It shows the LASG-based *PE* projections in 3rd in March quadrant can be appropriately adjusted in the 2nd quadrant by the quadrant transformation method.

Figure 7. Comparison of duration curve for evapotranspiration at each quadrant in March (LASG GCM).

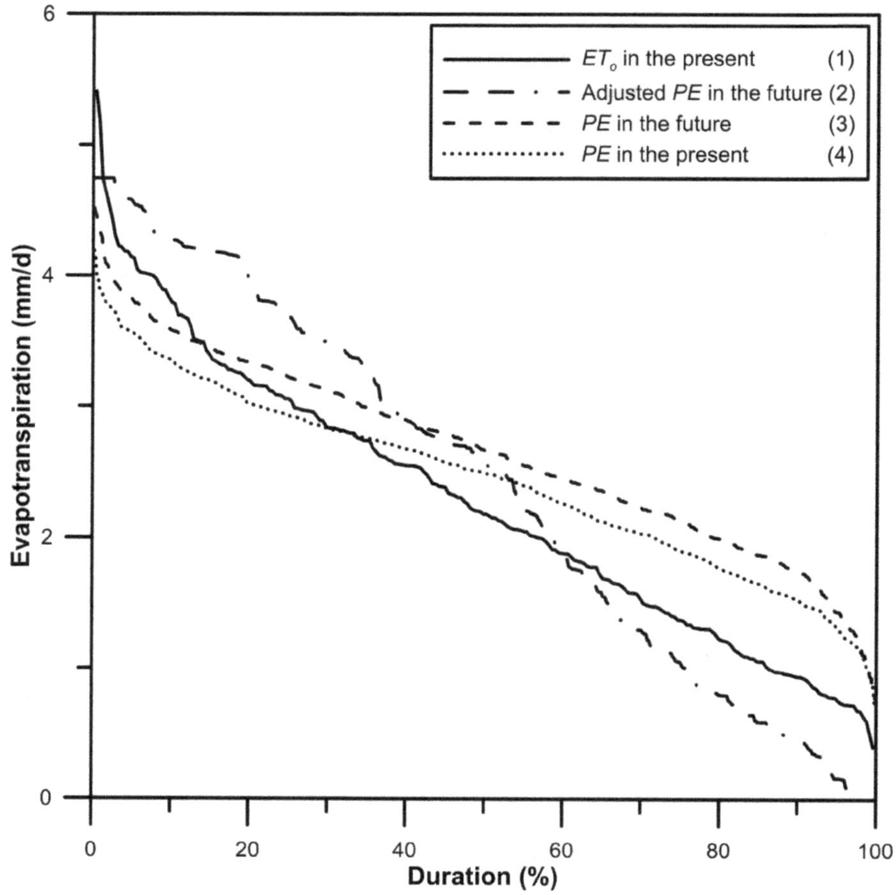

Here, Figure 8 shows the comparison with before/after bias correction for these GCMs. While focusing on the different between ET_o and PE in the present, we would find the PE calculated by Hamon method is always overestimated during January to September, especially June to August. After bias correction, most of the month will be adjusted to be lower, except some months for GCMs. The bias correction does not seem to work during June/July/August/September for some GCMs. The main reason is that the extreme high ET_o always happened in summer. If the extreme value of ET_o in the present is more than the value of PE in the future, and the future PE is always more than the present PE. Then, the adjusted evapotranspiration would be more than the non-adjusted probably.

3.2. Comparison of Model Estimation and Actual Investigation

Meteorological factors like temperature, wind speed and net radiation during 2004–2011 have been adopted for estimating irrigation demand by the model. Before starting the process to estimate future irrigation requirements, the methods mentioned in Section 2 have to be compared with the actual irrigation consumption, as shown in Table 1. For each year, the difference of model estimation and actual investigation is always over 20%, except for the first cropping season in 2005. The result seems to show that the model has failed. Why has this happened? It would be summarized by the following two reasons:

Figure 8. Comparison with before/after bias correction: (**a**) CCCma GCM; (**b**) CNRM GCM; (**c**) CSIRO GCM; (**d**) GFDL GCM; (**e**) LASG GCM.

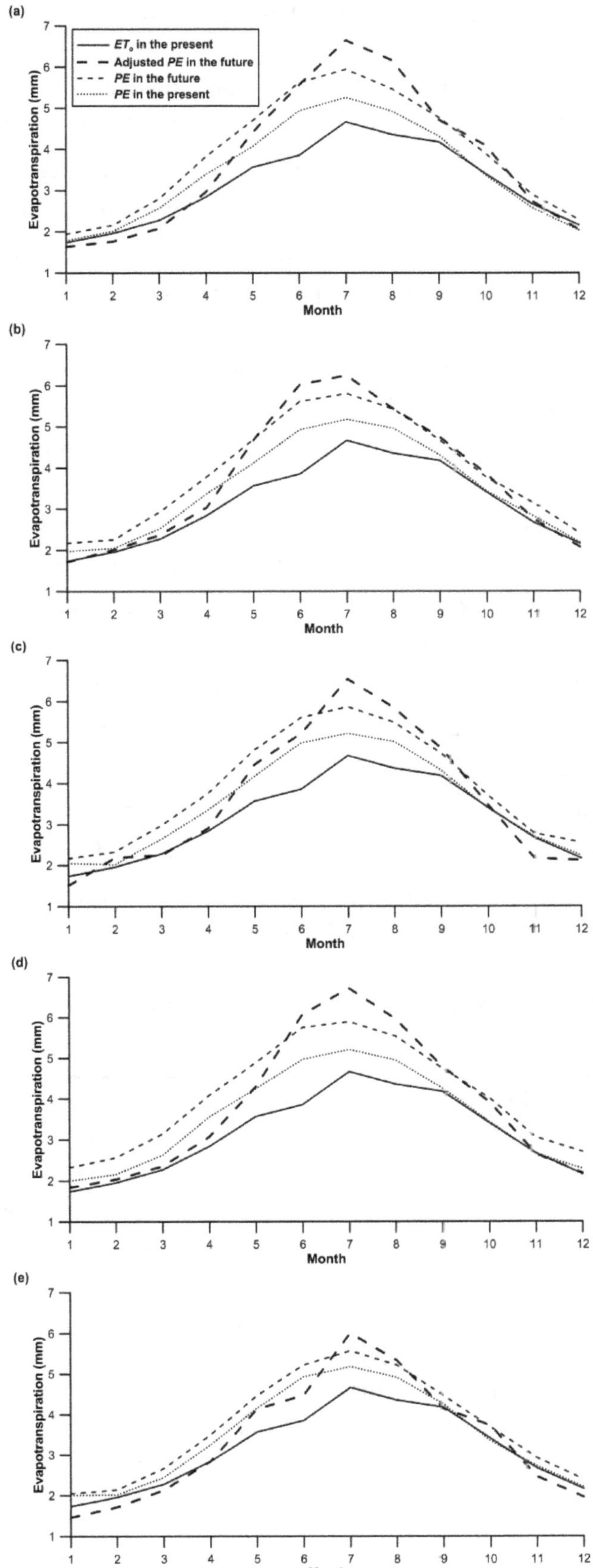

First, the actual investigation of irrigation consumption is according to the water received by canals, and it there would be interference from the discharge of the river, storage of reservoir, operation of canals, *etc.* However, the model estimation depends on the meteorology and the growth stages of paddy. There is a difference in the foundation between the actual investigation and model estimation.

Second, the strategies of the government would also influence the supply of irrigation water significantly. In January 2002, Taiwan became a member of the World Trade Organization. According to the membership commitment, Taiwan has to import 144,720 metric tons of rice per year. To achieve this goal, the cultivation area in TIA was reduced significantly (probably reduced from 20,000 to 6000 ha). After 2002, the amount of agricultural water consumption is not only for irrigation, it involves multi-purpose uses like water resources scheduling, groundwater recharge and environmental conservation.

Since comparing the difference between model estimation and actual investigation in the same years is not proper for evaluating the model, we try to use the data before 2002 for model validation. In this study, the two-sample t test is adopted because the data of validation is not the same as the period.

At the 0.01 level of significance, the absolute value of threshold value t for 16 sample size is 2.98 (a double-tailed test). The result of two-sample t test is shown in Table 2. In the comparison of the actual investigation between 1994–2001 and 2004–2011, the absolute values of t of the first and second seasons are 2.05 and 3.71, respectively. The t of the second season is greater than 2.98, and it implies that the difference in two means of 2nd season between 1994–2001 and 2004–2011 is significant, it proves the inference that irrigation consumption during 2004–2011 is not only for farming paddy and not proper for evaluating the model. In the comparison of actual investigation during 1994–2001 and model estimation during 2004–2011, the absolute value of t of 1st and 2nd seasons are 1.57 and 2.43 respectively, they are both smaller than 2.98. It implies that the difference in two means is insignificant between actual investigation and model estimation. That is, the proposed process could be accepted and applied to evaluating the impact on irrigation in the future.

Table 1. Comparison of irrigation water requirement between model estimation and actual investigation. (Unit: mm).

	Actual Investigation					Model Estimation		
Year	**First Season**	**Second Season**	**Year**	**First Season**	**Second Season**	**Year**	**First Season**	**Second Season**
1994	907	1145	2004	1512	1512	2004	1169	1204
1995	1191	1136	2005	995	1967	2005	983	1358
1996	839	1046	2006	1260	2672	2006	985	1457
1997	1242	1128	2007	840	828	2007	1029	1229
1998	1294	1108	2008	1762	2201	2008	1099	1362
1999	1343	1361	2009	1977	2219	2009	1130	1172
2000	1245	1218	2010	1741	1867	2010	867	1158
2001	1253	1152	2011	1949	1859	2011	1114	1337
Mean	1164	1162		1505	1891		1047	1285
Standard deviation	186	94		432	547		100	108

Table 2. The absolute value of tow sample t test for different periods.

Title	Actual Investigation (2004–2011)	Model Estimation (2004–2011)
Actual investigation	2.05	3.71 *
(1994–2001)	1.57	2.43

Note: *: Exceed the threshold value $t = 2.98$.

3.3. Projected Evapotranspiration Analysis

According to the projections of temperature from the five GCMs, the average increments of temperature during the first and second cropping seasons over TIA are 2.2 °C and 1.1 °C, respectively. This will cause evapotranspiration, ET_o, to increase in the future. As Table 3 shows, no matter whether the first or second season, evapotranspiration grows by all selected GCMs. The GFDL model yields higher projections of evapotranspiration, as LASG model gives lower ones. The average increments during the first and second are 133 mm and 95 mm, respectively, about 35.8% and 16.8% more than the present period (2004–2011).

Table 3. The assessment of water consumption and supply for paddy field in the present and future.

Assessment Factors	Cropping Seasons	Present	Future					
			CCCma	CNRM	CSIRO	GFDL	LASG	Average
Evapotranspiration (mm)	First season	372	504	506	511	532	472	505
	Second season	564	665	662	654	677	635	659
Crop water requirement (mm)	First season	383	521	523	525	549	487	521
	Second season	547	638	632	625	648	607	630
Rainfall (mm)	First season	793	1262	762	904	1105	1344	1075
	Second season	637	629	480	670	658	527	593
Effective rainfall (mm)	First season	465	769	503	579	659	788	660
	Second season	308	533	433	580	567	507	524
Effectiveness (%)	First season	62.6	64.2	70.7	67.7	61.0	62.0	65.1
	Second season	50.8	85.9	91.2	87.0	88.0	96.5	89.5
Irrigation water requirement (mm)	First season	923	803	1087	1000	945	742	915
	Second season	1,208	1179	1285	1110	1152	1172	1180

Furthermore, Figures 9 and 10 show the exceedance probability distribution of evapotranspiration distribution over time for the present and future (ensemble of GCMs). Since evapotranspiration is a cost factor in paddy fields, both figures give the representative values from an optimistic 90% to a pessimistic 10% for the first and second cropping seasons. Obviously, evapotranspiration increases in May and June of the plum period (first cropping season) and gradually decreases within the typhoon period from August through October (second cropping season). The comparison between 2004–2011 and 2046–2065 indicates the temporal distribution of the future is similar to the present, but the variance of the future is much lower than the present.

Figure 9. Exceedance probability distribution of evapotranspiration at each ten-day during the first cropping season. (**a**) present (2004–2011); (**b**) future (2046–2065).

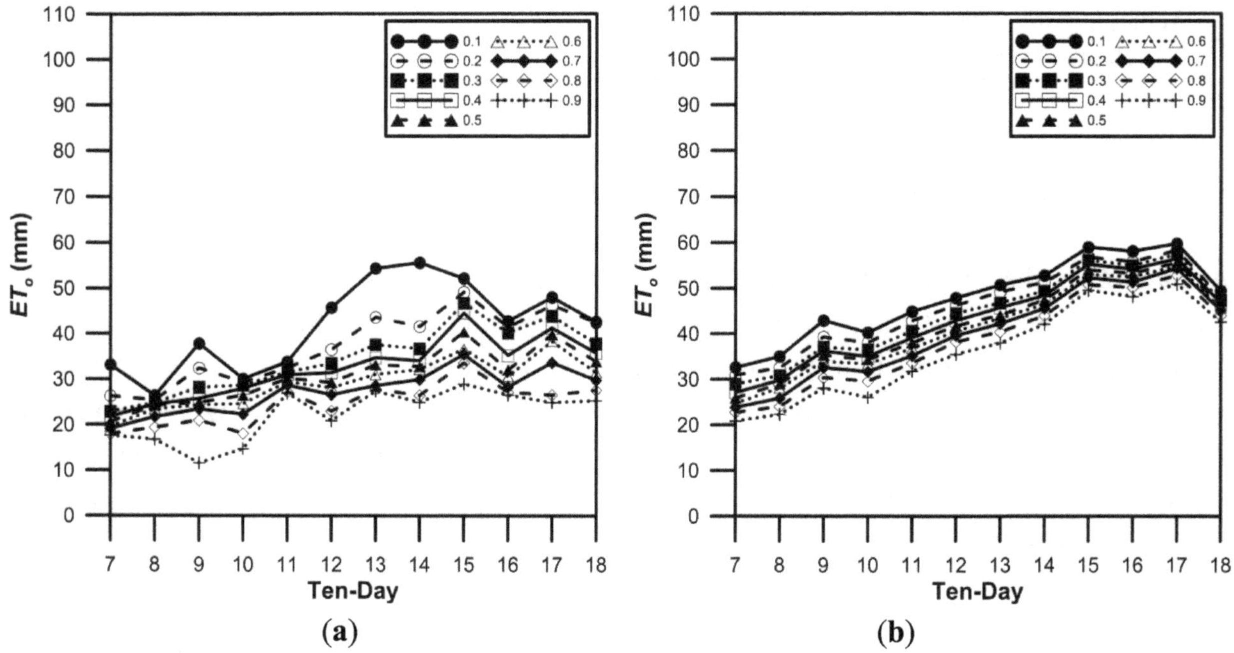

Figure 10. Exceedance probability distribution of evapotranspiration at each ten-day during the second cropping season. (**a**) present (2004–2011); (**b**) future (2046–2065).

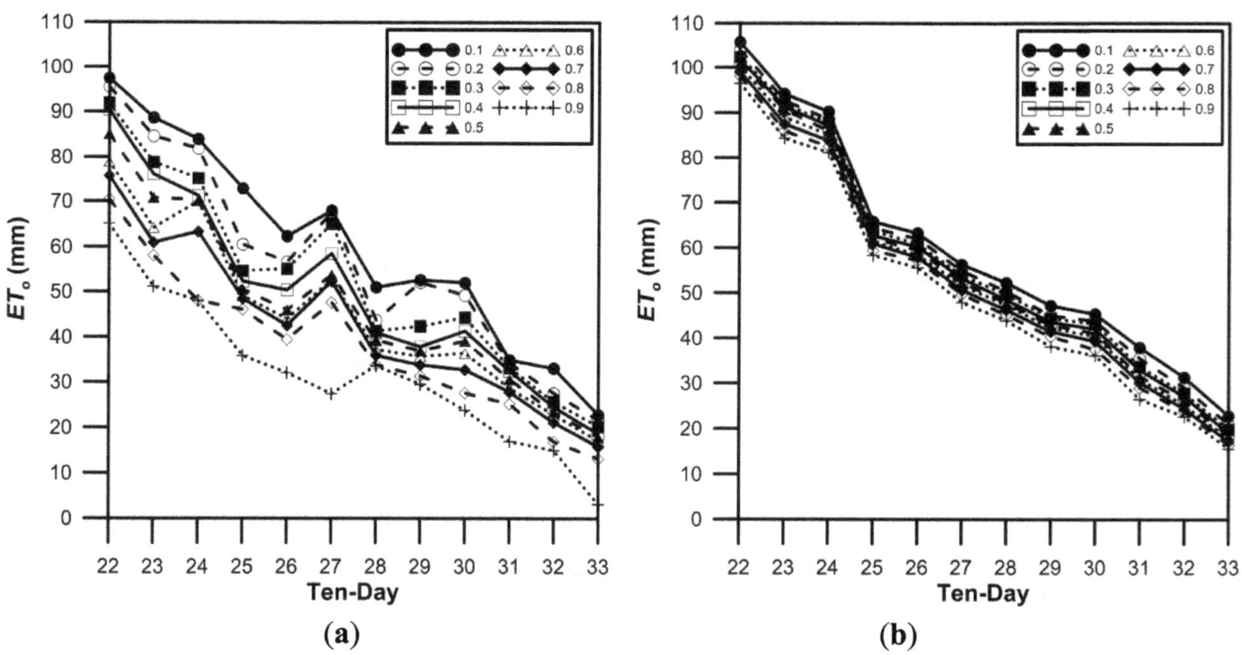

3.4. Crop Water Requirement Analysis

With evapotranspiration multiplied by the time-varying crop coefficient (K_c), as shown in Figure 3, the crop water requirement for rice cultivation could be obtained. Because of rising evapotranspiration in the future, it will result in an increase in the crop water requirement during the first and second cropping seasons. Table 3 shows the comparison of paddy water requirement between the present and future. Clearly, the second cropping season needs more water than the first season. Future also requests more water than the present, based on the GCMs' outputs. On average, 138 mm and 83 mm more water would be needed in the future during the first and second seasons, an increment of nearly 36.0% and 15.2%, respectively. This is due to the increase of temperature in the future, and would increase the possibility of water deficit if rainfall could not supply essential crop water requirement in the future. In addition, as illustrated in Figures 11 and 12, the patterns of exceedance probability distribution of crop water requirement seems to be a great difference between the present and the future in the second cropping season (ensemble of GCMs). The periods with maximum water requirements occur in May during the first cropping season and in September during the second cropping season. Notice that the representative values of exceedance probability distribution are from an optimistic 90% to a pessimistic 10%.

3.5. Projected Rainfall Analysis

Table 3 presents the comparison of the rainfall between observation (2004–2011) and projected (2046–2065). It shows, except for the CNRM model, that the GCMs project much more rainfall than the present in the first cropping season. The average rainfall increases 282 mm (35.6%) in the future. However, in the second cropping season, the average decreases 44 mm (6.9%). In particular, CNRM and LASG models produce much less rainfall than the present. Plus, Figures 13 and 14 give the representative values of rainfall exceedance probability distribution from a pessimistic 90% to an optimistic 10% for the first and second cropping seasons, because rainfall is a benefit factor in paddy fields. The figures show that the period in May and June in the first cropping season has more rainfall in the future (see Figure 13). However, a lower quantity of rainfall occurs in July to October in the second cropping season (see Figure 14).

Figure 11. Exceedance probability distribution of crop water requirement at each ten-day during the first cropping season. (**a**) present (2004–2011); (**b**) future (2046–2065).

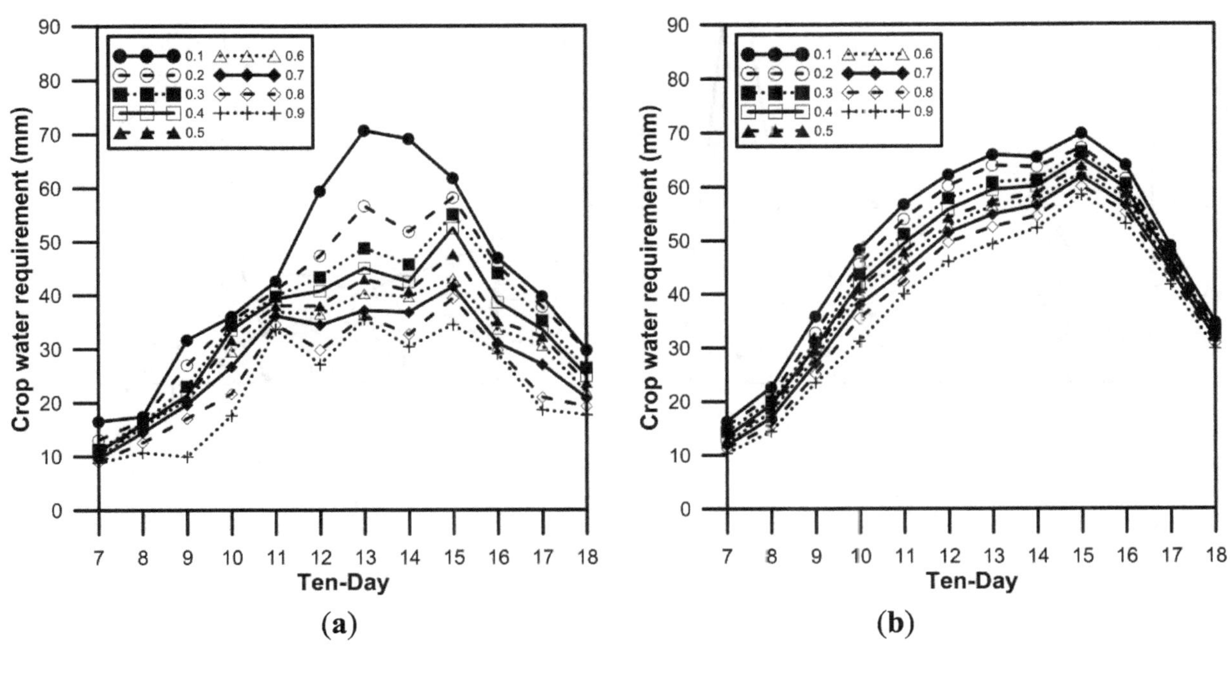

(a) (b)

Figure 12. Exceedance probability distribution of crop water requirement at each ten-day during the second cropping season. (**a**) present (2004–2011); (**b**) future (2046–2065).

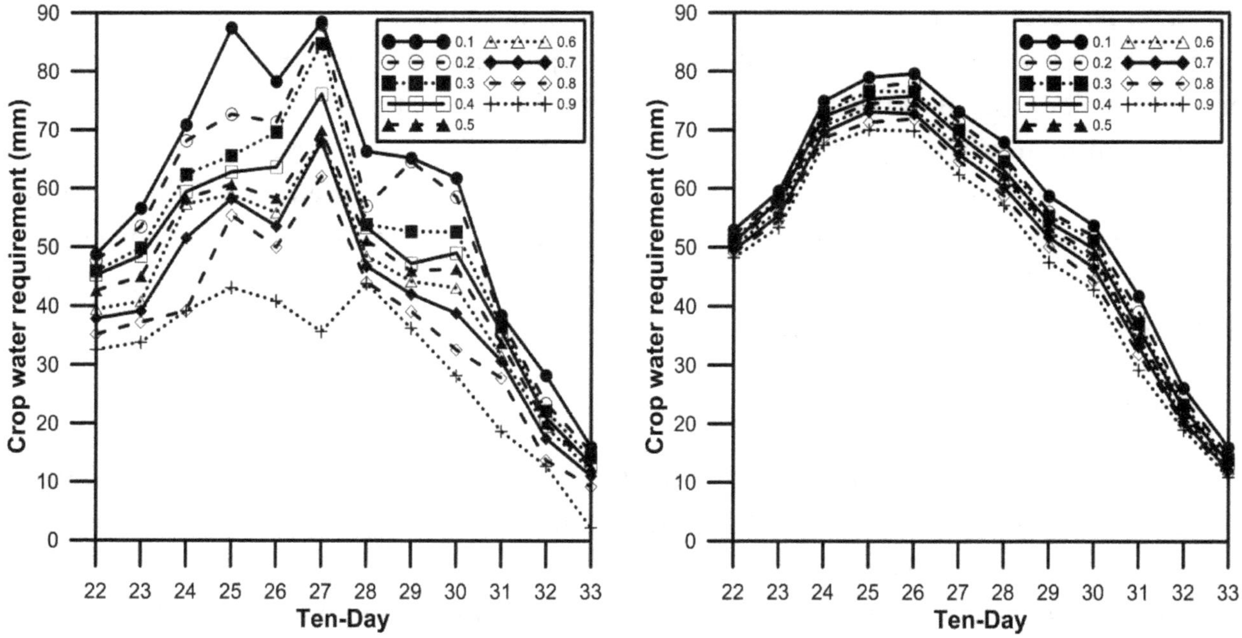

(a) (b)

Figure 13. Exceedance probability distribution of rainfall at each ten-day during the first cropping season. (**a**) present (2004–2011); (**b**) future (2046–2065).

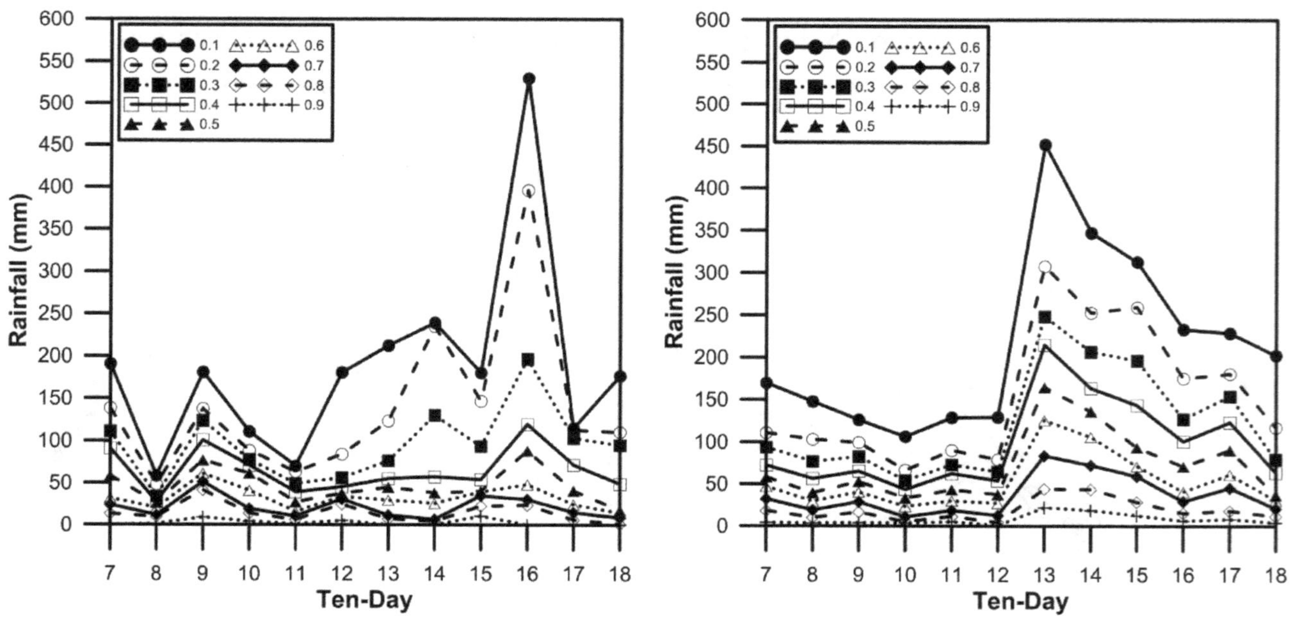

(a) (b)

Figure 14. Exceedance probability distribution of rainfall at each ten-day during the second cropping season. (**a**) present (2004–2011); (**b**) future (2046–2065).

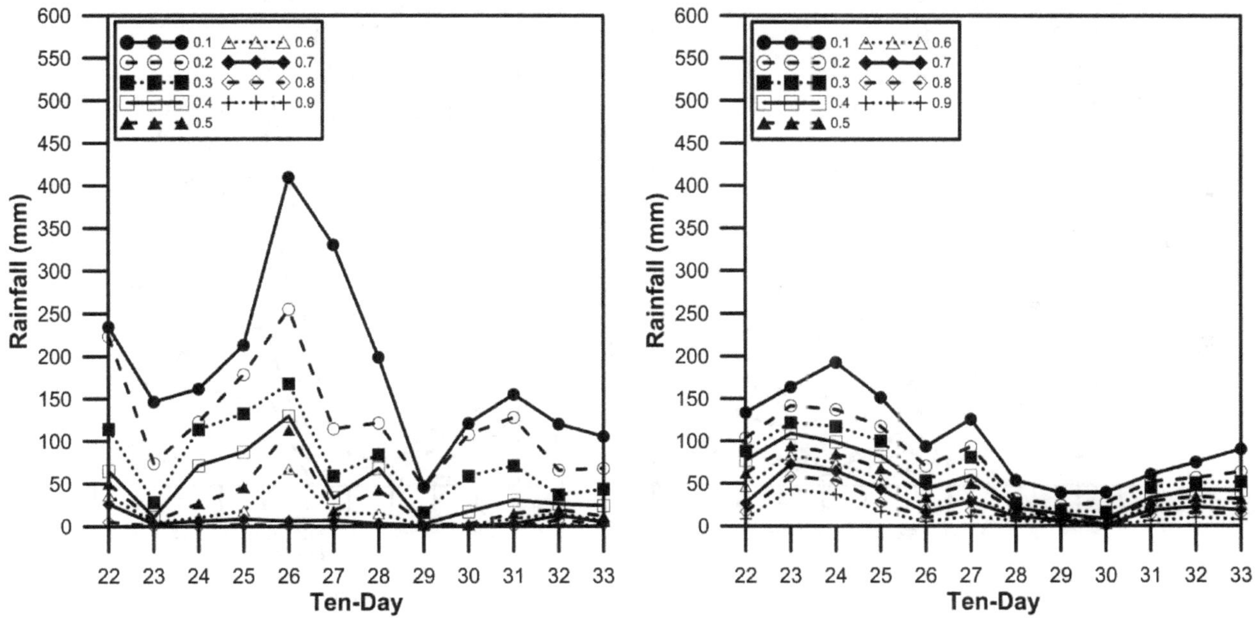

3.6. Effective Rainfall Analysis

As shown in Figures 13 and 14, rainfall distribution of the future at each ten-day period seems more even than the present. This may increase the occurrence of effective rainfall in the future. As a benefit factor in paddy fields, Figures 15 and 16 show representative values of exceedance probability distribution of effective rainfall from a pessimistic 90% to an optimistic 10% for the first and second cropping seasons. In the future, more effective rainfall is obtained in May–June during the first cropping season, and in August-September within the second cropping season. Moreover, from Table 3, we can find that all the five GCMs project more effective rainfall than the present during the cropping seasons. The average increments are 195 mm (41.9%) and 216 mm (70.1%), respectively, in the first and second cropping season. Certainly, this is helpful to paddy cultivation and reduces the irrigation requirement.

Effectiveness is defined by this study as a ratio of effective rainfall to total rainfall during cropping season. The difference of effectiveness between the observation and the projection is insignificant in the first cropping season. In contrast, although the projected rainfall is smaller in the second cropping season in the future, the effectiveness appears much better than the present because rainfall distribution become more even. Rainfall effectiveness in the future increases 2.6% and 38.7%, respectively, during the first and second cropping seasons. Apparently, rainfall is more effectively utilized during the second cropping season.

Figure 15. Exceedance probability distribution of effective rainfall at each ten-day during the first cropping season. (**a**) present (2004–2011); (**b**) future (2046–2065).

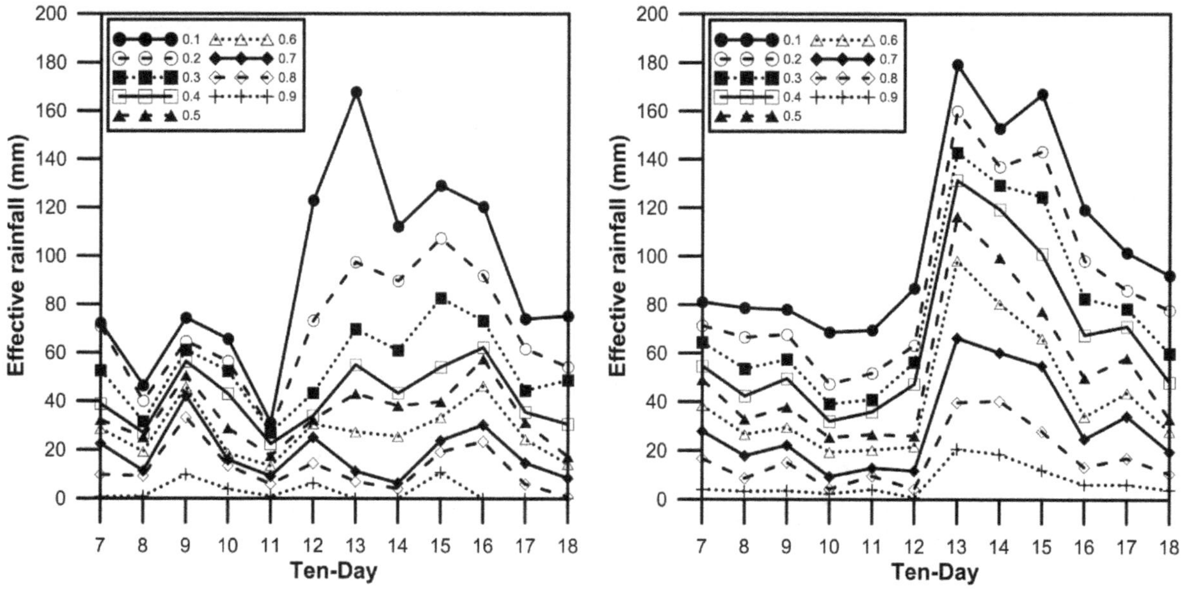

(a) (b)

Figure 16. Exceedance probability distribution of effective rainfall at each ten-day during the second cropping season. (**a**) present (2004–2011); (**b**) future (2046–2065).

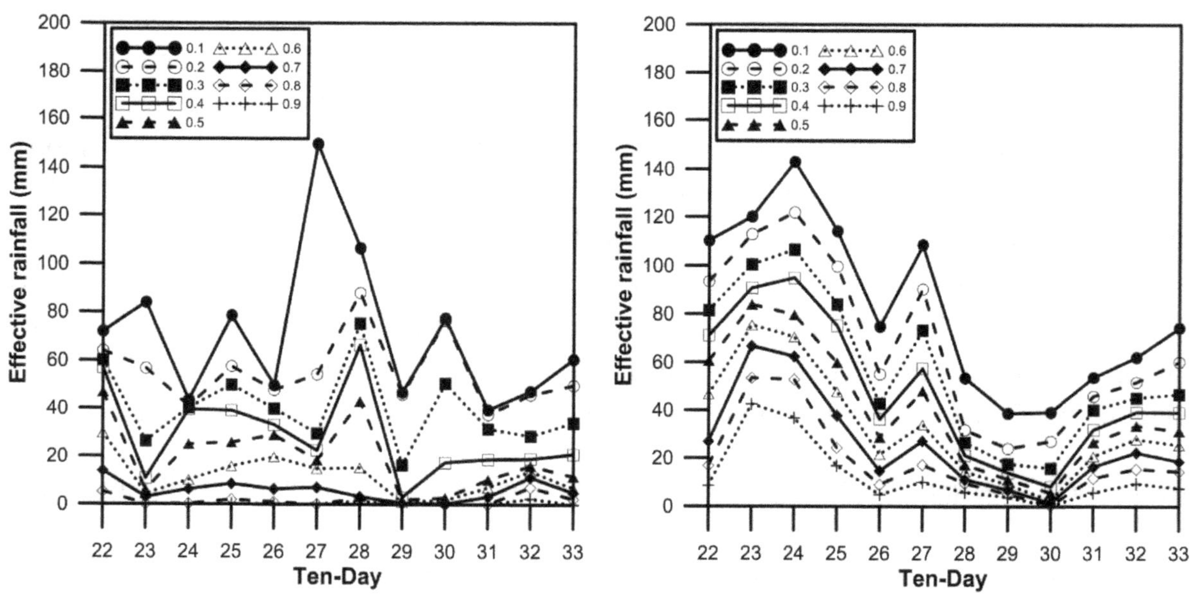

(a) (b)

3.7. Irrigation Water Requirement Analysis

Linking a higher crop water requirement with more effective rainfall in the future, the impact on irrigation water requirement would probably be neutralized. The process for estimating irrigation requirement can be done by Equations (6). As seen in Figures 17 and 18, the pattern of exceedance probability distribution on irrigation water requirement, by comparing the present with the future, is similar. Notice that the exceedance probability distribution from 90% to 10% represents optimistic to pessimistic, because irrigation requirement is a cost factor. Overall, in the present, the agricultural sector needs to supply 923 mm and 1208 mm water, respectively, for irrigation. That is, 9230 m^3/ha and 12,080 m^3/ha during the first and second cropping seasons. In fact, the estimation of the future irrigation requirement depends on a chosen GCM. For example, in the first cropping season, CCCma and LASG produce less requirement, but CNRM, CSIRO and GFDL models request more irrigation water (see Table 3). On average, future requirements, respectively, reach 915 mm and 1180 mm in the first and second cropping seasons. In contrast to the present, future needs less irrigation water, though the difference is not significant.

Figure 17. Exceedance probability distribution of irrigation water requirement at each ten-day during the first cropping season. (**a**) present (2004–2011); (**b**) future (2046–2065).

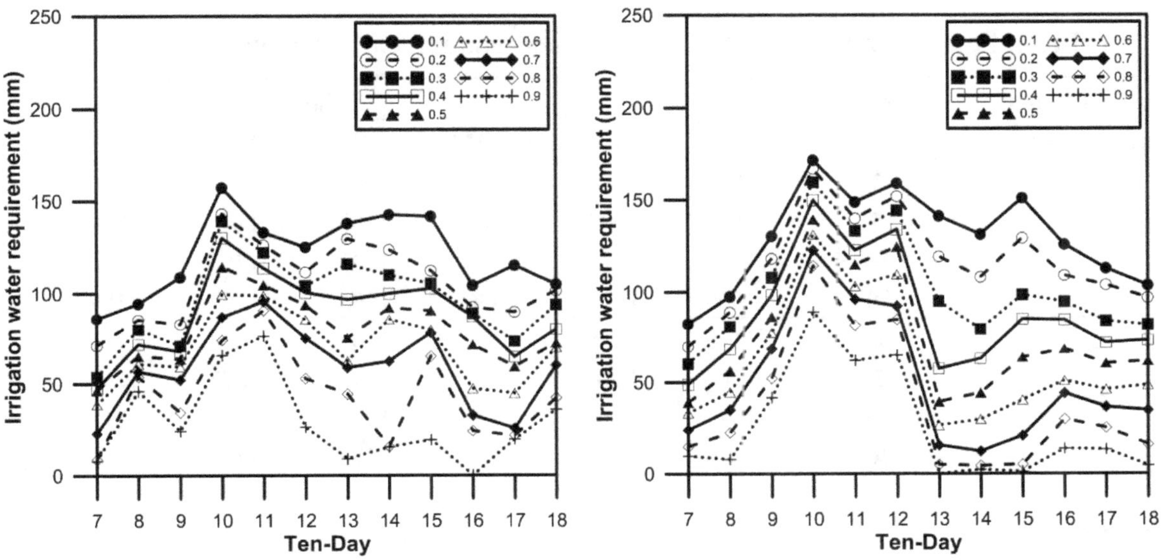

(**a**) (**b**)

Figure 18. Exceedance probability distribution of irrigation water requirement at each ten-day during the second cropping season. (**a**) present (2004–2011); (**b**) future (2046–2065).

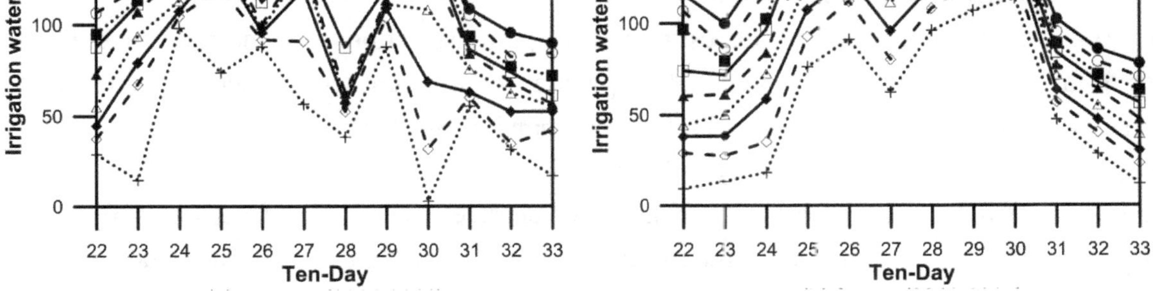

(**a**) (**b**)

Generally, the threshold value for the water requirement of an irrigation plan can be determined on the basis of a 5-year deficit, a deficit event that has a 20% probability of occurring in any given year. That is, the exceedance probability of greater than the threshold value is 20%. By using Log-Pearson Type III frequency analysis [28,29], Table 4 shows the required irrigation water in accordance with a different return period. For the first cropping season, with the exception of a 2-year event, the future would need more irrigation water. In contrast, the required irrigation water at the second cropping season is less in the future. By choosing a 5-year return period event as the standard for the irrigation plan, future planning irrigation water requirement in the first cropping season would increase 73 mm (730 m^3/ha), and decrease 28 mm (280 m^3/ha) in the second cropping season. Multiplying the planned farming area 24,236.93 ha, extra 17.7 million m^3 of water would be needed in the first cropping season. In contrast, water would be 6.8 million m^3 less than the present in the second cropping season.

Table 4. Frequency analysis of irrigation water requirement in the present and future.

Irrigation Water Requirement (mm)	Return Period (year)	Present	Future					
			CCCma	CNRM	CSIRO	GFDL	LASG	Average
First season	2	933	824	1072	1000	931	721	910
	5	1021	1004	1220	1153	1146	946	1094
	10	1060	1078	1308	1231	1266	1076	1192
	20	1089	1127	1386	1295	1369	1189	1273
Second season	2	1199	1215	1299	1109	1134	1161	1183
	5	1307	1296	1373	1192	1277	1260	1279
	10	1371	1319	1404	1236	1365	1318	1328
	20	1426	1330	1426	1272	1445	1371	1369

4. Conclusions

This study investigates the impact on the irrigation water requirement under climate change between the present (2004–2011) and future (2046–2065). Impacts in terms of five selected GCMs under the SRES A1B scenario were assessed for the paddy fields of the Taoyuan Irrigation Association (TIA) in northern Taiwan. Projected meteorology in the future would be different because of GCM features and downscaling methods; therefore, considering several GCMs to reduce the uncertainty of models is necessary.

The FAO-PM equation is mostly used for evapotranspiration assessment, but it would not be suitable for evaluating the projection of evapotranspiration. This paper tries to combine the Hamon method and the Quadrant transformation method for assessing evapotranspiration in the future, and it is a possible and effective way to solve the problem.

Due to the rising temperature, the estimated evapotranspiration will increase in both cropping seasons in the future. Meanwhile, estimated crop water requirement would increase 36.0% and 15.2% in the 1st and 2nd seasons respectively. On the other hand, projected rainfall increases 35.6% in the 1st cropping season, but decreases 6.9% in the 2nd cropping season.

The impact of irrigation water requirement under climate change would not be easily assessed by crop water requirement and rainfall, although they play an important role. In the paddy field, storage, percolation and conveyance loss also influence the magnitude of irrigation water requirement. For evaluating the impact, this study simulates the water balance in paddy fields day by day.

As mentioned above, the projected rainfall decreases in the second cropping season, but estimated effective rainfall augments by 41.9% and 70.1% during the first and second seasons, respectively. This is because of the rainfall distribution, which becomes more even in the future.

Increased effective rainfall neutralizes the augmented crop water requirement, and causes the difference of irrigation requirement between the future and present to be insignificant. Estimated irrigation water requirements decrease 0.9% and 2.3% in the 1st and 2nd seasons, respectively. The decrements are equal to 1.9 million m^3 and 6.8 million m^3 by multiplying the planned farming area 24,236.93 ha.

In addition, this study uses frequency analysis to analyze the change of irrigation requirement. Based on the 5-yr threshold value, the estimated irrigation water requirement would increase by 7.1% in the first cropping season, and decrease by 2.1% in the second cropping season. By multiplying the planned farming area, the difference of irrigation requirement based on 5-yr return period between the future and present would increase by 17.7 million m^3 in the first cropping season but decrease by 6.8 million m^3 in the second cropping season.

The variation of the irrigation water requirement of TIA in the future would be insignificant. Nevertheless, since the projected meteorology of the basin of Shihmen reservoir would probably change [2,5,30], which is the main facility to supply the irrigation water for the TIA. This could be crucial for water resource planning of the Taoyuan area. The risk of water shortage for future irrigation demand needs further study. In addition, some possible adaptations to changing conditions either on the supply side or demand side is worthy of concern.

Acknowledgments

The authors appreciate the National Taiwan University Global Change Center and the Agricultural Engineering Research Center for providing relevant GCMs and meteorological data. This study is sponsored by the project of the Council of Agriculture under Executive Yuan of Taiwan (101AS-8.2.5-IE-b1).

Author Contributions

Jyun-Long Lee analyzed the data and drafted the manuscript; Wen-Cheng Huang revised the manuscript. The study concept and design was to come up with both authors.

References

1. IPCC Fourth Assessment Report: Climate Change 2007. Available online: http://www.ipcc.ch/publications_and_data/ar4/syr/en/contents.html (accessed on 4 November 2014).
2. Yang, T.C.; Yu, P.S.; Wei, C.M.; Chen, S.T. Projection of climate change for daily precipitation: A case study in Shih-Men reservoir catchment in Taiwan. *Hydrol. Process.* **2011**, *25*, 1324–1354.

3. Hsu, H.H.; Chen, C.T.; Lu, M.M.; Chen, Y.M.; Chou, C.; Wu, Y.C. *Climate Change in Taiwan: Scientific Report*; National Science Council, Executive Yuan: Taipei, Taiwan, 2011.
4. Liu, T.M.; Tung, C.P.; Ke, K.Y.; Chuang, L.G.; Lin, C.Y. Application and development of a decision-support system for assessing water shortage and allocation with climate change. *Paddy Water Environ.* **2009**, *7*, 301–311.
5. Huang, W.C.; Chiang, Y.; Wu, R.Y.; Lee, J.L.; Lin, S.H. The impact of climate change on rainfall frequency in Taiwan. *Terr. Atmos. Ocean. Sci.* **2012**, *23*, 553–564.
6. Yu, P.S.; Yang, T.C.; Kuo, C.C. Evaluating long-term trends in annual and seasonal precipitation in Taiwan. *Water Resour. Manag.* **2006**, *20*, 1007–1023.
7. Chang, L.N.; Lin, P.L.; Chang, C.M.; Shen, H.S. *Assessment of Climate Change in Taiwan (II)*; Environmental Protection Administration, Executive Yuan: Taipei, Taiwan, 1998. (In Chinese)
8. Williams, J.R.; Jones, C.A.; Dyke, P.T. A modeling approach to determining the relationship between erosion and soil productivity. *Trans. Am. Soc. Agric. Eng.* **1984**, *27*, 129–144.
9. Eheart, J.W.; Tornil, D.W. Low-flow frequency exacerbation by irrigation withdrawals in the agricultural midwest under various climate change scenario. *Water Resour. Res.* **1999**, *35*, 2237–2246.
10. Döll, P. Impact of climate change and variability on irrigation requirements: A global perspective. *Clim. Chang.* **2002**, *54*, 269–293.
11. Döll, P.; Siebert, S. Global modeling of irrigation water requirements. *Water Resour. Res.* **2002**, *38*, 8-1–8-10.
12. Smith, M. *CROPWAT: A Computer Program for Irrigation Planning and Management*; Food and Agriculture Organization of the United Nations: Rome, Italy, 1992.
13. Chung, S.O.; Nkomozepi, T. Uncertainty of paddy irrigation requirement estimated from climate change projections in the Geumho river basin, Korea. *Paddy Water Environ.* **2012**, *10*, 175–185.
14. Kloss, S.; Pushpalatha, R.; Kamoyo, K.J.; Schütze, N. Evaluation of crop models for simulating and optimizing deficit irrigation systems in arid and semi-arid countries under climate variability. *Water Resour. Manag.* **2012**, *26*, 997–1014.
15. Taoyuan Irrigation Association. *Irrigation Plan of Taoyuan Irrigation Association in 2005*; Council of Agriculture, Executive Yuan: Taipei, Taiwan, 2005. (In Chinese)
16. Contribution of Working Group I to the Fourth Assessment Report of the Intergovernmental Panel on Climate Change, 2007 (WG1). Available online: http://www.ipcc.ch/ipccreports/tar/wg1/029.htm (accessed on 4 November 2014).
17. Lin, S.H.; Liu, C.M.; Huang, W.C.; Lin, S.S.; Yen, T.H.; Wang, H.R.; Kou, J.T.; Lee, Y.C. Developing a yearly warning index to assess the climatic impact on the water resources of Taiwan, a complex-terrain island. *J. Hydrol.* **2010**, *309*, 13–22.
18. Kalnay, E.; Kanamitsu, M.; Kistler, R.; Collins, W.; Deaven, D.; Gandin, L.; Iredell, M.; Saha, S.; White, G.; Woollen, J.; *et al.* The NCEP/NCAR 40-year reanalysis project. *Bull. Amer. Meteor. Soc.* **1996**, *77*, 437–470.
19. Lin, S.H.; Chen, Y.C.; Lin, W.S.; Liu, C.M. Climate Change Projection for Taiwan based on Statistical Downscaling on Daily Temperature and Precipitation. In Proceedings of the 15th International Joint Seminar on the Regional Deposition Processes in the Atmosphere and Climate Change, Taipei, Taiwan, 12–14 November 2009.

20. Allen, R.G.; Pereira, L.S.; Raes, D.; Smith, M. *Crop Evapotranspiration*; FAO irrigation and drainage paper No.56; Food and Agriculture Organization of the United Nations: Rome, Italy, 2006.
21. Kan, C.E. *Studies on Water Distribution for Irrigation System*; Water Resources Agency, Ministry of Economic Affairs: Taipei, Taiwan, 1979.
22. Hamon, W.R. Estimating potential evapotranspiration. *J. Hydraul. Div.* **1961**, *87*, 107–120.
23. Shaw, S.B.; Riha, S.J. Assessing temperature-based PET equations under a changing climate in temperate, deciduous forests. *Hydrol. Process.* **2011**, *25*, 1466–1478.
24. Tsai, A.Y.; Huang, W.C. Estimation of regional renewable water resources under the impact of climate change. *Paddy Water Environ.* **2012**, *10*, 129–138.
25. Searcy, J.K. Flow-Duration Curves. *Manual of Hydrology: Part 2. Low-Flow Techniques*; United States Department of the Interior: Washington, DC, USA, 1959.
26. Cheng, M. *Irrigation and Drainage Management for Paddy*; Council of Agriculture, Executive Yuan: Taipei, Taiwan, 1977. (In Chinese)
27. Kan, C.E. *Water Conveyance Loss and Irrigation Efficiency in Agriculture*; Tsao-Jin Memorial Foundation for R&D for Agriculture and Irrigation: Kaohsiung, Taiwan, 2000. (In Chinese)
28. Haan, C.T. *Statistical Methods in Hydrology*; Iowa State University Press: Hertfordshire, UK, 1982.
29. Linsley, R.K.; Kohler, M.A.; Paulhus, J.H.L. *Hydrology for Engineers*; McGraw-Hill book Co.: New York, NY, USA, 1988.
30. Yu, P.S.; Wang, Y.C. Impact of climate change on hydrological processes over a basin scale in northern Taiwan. *Hydrol. Process.* **2009**, *23*, 3556–3568.

10

Water Discharge and Sediment Load Changes in China: Change Patterns, Causes and Implications

Chong Jiang [1,2,3,4,*], Linbo Zhang [1,4], Daiqing Li [1,4] and Fen Li [1,4,*]

1. State Key Laboratory of Environmental Criteria and Risk Assessment, Chinese Research Academy of Environmental Sciences, Beijing 100012, China; E-Mails: zhanglb@craes.org.cn (L.Z.); lidq@craes.org.cn (D.L.)
2. College of Global Change and Earth System Science, Beijing Normal University, Beijing 100875, China
3. Joint Center for Global Change Studies, Beijing 100875, China
4. Key Laboratory of Regional Eco-Process and Function Assessment and State Environment Protection, Chinese Academy of Environmental Sciences, Beijing 100012, China

* Authors to whom correspondence should be addressed; E-Mails: jiangchong1987@gmail.com (C.J.); lifen@craes.org.cn (F.L.).

Academic Editor: Yingkui Li

Abstract: In this research, monthly hydrological and daily meteorological data were collected across China for the period 1956–2012. Modified Mann–Kendall tests, double mass curve analysis, and correlation statistics were performed to identify the long-term trends and interrelation of the hydrometeorological variables and to examine the influencing factors of streamflow and sediment. The results are as follows: (1) In the last 60 years, the streamflow in northern China has shown different decreasing trends. For the southern rivers, the streamflow presented severe fluctuations, but the declining trend was insignificant. For the streamflow in western China, an increasing trend was shown. (2) In the northern rivers, the streamflow was jointly controlled by the East Asian monsoon and westerlies. In the southern rivers, the runoff was mainly influenced by the Tibet–Qinghai monsoon, the South Asian monsoon, and westerlies. (3) Sediment loads in the LCRB (Lancang River Basin) and YZRB (Yarlung Zangbo River Basin) did not present significant change trends, although other rivers showed different degrees of gradual reduction, particularly in the 2000s. (4) Underlying surface and precipitation changes jointly influenced the streamflow in eastern rivers. The water consumption for industrial and residential purposes, soil and water conservation engineering, hydraulic engineering, and underlying surface changes induced by other factors were the main causes of streamflow and sediment reduction.

Keywords: streamflow; sediment load; precipitation; monsoon; soil and water conservation engineering

---

1. Introduction

Global warming caused by human-induced emissions of greenhouse gases is accelerating the global hydrological cycle [1]. The accelerated hydrological cycle is in turn altering the spatial-temporal patterns of precipitation, resulting in increased occurrences of precipitation extremes that cause increased occurrences of floods and droughts in many regions of the world [2], including China [3–5]. As a vital natural resource, water is fundamental for the sustainable development of the economy, ecosystem, and biodiversity. Therefore, water security and related implications for ecosystem and river diversity, particularly the variability and availability of regional water resources under the influences of climatic change and human activities, have been discussed in recent years [6–8]. Much attention has been given to water resource changes and their effects on the economic society by the international community. For example, the Intergovernmental Panel on Climate Change (IPCC) has reviewed changes in the global hydrological cycle and has assessed the impacts of climatic change on water resources [9]. Many countries, such as the United Kingdom, have addressed the impact of climatic change on water resource variation [10]. The climatic changes of China are controlled mainly by winter and summer monsoons [11]. Generally, precipitation in southwest China is greater than that in northwest China; these patterns are controlled mainly by the monsoon system and the effects of topography [12]. Rainy seasons in eastern China hinge on progress and retreat of the East Asian monsoon. Detailed information on the evolution of summer Asian monsoons and the associated propagation of rain belts has been reported by Ding [13].

River sediment is an important aspect of land surface processes and global change research. River sediment generation, transportation, and river delta response have become important aspects of Earth system science. In 1968, Holeman [14] investigated global sediment discharge by using global hydrological data; further research was conducted by Holland [15]. Walling and Fang [16] investigated the temporal variation of 145 rivers by using long-term data (longer than 25 years) in Asia, Europe, and North America. They reported that the sediment discharge in more than 50% of analyzed rivers presented upward or downward trends, the latter of which was dominant. However, in the remaining 50%, the sediment flux essentially remained stable [17]. A study of the sediment load in Russia showed that of the 20 rivers flowing into the Arctic Ocean, 35% showed increasing trends, 60% presented declining trends; only 5% remained stable [18]. Similar research was conducted by Liu [19], Subramainian [20], and Siakeu [21] for major rivers of Asia, India, and Japan, respectively. This research revealed that human activities, particularly reservoir and dam construction, were the main causes of sediment flux reduction.

Many researchers investigated the sediment and streamflow change in major rivers of China. The Yangtze River Basin (YARB) [22], Yellow River Basin (YRB) [23], Huai River Basin (HURB) [24], Liao River Basin (LRB) [25], and Songhua River Basin (SRB) [26] showed different degrees of decreasing trends. However, the sediment flux in western rivers such as the Yarlung Zangbo River Basin (YZRB) and Lancang River Basin (LCRB) remained stable or increased slightly. The Yellow and Yangtze rivers are two of the largest rivers in China and therefore receive more attention. Yang [27] considered that reservoir and dam construction was the main reason of sediment reduction in the Yangtze River. Miao [23] reported that reservoir construction, reduced precipitation, soil and water conservation projects jointly induced sediment reduction in the Yellow River. So far, the focus in China has been mostly on regional water and sediment resources. On the basis of instrument records of streamflow and sediment, many scholars conducted research in different river basins in China to reveal different changing patterns of runoff and sediment. Most of these studies are based on the monitoring data of individual rivers without considering the impacts of water and sand resource consumption of the economic society within the river basin. Regarding spatial scale, most studies completed thus far are based on one river basin; few are based on a national scale. An acceptable evaluation of water and sediment resources requires sufficient hydrometeorological data and extensive data-driven analysis, which is the motivation for the current study. Further, possible causes of precipitation and streamflow and sediment resource variation need to be investigated, and the related implications should be discussed.

Therefore, the objectives of this study are to (1) investigate streamflow and sediment changes in major rivers in China; (2) determine the streamflow changes and their relationship with precipitation, monsoons, and water consumption for industrial and residential purposes; (3) determine regional sediment load changes and their relationship with hydraulic engineering, soil and water conservation engineering, and underlying surface changes induced by human activities; and (4) discuss the relationship between sediment and runoff changes and their relationship with specific events, the implications of which will also be discussed. The primary goal of this study is to evaluate the impact of climate change and human activities on streamflow and sediment load and to provide basic information for water and soil resources management in this region.

2. Data and Methodology

2.1. Data Collection and Processing

In this study, annual precipitation data from 725 rain gauge stations for the period 1951–2012 were obtained from the National Climate Center (NCC) of the China Meteorological Administration (CMA). The quality of meteorological data was firmly controlled [28]. To guarantee the accuracy of the results, the data was preprocessed as follow before the analysis. The observational data of missing data years of more than 5 years (including 5 years) were excluded. The time series data of partial relocation stations were unified, and the remaining missing observation data were completed with a linear regression method and adjacent station interpolation to ensure the integrity of the time series. The missing data in 725 stations only accounted for less than 5% of total data amount. The regional averages refer to the arithmetic mean value of the stations within a region. Annual precipitation average was, thus, calculated from these records using the Thiessen Polygon method for each river basin.

The consecutive monthly data of streamflow and sediment yields from the 30 gauge stations were collected for the same period from Ministry of Water Resources (MWR). The hydrologic data from the 30 gauge stations listed in Table 1 were used to analyze changes in streamflow and sediment load.

Figure 1 and Table 1 provide information on the station location and associated drainage area. It is worth to address is that, the Xinjiang Inland River Basin and the Hexi Inland River Basin (HIRB) in northwestern China are composed of many tributaries. Therefore, to reflect the overall change in streamflow, we summed the streamflow in tributary data to represent the runoff of the entire basin. In southwestern China, although many large rivers are present, we selected as study objects only two river systems, the LCRB and YZRB, considering data availability.

The daily discharge was computed from the water level by using previously calibrated discharge-water level curves. Water was sampled at fixed intervals, and suspended sediment concentration was obtained by measuring water samples in the laboratory. All the measurements of water level, discharge, SSC followed national standards issued by the Ministry of Water Conservancy, and were printed in the China Gazette of River Sedimentation [29]. Sediment loads refers to the suspended fraction only, whereas bedload was excluded due to its difficulty in field sampling. Measurement of the sediment loads was on the basis of standard procedures [30,31]. Errors in calculating sediment load were introduced through the low frequency of sampling, rather than continuous monitoring, which is likely to underestimate sediment load during peak hours. The monthly and annual streamflow and sediment load at the gauging stations were derived from the daily measured data. The accuracy and consistency of all the data used in this study have been checked out by the corresponding agencies before their release. In instances when discharge data were missing, we used discharge data from similar rainfall conditions at other times as a replacement, and the missing data in 30 stations only accounted for less than 3% of total data amount.

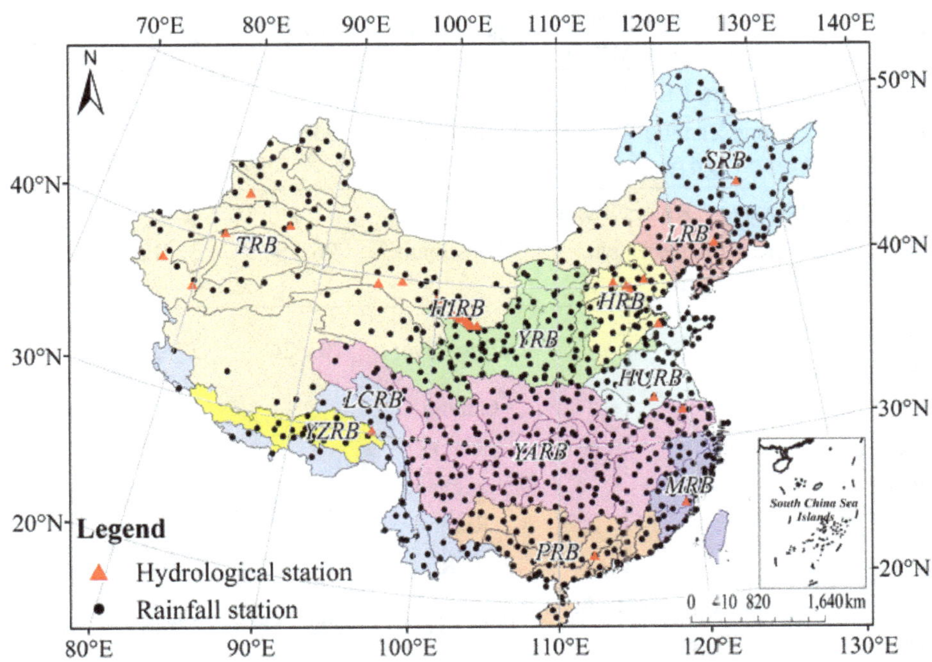

Figure 1. Distribution of 12 major basins in China showing gauge stations (red triangles) and meteorological stations (black dots) used in this study.

1:100,000-scale land use maps in 1985 and 2010 were respectively obtained from the Earth System Science Data Sharing Platform and The Remote Sensing Monitoring and Assessment of Decadal Changes of National Eco-environment (2000–2010) project group. The digital elevation model (DEM), the Monitoring Report of Soil and Water Loss in China, and other maps were obtained from the Earth System Science Data Sharing Platform.

Table 1. Summary of gauging stations and hydrological characteristics in the 12 major basins in China. DA is the drainage area.

ID	Gauge Station	Basin (Abbreviation)	Location	Longitude (N°)	Latitude (E°)	DA (km^2)	Time Span
1	Lijin	Yellow River Basin (YRB)	Mainstream	118°15′	37°29′	752,032	1952–2012
2	Datong	Yangtze River Basin (YARB)	Mainstream	117°03′	32°37′	1,705,383	1950–2012
3	Wujiadu	Huai River Basin (HURB)	Mainstream	117°23′	32°54′	121,330	1950–2011
4	Zhangjiafen	Hai River Basin (HRB)	Bai River	116°10′	39°48′	8506	1954–2011
5	Xiahui		Chao River	117°18′	40°22′	5340	1961–2011
6	Shixiali		Sanggan River	114°43′	40°16′	23,944	1952–2011
7	Xiangshuipu		Yang River	109°40′	38°01′	14,507	1952–2011
8	Yanling		Yongding River	115°49′	40°01′	43,674	1952–2011
9	Haerbin	Songhua River Basin (SRB)	Mainstream	126°32′	45°48′	389,769	1955–2012
10	Tieling	Liao River Basin (LRB)	Mainstream	123°43′	42°13′	120,764	1954–2012
11	Gaoyao	Pearl River Basin (PRB)	Xi River	112°27′	23°01′	351,535	1957–2011
12	Zhuqi	Southeast Rivers Basin (Min River Basin, MRB)	Mainstream	119°06′	26°08′	54,500	1950–2011
13	Nuxia	Southwestern Rivers Basin (Yarlung Zangbo River, YZRB)	Mainstream	95°05′	31°17′	191,235	1956–2009
14	Xiangda	Southwestern Rivers Basin (Lancang River Basin, LCRB)	Mainstream	96°28′	32°12′	17,909	1956–2012
15	Dajingxia Reservoir	Hexi Inland River Basin (HIRB)	Dajing River	103°24′	37°28′	68,300	1961–2010
16	Gulang		Gulang River	102°52′	37°27′		1961–2010
17	Huangyanghe Reservoir		Huangyang River	102°44′	37°35′		1961–2010
18	Zamusi		Zamu River	102°34′	37°42′		1961–2010
19	Nanying Reservoir		Jinta River	102°31′	37°48′		1961–2010
20	Jiutiaoling		Xiying River	102°03′	37°52′		1961–2010
21	Shagousi		Dongda River	101°55′	37°58′		1961–2010
22	Xidahe Reservoir		Xida River	101°23′	38°03′		1961–2010
23	Changmapu		Shule River	96°51′	39°49′		1961–2010

Table 1. *Cont.*

ID	Gauge Station	Basin (Abbreviation)	Location	Longitude (N°)	Latitude (E°)	DA (km²)	Time Span
24	Dangchengwan	Hexi Inland River Basin (HIRB)	Dang River	94°53′	39°30′	68,300	1961–2010
25	Yingluoxia		Hei River	99°55′	38°57′		1961–2010
26	Binggou		Beida River	101°56′	37°54′		1961–2010
27	Kaqun	Xinjiang Inland River Basin (Tarim River Basin, TRB)	Yeerqiang River	76°54′	37°59′	50,248	1957–2011
28	Tongguziluoke		Yulongkashi River	79°55′	36°49′	14,575	1957–2011
29	Yanqi		Kaidu River	86°34′	42°02′	22,516	1957–2011
30	Alaer		Mainstream	81°19′	40°32′	127,900	1957–2011

2.2. Methodology

2.2.1. Mann–Kendall Test for Monotonic Trend

To analyze the long-term trends of hydrometeorological variables, the non-parametric Mann–Kendall test [32,33] was applied. This method has been widely used to detect trends in climate and streamflow time series [34]. In the Mann–Kendall test, the null hypothesis H_0 states that $x_1,..., x_n$ are samples of n independent and identically distributed random variables with no seasonal change. The alternative hypothesis H_1 for a two-sided test defines the distributions of x_k and x_j as non-identical for all $k, j \leq n$; with $k \neq j$. The test statistic S is given as

$$S = \sum_{i=1}^{n-1} \sum_{k=i+1}^{n} \operatorname{sgn}(x_k - x_i) \tag{1}$$

$$\operatorname{sgn} = \begin{cases} +1 & \\ 0 & if \\ -1 & \end{cases} \begin{matrix} \theta > 0 \\ \theta = 0 \\ \theta < 0 \end{matrix} \tag{2}$$

If the dataset is independent and identically distributed, the mean of S will be zero, and the variance of S will be:

$$\operatorname{var}(S) = \left[n(n-1)(2n+5) - \sum_{j=1}^{m} t_j (t_j - 1)(2t_j + 5) \right] / 18 \tag{3}$$

where n is the number of data points, t is the extent of a given time, m is the number of tied groups, and t_j is the number of data points in the j-th group. A tied group is a set of data points having the same value. A normalized test statistic Z can be computed on the basis of S as:

$$Z = \begin{cases} \dfrac{S-1}{\sqrt{\operatorname{var}(S)}} & S > 0 \\ 0 & S = 0 \\ \dfrac{S+1}{\sqrt{\operatorname{var}(S)}} & S < 0 \end{cases} \tag{4}$$

When the significance levels are set at 0.01, 0.05, and 0.1, $|Z_\alpha|$ is 2.58, 1.96, and 1.65, respectively. At a certain significance level, if $|Z| > |Z_\alpha|$, the null hypothesis H_0 is rejected. That is, the trend is significant at the set level of significance. Otherwise, no significant trend exists.

In the Mann–Kendall test, the slope estimated by using the Theil–Sen estimator [35,36] is usually considered to detect the monotonic trend and to indicate the variable quantity in the unit time. It is a robust estimate of the magnitude of a trend and has been widely used to identify the slope of a trend line in a hydrological or climatic time series [37]. The estimator is given as:

$$\beta = Median\left(\frac{x_j - x_l}{j - l}\right) \forall 1 < l < j \tag{5}$$

where $1 < l < j < n$, β is the median overall combination of record pairs for the entire dataset and is resistant to extreme observations. A positive β denotes an increasing trend, and a negative β indicates a decreasing trend.

2.2.2. Modified Mann–Kendall Test (Mann–Kendall Test with Trend-Free Pre-whitening)

The Mann–Kendall test assumes that the series is independent and the series is not robust against autocorrelation. However, certain hydrological time series may frequently display statistically significant serial correlation. This may lead to a disproportionate rejection of the null hypothesis of no trend, whereas the null hypothesis is actually true. Therefore, the effect of serial correlation is a major source of uncertainty in testing and interpretation trends. To eliminate the influence of serial correlation, "pre-whitening" was proposed by Von Storch [38] to remove the lag one serial correlation (r_1) from the time series. This method has been applied in an increasing number of studies [23,26,39].

To determine whether the observed dataset is serially correlated, the significance of the lag-1 serial correlation (r_1) should be tested at the 0.10 significance level. r_1 is calculated by using the following Equation [23,26]:

$$r_k = \frac{\frac{1}{n-k}\sum_{i=1}^{n-k}\left(x_i - \bar{x}\right)\left(x_{i+k} - \bar{x}\right)}{\frac{1}{n}\sum_{i=1}^{n}\left(x_i - \bar{x}\right)^2} \tag{6}$$

If $\frac{-1-1.645\sqrt{n-2}}{n-1} \le r_1 \le \frac{-1+1.645\sqrt{n-2}}{n-1}$, the time series is assumed to be independent at the 0.10 significance level and can be subjected to the original Mann–Kendall test. Otherwise, the effect of serial correlation should be removed from the time series by pre-whitening prior to application of the Mann–Kendall test. The Mann–Kendall test is then used to detect trends in the residual series. The new time series is obtained as [40].

$$x_i' = x_i - (\beta \times i) \tag{7}$$

The r_1 value of this new time dataset is calculated and used to determine the residual series as:

$$y_i' = x_i' - r_1 \times - x_{i-1}' \tag{8}$$

The value of $\beta \times i$ is added again to the residual dataset as:

$$y_i = y_i' + (\beta \times i) \tag{9}$$

The y_i series is then subjected to trend analysis.

2.2.3. Double Mass Curve

Double mass curve analysis is a simple and practical visual method widely used in the study of the consistency and long-term trend test of hydrometeorological data [41]. This method was first used to analyze the consistency of precipitation data in Susquehanna watershed, Pennsylvania, USA [42]; a theoretical explanation was later reported [43]. The theory of the double mass curve is based on the fact that a plot of two cumulative quantities during the same period exhibits a straight line if the proportionality between the two remains unchanged; the slope of the line represents the proportionality. This method can smooth a time series and suppress random elements in the series; thus, it can show the main trends of the time series. In the last 30 years, Chinese scholars analyzed the effects of soil and water conservation measures and land use/cover changes on streamflow and sediment by using this method and have achieved good results [44]. In the present study, double mass curves of sediment *versus* streamflow were plotted for the different periods to detect the relationship change before and after transition years. The appearance of the inflection point denotes that the relationship between the sediment and streamflow begins to change significantly [44].

3. Variation of Streamflow and Sediment Load

3.1. Overall Change of Streamflow

In this study, we defined the northern rivers and southern rivers as those north and south of the eastern monsoon zone in China, respectively. Therefore, the northern rivers include the YRB, HRB, LRB, SRB, and HURB; the southern rivers include the Pearl River Basin (PRB), YARB, and Min River Basin (MRB); and the western rivers include the Tarim River Basin (TRB), HIRB, LCRB, and YZRB. Based upon the years of average runoff (Table 2), the order of runoff in the southern rivers was YARB (8944.8×10^8 m^3) > PRB (2166.4×10^8 m^3) > MRB (532.7×10^8 m^3); that in the northern rivers was SRB (404.7×10^8 m^3) > YRB (299.2×10^8 m^3) > HURB (266.9×10^8 m^3/a) > LRB (29.3×10^8 m^3) > HRB (18.1×10^8 m^3); and that in the western rivers was YZRB (312.5×10^8 m^3) > LCRB (247.3×10^8 m^3) > TRB (157.3×10^8 m^3) > HIRB (44.0×10^8 m^3). It should be noted that Xiangda Station, which represents the LCRB, is located at the source area of LCRB; therefore, the streamflow was smaller than that in the entire basin. Actually, the average streamflow in the downstream region of the LCRB was 740.5×10^8 m^3 [45].

Figure 2a,b show the cumulative curve of the streamflow in the major rivers. The cumulative curves of YARB, PRB, MRB, LCRB, YZRB, TRB, and HIRB, presented linear increasing trends with essentially no fluctuation or inflection point. Among them, the runoff in the HIRB showed a significant increasing trend at 0.16×10^8 m^3/a, $P < 0.001$. That of other basins fluctuated near the mean level, as shown in Figure 3; the interdecadal anomalies are shown in Table 3. For the northern rivers, the cumulative curve of streamflow in HURB presented complicated changes. The overall direction was a straight line, indicating severe fluctuation on the interannual scale, and no obvious trend was found. The other northern rivers generally presented convex curves or lines, indicating that runoff in these rivers showed decreasing trends (Figure 3). As shown in Table 4, the order of decrease rate was YRB (-7.25×10^8 m^3/a, $P < 0.001$) > SRB (-3.32×10^8 m^3/a, $P < 0.001$) > HRB (-0.69×10^8 m^3/a, $P < 0.001$) > LRB (-0.48×10^8 m^3/a, $P < 0.001$).

Table 2. Average annual streamflow and sediment load.

Basins	Water Discharge (10^8 m^3)	Sediment Load (10^4 t)
HRB	18.1	795.1
HURB	266.9	881.4
YRB	299.2	76,655
LRB	29.3	1112.9
SRB	404.7	598.8
MRB	532.7	573.3
PRB	2166.4	6274.6
YARB	8944.8	40791
HIRB	44.0	–
TRB	157.3	–
YZRB	312.5	1710.4
LCRB	247.3	341.0

3.2. Overall Change in Sediment Load

Figure 2c,d show the cumulative curve of the sediment load in the major rivers. No obvious convex state was presented in those of LCRB and YZRB, which means the sediment discharge variation had no significant trend. The cumulative curve of sediment load in other rivers showed obvious convex shapes, which denote that the sediment had different degrees of gradual reduction. In particular, after 2000, the decrease was between 59.1% and 98.7% (Table 3). In the southern rivers, the decrease in the 2000s was between 59.1% and 63.4%, with YARB showing the largest value. In the northern rivers, the decrease was between 59.4% and 98.7%; LRB and HRB were reduced by 97.6% and 98.7%, respectively (Table 3).

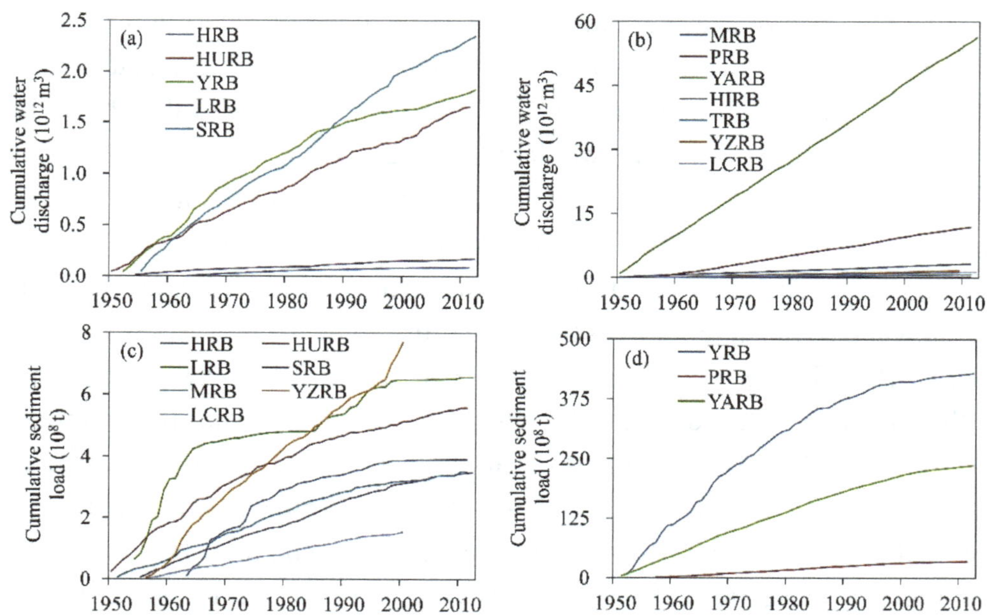

Figure 2. Cumulative curves of (**a**), (**b**) water discharge and (**c**), (**d**) sediment discharge in major basins in China, during 1950–2012.

Table 3. Interdecadal anomalies of water discharge and sediment load. The reference value is the average value during the 1950s to 1960s.

Basins	Change in Water Discharge (%)				Change in Sediment Load (%)			
	1970s	1980s	1990s	2000s	1970s	1980s	1990s	2000s
HRB	–18.6	–55.9	–56.4	–84.6	–35.0	–77.5	–77.2	–98.7
HURB	–27.8	0.1	–45.5	–6.5	–43.4	–46.8	–73.7	–69.6
YRB	–36.4	–41.5	–71.2	–67.5	–26.0	–47.3	–67.8	–88.7
LRB	–56.9	–36.8	–27.3	–62.8	–90.2	–80.3	–59.9	–97.6
SRB	–31.6	–2.9	–4.6	–43.4	–21.3	7.7	–19.6	–59.4
MRB	–1.7	–2.6	4.2	–5.4	–1.6	–13.4	–49.0	–63.4
PRB	11.2	–4.0	13.3	–6.4	10.9	14.7	4.4	–59.1
YARB	–6.1	–0.9	5.8	–5.7	–13.0	–10.9	–29.7	–63.1
YZRB	–6.6	–14.2	–4.5	5.2	–14.9	–24.0	–5.2	–
LCRB	–1.4	19.3	1.7	14.1	5.7	39.4	–14.4	–

Table 4. Results of Sen's slope estimator and the Z value by using linear regression and the Mann–Kendall test, respectively.

Basins	Water Discharge (10^8 m^3)			Sediment Load (10^4 t)		
	Slope	Z	Significance	Slope	Z	Significance
HRB	–0.69	–6.96 **	$P < 0.001$	–46.96	–6.73 **	$P < 0.001$
HURB	–1.50	–1.49	$P > 0.1$	–25.35	–5.42 **	$P < 0.001$
YRB	–7.25	–6.01 **	$P < 0.001$	–2300	–6.58 **	$P < 0.001$
LRB	–0.48	–3.2 *	$P < 0.01$	–55.33	–4.90 **	$P < 0.001$
SRB	–3.32	–3.37 **	$P < 0.001$	–8.44	–4.11 **	$P < 0.001$
MRB	–0.53	–0.58	$P > 0.1$	–9.74	–4.10 **	$P < 0.001$
PRB	–2.04	–0.98	$P > 0.1$	–77.49	–3.31 **	$P < 0.001$
YARB	–5.72	–0.36	$P > 0.1$	–6000	–6.73 **	$P < 0.001$
HIRB	0.16	2.81 *	$P < 0.01$	–	–	–
TRB	0.10	0.80	$P > 0.1$	–	–	–
YZRB	0.20	0.40	$P > 0.1$	6.12	0.47	$P > 0.1$
LCRB	0.61	1.58	$P > 0.1$	–0.96	–0.67	$P > 0.1$

Note: "*" and "**" mean the correlation coefficient reach the significance level of 0.01 and 0.001, respectively.

3.3. Pattern of Changes in Streamflow and Sediment Load

Figures 3 and 4 show respectively the temporal variation and double mass curves of sediment load and streamflow in the 12 major rivers in China. On the whole, the variation of streamflow *versus* sediment can be divided into three categories. In the first, the streamflow was stably maintained, but the sediment load was reduced (HURB, YARB, MRB, PRB). In the second, the streamflow and sediment load were both reduced (YRB, LRB, HRB, SRB). In the third, both water and sediment discharge remained stable (LCRB, YZRB).

3.3.1. Streamflow Remained Stable and Sediment Load Reduced

In the HURB, the streamflow and sediment discharge experienced continuous declines before 1978 (Figures 3a and 4a). Since the early 1980s, and particularly in the 1990s, an obvious decrease in sediment discharge occurred. This result was due mainly to the effects of soil and water conservation projects in the HURB with areas of 1.53 × 10^4 km^2 and 1.04 × 10^4 km^2 in the 1980s and 1990s, respectively. Shi [24] reported that the soil erosion amount of the small watershed in the upper reaches of the Huai River was reduced 77%–85% after implementation of engineering projects.

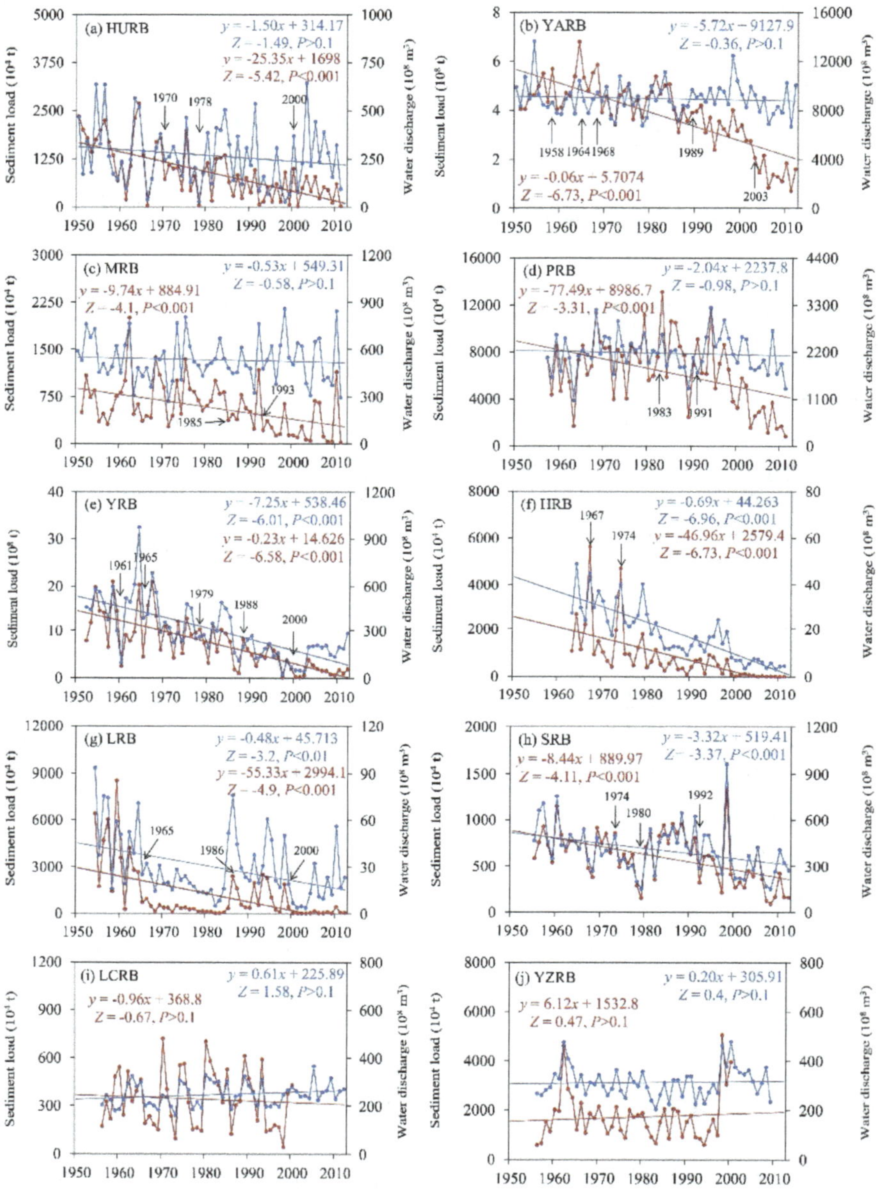

Figure 3. Interannual variability of streamflow and sediment load in major basins in China during 1950–2012. Blue and red lines represent changes in water discharge and sediment discharge, respectively.

In the YARB, the streamflow variation was relatively stable; however, the sediment load decreased significantly (Figures 3b and 4b). The sediment yields in 1956, 1958, and 1963–1968 were obviously high, which is likely attributed to large-scal edeforestation activities such as "Devastating Forests for Arable Land". After 1969, the sediment load value suddenly dropped. This result is likely attributed to the Danjiangkou Reservoir operation that began in 1968, which intercepted a large amount of sediment. After 1989, the sediment yield decreased again, which could have been caused by three aspects. Firstly, a series of ecological engineering projects was conducted since 1982, particularly the Changzhi Project launched in 1989. Secondly, the reservoirs were constructed in the mainstream and tributaries of the Yangtze River. Thirdly, the amount of sand dredging increased annually. The total amount in 1990–2002 was 5 × 10^7 t. After 2003, the sediment load decreased again, which was mainly caused by operation of the Sanxia Reservoir in 2003. In 2012, sedimentation in Sanxia Reservoir reached 14.37 × 10^8 t.

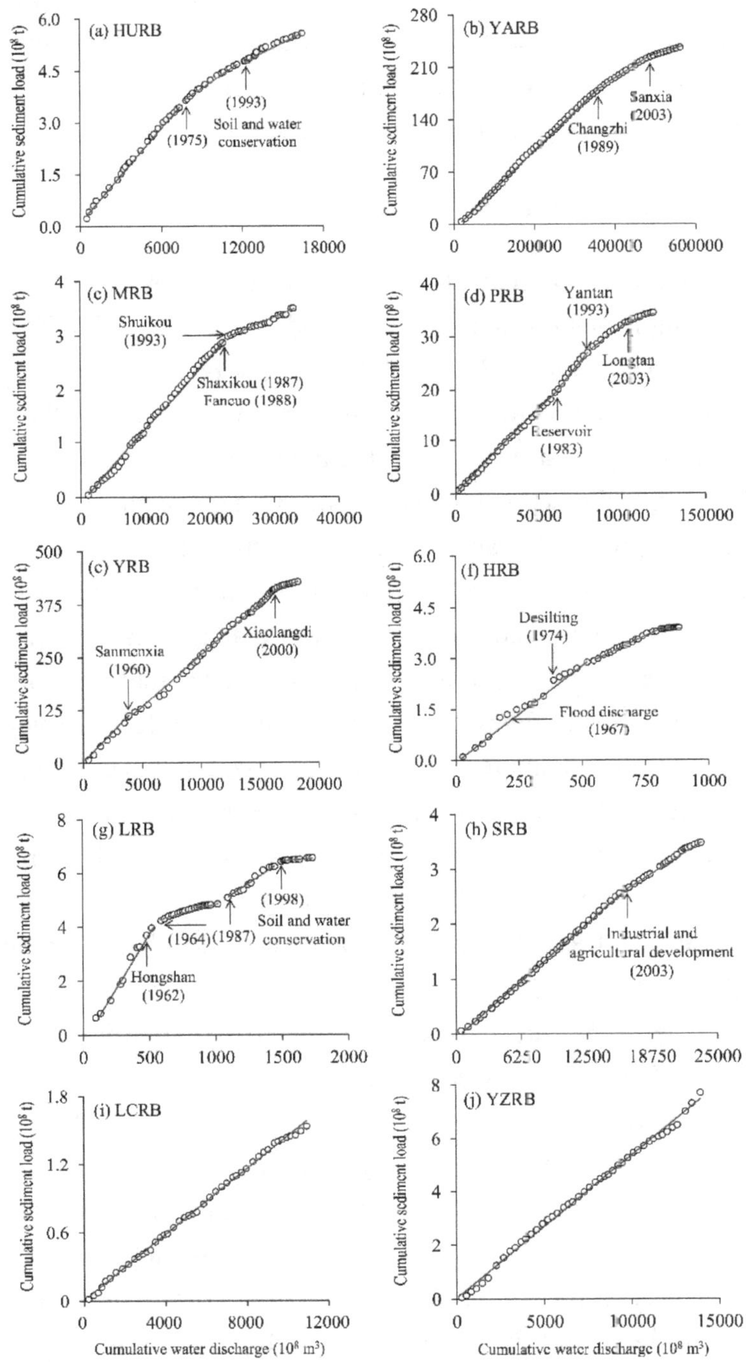

Figure 4. Double mass curve of streamflow *versus* sediment load.

For MRB, as observed by Zhuqi Station, sediment load changed with streamflow variation before 1985, although the amplitude of the sediment was larger than that of the streamflow (Figures 3c and 4c). After 1985 and 1993, we detected two distinct decreasing processes obviously related to reservoir construction such as the Shaxikou hydropower station in 1987, Fancuo Dam in 1988, and the Shuikou hydropower station in 1993. In particular, the sedimentation of Shuikou Dam accounted for 86% of the total amount of sediment in the MRB. An additional factor is that the annual sand dredging amount in Mawei near Shuikou Dam was 1×10^7 t after the late 1980s. The streamflow in Gaoyao Station of PRB was stable, although the sediment load decreased significantly (Figures 3d and 4d). Before 1983, the sediment load fluctuated with streamflow change and maintained a consistent pace. In 1983–1991, the sediment load increased obviously, which may related to the construction of multiple reservoirs. Since 1994, the sediment load presented a sharp decline, which is mainly attributed to the sediment-retaining functions of reservoirs and dams such as Yantan Reservoir in the Hongshui River (operated in 1993) and Longtan Reservoir (operated in 2003).

3.3.2. Streamflow and Sediment Load Reduced Together

In the YRB, HRB, LRB, and SRB, the water and sediment discharge showed clear downward trends. In YRB, as observed by Linjin Station, the sediment loads in 1961–1965, 1980–1988, and 2000–2012 were lower than the mean level (Figures 3e and 4e). The low point in 1961 reflected the operation of Sanmenxia Reservoir, which began to retain water and sediment. In 1965, the operation mode of this reservoir changed to begin storing clear water and releasing muddy sediment; therefore, the sediment load recovered slightly. After 1980, the sediment load reduced again, with the following possible causes: Firstly, the water diversion project began operation, which reduced the streamflow and sediment discharge. Secondly, the soil and water conservation projects in the upstream function are functioning efficiently; thus, the water discharge is reduced in the lower reaches. Thirdly, the rainfall is concentrated mainly in the midstream region with less soil loss area; thus, the sediment is reduced downstream. In the early 2000s, the sediment and streamflow reduced again, mainly due to the water diversion projects and water reservoirs operating in the upstream region, such as Xiaolangdi Reservoir in 1999.

In the HRB, the sediment and streamflow after the 1980s was maintained at relatively low levels, the sediment load approached zero since 2000 (Figures 3f and 4f). The abnormally high sediment load in 1967 and 1974 was due to the desilting effect of Guanting Reservoir during the flood season through high rainfall and runoff. In the LRB, the sediment and water discharge presented periodic changes (Figures 3g and 4g). Before the 1960s, strong rainfall caused severe soil erosion, thereby inducing sediment load increases. After 1964, the sediment load decreased significantly due to the operation of Hongshan Reservoir in 1962. Until 1999, the sedimentation in Hongshan Reservoir reached 9.41×10^8 m^3, accounting for 58% of the total storage capacity. Severe fluctuation of sediment and streamflow occurred in 1985–2000, which was caused mainly by rainfall.

In the SRB, the streamflow and sediment discharge decreased at the same pace (Figures 3h and 4h). On the interdecadal scale, the streamflow and sediment discharge experienced "low-high-low" alternating variation processes in the 1970s to 1990s, which were mainly affected by precipitation. In addition, agricultural and industrial development since the beginning of the 1960s also accelerated the water consumption for industrial and residential purposes. Although some large- and medium-sized reservoirs are located in the upstream regions of Songhua River, the reciprocal relationship between runoff and sediment has been relatively good, and no obvious anomalies have been detected. This result occurred essentially because the vegetation coverage in the source area is relatively high, and density of population is low; thus, the river is seldom disturbed by human activities.

3.3.3. Both Streamflow and Sediment Discharge Remained Stable

Both streamflow and sediment discharge in the LCRB and YZRB maintained stability. No significant upward or downward trends were detected, and fluctuations of streamflow *versus* sediment have been essentially consistent (Figures 3i,j and 4i,j). Actually, the vegetation coverage is good, the population density in the YZRB and LCRB is relatively low, and the level of economic development and social construction is also relatively low. Thus; the streamflow and sediment are seldom affected by human activities, and presented a good correlation among rainfall, streamflow, and sediment.

4. Influencing Factors of Streamflow Variation

4.1. Precipitation

In general, climate change is mainly characterized by temperature and precipitation variability. Precipitation drives runoff and hence directly influences the discharge of a river. Figure 5 shows precipitation changes in the major rivers and in the entire country that occurred during the past 60 years. China's precipitation experienced a "decrease, increase, decrease, increase, decrease" pattern. The 1950s was a rainy decade; the average rainfall reached 820.3 mm, which was the highest value recorded in several decades. After 1960, precipitation began to decrease. Levels were low in the 1980s and increased slightly in the 1990s. After 2000, the overall trend obviously reduced. The SRB and LRB experienced a rainy period in the 1950s that decreased in the 1960s and 1970s. The 1980s and 1990s presented slight increases that decreased again in the 2000s. Precipitation in the YRB essentially showed a decreasing trend by decade with lowest values occurring in the 1990s. After 2000, however, the level increased slightly. Rainfall in the TRB and HIRB increased by decade; however, the YZRB and LCRB presented decreasing trends since the 1960s. For the southern rivers including YARB, MRB, and PRB, changes in precipitation were consistent. Levels were lowest in the 1950s, increased in the 1960s and 1970s, and fell slightly in the 1980s before returning to the less-rainfall stage in 2000.

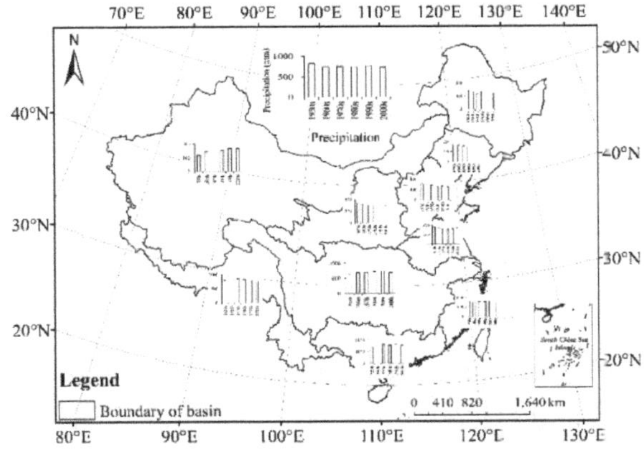

Figure 5. Interdecadal rainfall changes in the main basins in China during the 1950–2012.

Since the Reform and Opening-up policy was implemented in 1979, the demand for water resources increased substantially, and the streamflow in some rivers showed different degrees of decreasing trend (Figure 3). Therefore, we divided the streamflow sequence into two periods of before and after 1980 to explore the relationship between streamflow and precipitation (Figure 6). The subsection fitting lines of precipitation *versus* runoff represent the runoff depth produced by rainfall in 1950–1979 and 1980–2010, respectively. Assuming that the precipitation had no significant change on the basin scale, changes in the relationship of precipitation and streamflow can reflect the influence of the underlying surface on the original hydrological process. In the northern rivers, the fitting lines of rainfall-runoff in the HRB (Figure 6a), YRB (Figure 6c), and SRB (Figure 6e) moved downward during 1980–2010 comparing with that occurring in 1950–1979.

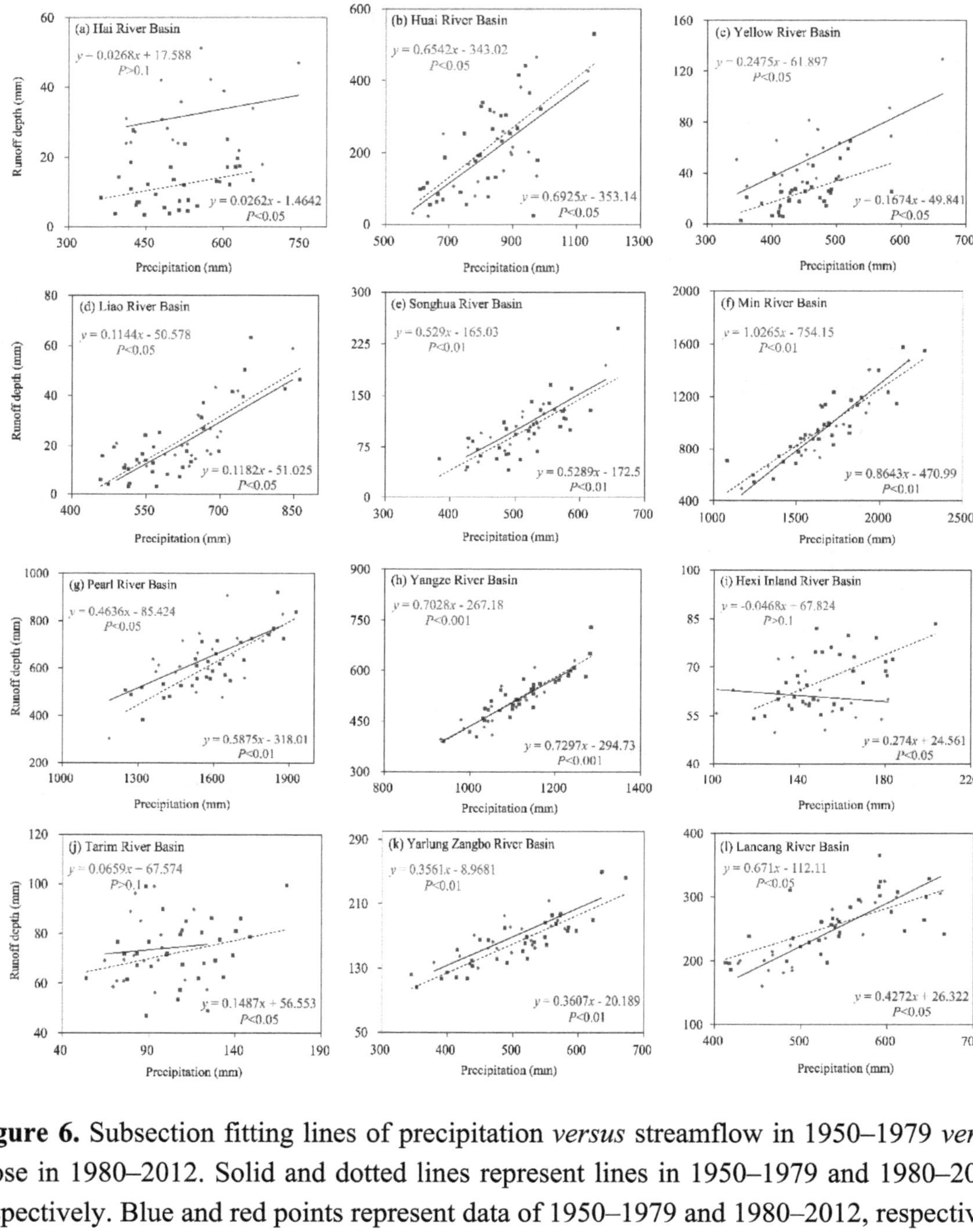

Figure 6. Subsection fitting lines of precipitation *versus* streamflow in 1950–1979 *versus* those in 1980–2012. Solid and dotted lines represent lines in 1950–1979 and 1980–2012, respectively. Blue and red points represent data of 1950–1979 and 1980–2012, respectively.

This downward movement denote the capacity of runoff yield decline. For the HURB (Figure 6b) and LRB (Figure 6d), the fitting lines moved upward, which illustrates that the capacity of runoff yield was enhanced. In the southern rivers, the runoff capacity of the PRB declined after 1980 (Figure 6g), whereas the change of MRB was not obvious (Figure 6f). The YARB also remained stable (Figure 6h). The streamflows of the TRB and HIRB in the northwest and the LCRB and YZRB in the southwest, originating from Tibet–Qinghai Plateau, were recharged by rainfall in addition to runoff from glacier and snow melt water. In this condition, the rainfall-runoff fitting lines could not fully reflect the changes in runoff yield ability. Nevertheless, the fitting lines can still be used to determine the role of rainfall change on runoff. The slope of the regression line in the HIRB during 1950–1979 was a negative value (Figure 6i), which shows that the water supply in addition to rainfall decreased. In 1980–2010, however, this part of the water supply increased. Similarly, the runoff yield in the TRB (Figure 6j) and the YZRB (Figure 6k) decreased, whereas the change in the LCRB was not obvious (Figure 6l). Changes in the rainfall–runoff relationship showed that the underlying surface affected the hydrological process, which will be further discussed in subsequent sections.

4.2. Monsoons

To determine the possible causes of streamflow variation, we examined Asian monsoon indices, including the East Asian monsoon index (EAMI), South Asian monsoon index (SAMI), Tibet–Qinghai Plateau monsoon index (TPMI), West Pacific subtropical high (WPSH), and westerly index (WI). The Asian monsoon indices were created by Li [46,47], and WI was derived from National Centers for Environmental Prediction (NCEP) reanalysis data. The TPMI was calculated from NCEP 600 mb height reanalysis data based on the method given by Wang [48].

Figure 7 shows the cumulative anomaly curves of runoff *versus* climatic indexes and their correlation coefficients. The streamflow in the northern rivers correlated positively with the EAMI and negatively with the WI, including the HRB (Figure 7a,b), HURB (Figure 7c,d), YRB (Figure 7e,f), LRB (Figure 7g,h), and SRB (Figure 7i,j). Among the northern rivers, the correlation coefficient between the YRB and EAMI streamflow was the highest ($R = 0.82$, $P < 0.001$). However, the cumulative anomaly curves of streamflow against the WI presented an opposite phase ($R = -0.83$, $P < 0.001$). The results showed that streamflow in the YRB was jointly affected by the East Asian monsoon and westerlies, which also reflects the interaction of westerlies and the East Asian monsoon. The relationship of streamflow in the HRB, HURB, LRB, and SRB and climatic indices was essentially the same as that in the YRB. However, the correlation was not as strong as that with the YRB, which may be related to the locations of the East Asia monsoon and westerly belt, as well as the influence of human activities on natural runoff. In the southeastern rivers, the runoff of the MRB was affected mainly by the West Pacific subtropical high ($R = 0.46$, $P < 0.01$) and the Tibet–Qinghai Plateau Monsoon ($R = 0.38$, $P < 0.01$) (Figure 7k,l). The PRB (Figure 7m,n) is adjacent to the South China Sea and is also located in the westerly belt and the East Asian monsoon zone. Therefore, its runoff was influenced by several monsoon systems and therefore lacked direct correlation with a specific monsoon index; the closest relationships were with the SAMI ($R = 0.24$, $P < 0.1$) and TPMI ($R = -0.55$, $P < 0.01$). The case of the YARB (Figure 7o,p) was similar to that of the PRB. The runoff was influenced by several monsoon systems dominated by the WI ($R = 0.64$, $P < 0.001$) and TPMI ($R = 0.39$, $P < 0.01$). Ma [49] also found a similar phenomenon, although deeper indications require further investigation. In northwestern China, runoff in the HIRB (Figure 7q,r) and TRB (Figure 7s,t) correlated positively with that in the TPMI and WI, indicating that streamflow changes in these areas were controlled mainly by the Tibet–Qinghai Plateau monsoon and westerlies.

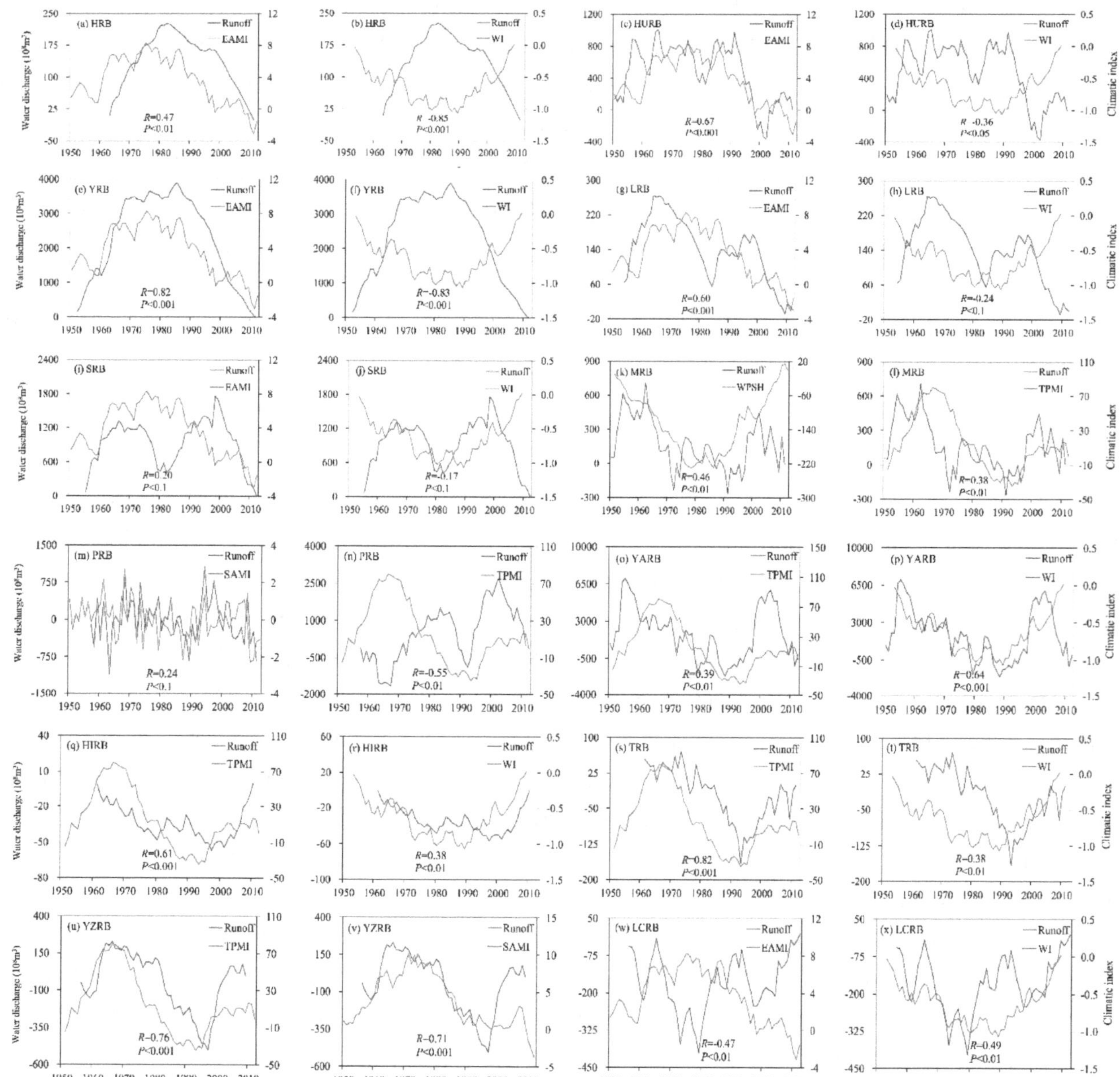

Figure 7. Cumulative anomaly curves of streamflow *versus* monsoon intensity indices in the main basins of China. EAMI, SAMI, WI, WPSH and TPMI represent the East Asian monsoon index, westerly index, West Pacific subtropical high, and Tibetan Plateau monsoon index, respectively. *R* and *P* represent the correlation coefficient and significance level, respectively.

A deeper indication is that the TRB and HIRB were located in the central area of Eurasia, which could have hardly been reached by the Northwest Pacific summer monsoon. Therefore, the westerlies and Tibet–Qinghai Plateau monsoon became the dominant factors. In southwestern China, the YZRB originated from the Tibet–Qinghai Plateau hinterland. Its runoff was therefore subjected to the Tibet–Qinghai plateau monsoon, and the correlation coefficient between them was as high as 0.76 ($p < 0.001$) (Figure 7u,v). In contrast, because the YZRB was near the India Ocean, the runoff and particularly the midstream was influenced by the South Asia monsoon. In the LCRB (Figure 7w,x), although its headwater was in the Tibet–Qinghai Plateau monsoon zone, the runoff was mainly affected by the westerlies and East Asian monsoon.

4.3. Water Consumption for Industrial and Residential Purposes

Although climate change is important, particularly the impact of precipitation change on runoff, water consumption for industrial and residential purposes also plays important roles in runoff change [50–52]. Long-term statistics of water consumption are not currently available, which makes detailed analysis difficult. The "China Water Resources Bulletin (2004–2012)", reported that the YARB had the largest annual water consumption (Figure 8a). Aside from a slight reduction in water consumption in the HRB, the other basins all presented increasing trends in 2004–2012 (Figure 8a). Water for agriculture and industry accounted for the largest proportion at more than 80% of the total water consumption (Figure 8b,c). For long term trends, the "China Statistical Yearbook (1950–2009)" reported that the population, gross domestic product (GDP), and gross agricultural products showed significantly increasing trends (Figure 9). In particular, definite increases in water consumption are expected by the former two sectors. Thus, it can be inferred that the water consumption during the past 60 years also showed sharp increasing trends.

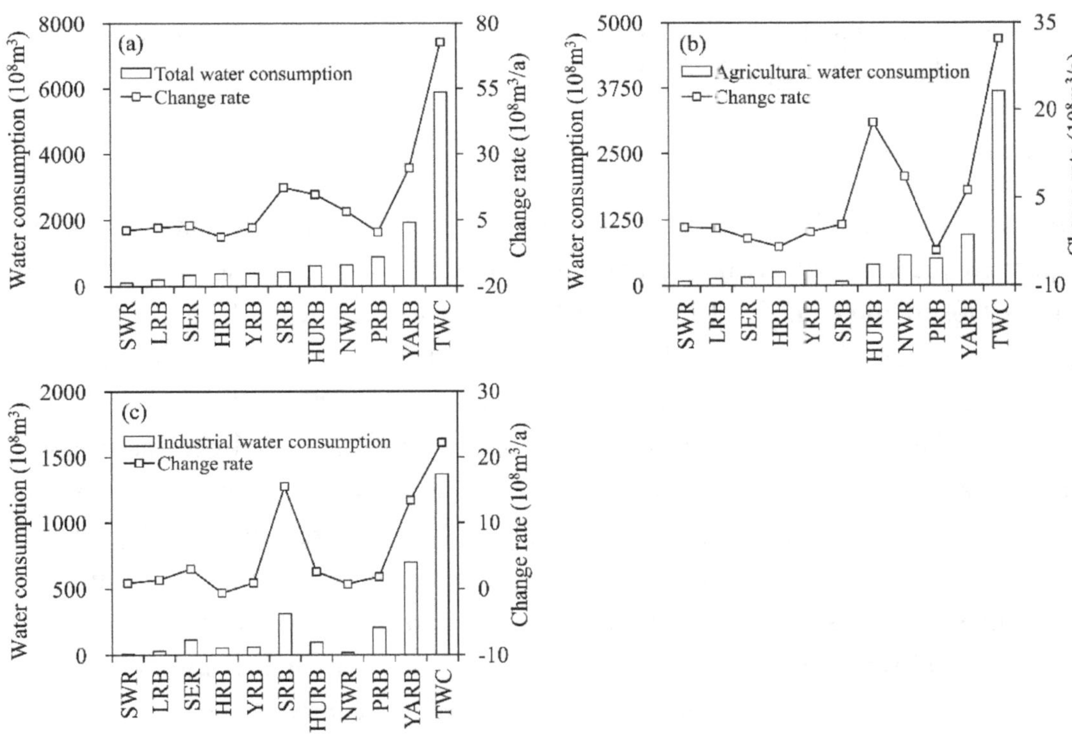

Figure 8. Interannual variability of water consumption and its change rate in the 10 major basins and in the whole country (TWC). (**a**), (**b**), and (**c**) are respectively gross water consumption, agricultural water consumption, and industrial water consumption.

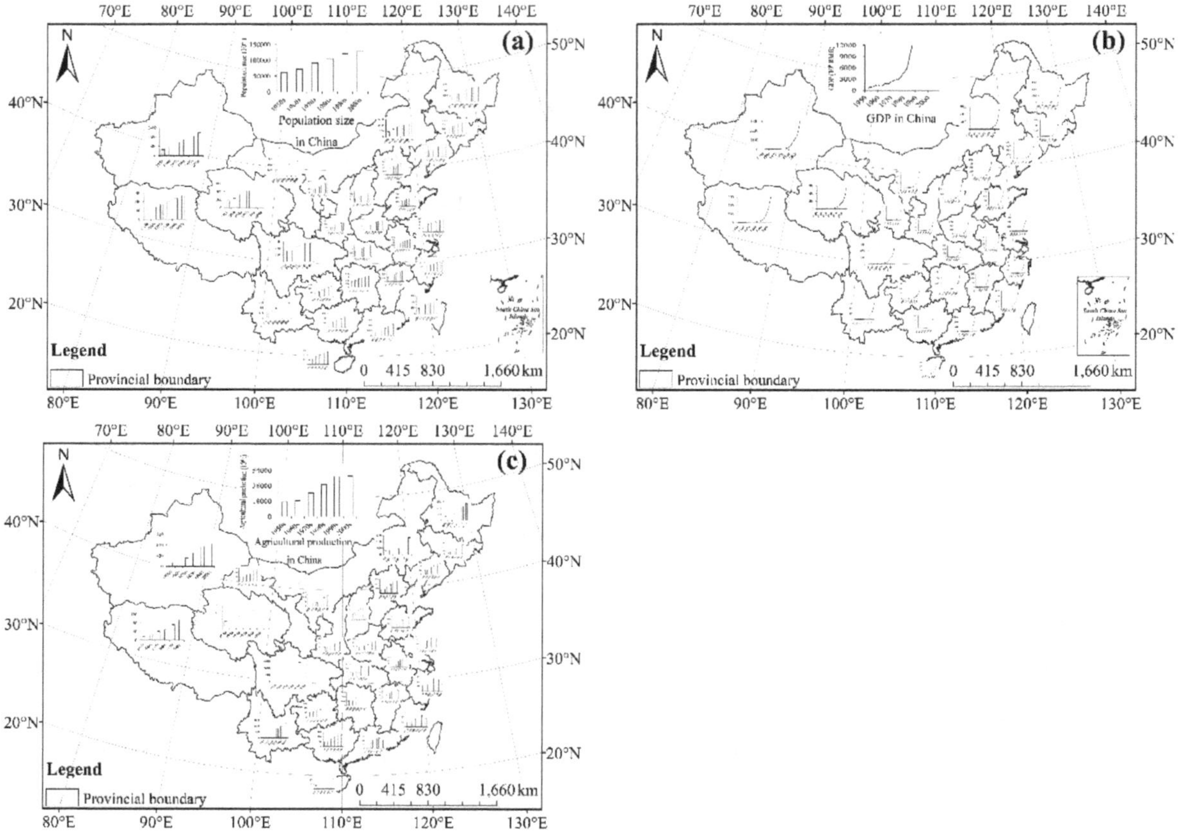

Figure 9. (**a**) Interannual variability of population; (**b**) gross domestic product (GDP); and (**c**) agricultural products in 1950–2012 in China.

4.4. Land Use/Land Coverage Change

The statistics of land use change from 1985 to 2010 are listed in Table 5. National land use change during the past 30 years shows obvious spatial differences and can be divided into two periods: from the end of the 1980s to 2000 and from 2000 to 2010. The characteristics of land use change during the first period indicate that farmland, residential land, industrial land, paddy fields, and water areas increased rapidly, and ecological land was greatly reduced. In 2000–2010, growth in farmland, residential land, paddy fields, and water areas slowed, and ecological land decreased slightly. Three possible causes can be considered. Firstly, construction land expansion was the main cause of the farmland decrease in traditional agricultural areas. In addition, large-scale ecological engineering promoted decreases in farmland. Another aspect, annual accumulated temperature increases in the northern arid zone, created suitable climatic conditions for reclamation in some forest zones, and a large area of dry land was converted to paddy fields. Secondly, implementation of the Reform and Opening-up policy resulted in rapid urbanization processes that occupied farmland and ecological land, particularly in eastern China. Thirdly, before 2000, the reclamation of grassland and forests caused soil degradation, which induced severe soil erosion. After 2000, the 'Grain for Green' program was implemented, which significantly increased the areas of grassland and forests in the ecologically fragile areas of western China.

Table 5. Area of land use conversion in China during the two periods of 1980s–2000 and 2000–2010.

Transfer Types	Dryland–Paddy Field	Grassland–Forest	Others–Waters	Others–ConstructionLand	Forest–Farmland	Forest–Grassland	Grassland–Farmland	Grassland–Forest	Waters–Others
1980–2000	173.15	115.80	80.81	177.63	174.73	31.14	345.76	104.73	62.40
2000–2010	138.12	142.36	114.55	378.24	37.54	41.85	197.38	88.79	83.27

The potential hydrological effect of land use/land coverage change has been a highly contested issue. Firstly, the high intensity of human activities in the city have resulted in accelerated soil compaction and crusting processes, thereby decreasing the infiltration rate and capacity of soil water storage and resulting in surface runoff increases [22,53]. Secondly, the runoff coefficients of forest areas and grassland are relatively small. When the forests and grassland converted to arable land, the flow yield will therefore increase. Thirdly, in southern China, a certain degree of lake reclamation occurred, which inhibited the process of evaporation and infiltration and resulted in runoff increases.

5. Influencing Factors of Sediment Load Variation

5.1. Hydraulic Engineering and Soil and Water Conservation Engineering

5.1.1. Soil and Water Conservation Engineering

China began to conduct soil and water conservation projects in the early 1950s, which included the construction of silt dams and terraced fields and the planting of trees and grass [54,55]. In the middle reaches of the Yellow River, for example, the most severe soil erosion in China occurred. The statistics of soil and water conservation engineering projects conducted in this area in 1979, 1989, 1996, and 2011 are plotted in Figure 10a.

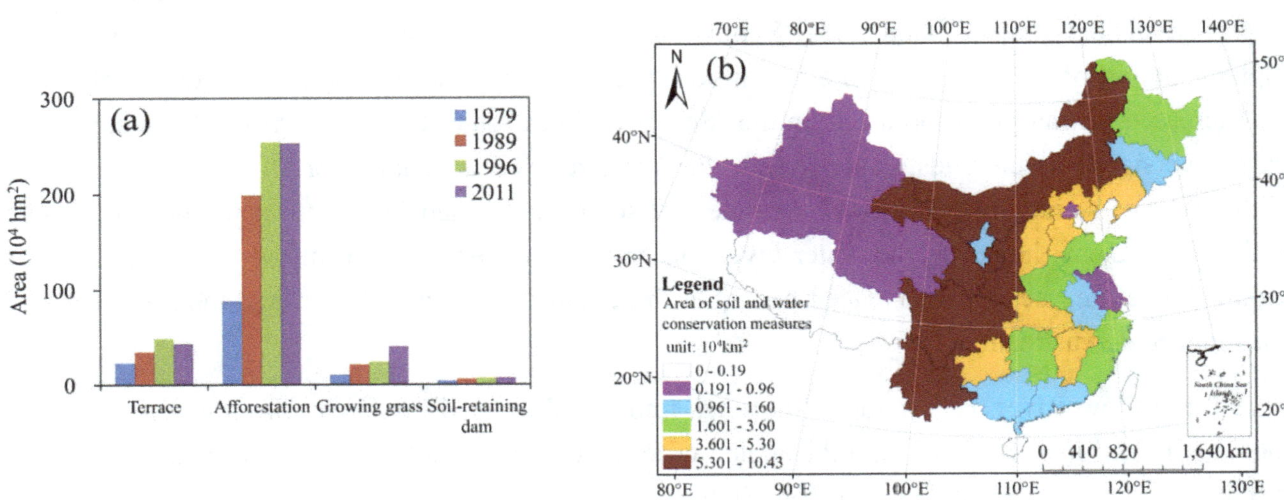

Figure 10. Statistics of soil and water conservation engineering areas in (**a**) the middle stream of the Yellow River basin (YRB) and (**b**) the national distribution.

All types of engineering showed an increasing trend, with the largest area in afforestation. On the national scale, the total area of present soil and water conservation engineering projects is 99.2×10^4 km^2, and the area of silt dams is 925.6 km^2 [56]. The largest distribution is in western China, which incurred severe soil erosion (Figure 10b). During the past 60 years, the runoff in some large rivers declined significantly (Figure 3), largely due to the impact of soil and water conservation engineering.

5.1.2. Reservoir and Dam Construction

By the end of 2012, the total number of large, medium, and small reservoirs in mainstreams and tributaries was 97,543, and a capacity of 8255×10^8 m^3 [56]. These reservoirs regulated the runoff and sediment change processes and reduced the sediment load in the downstream regions. The statistics of the "Water Resources Yearbook in 2006" [57] indicate that by the end of 2006, the numbers of large and medium reservoirs in the YARB were 149 and 1115, respectively, which ranked as highest. For the southern rivers including the YARB, PRB, HURB, and SERB, water resources were abundant. Thus, there were many reservoirs in these basins, and the ratios of total reservoir capacity (TRC) to average annual flow (AAF) was below 70%. However, in the northern rivers including the LRB, SRB, and HRB, the water resources are relatively limited; thus, the TRC/AAF was above 100%. In the two largest basins (YRB and YARB), the total amount of sedimentation in all of the reservoirs in the YARB during 1991–2005 was 17958×10^4 t. For the Sanmenxia and Xiaolangdi reservoirs, sedimentation in 1960–2012 and 1997–2012 was 64.108×10^8 m^3 and 27.625×10^8 m^3, respectively [56].

5.1.3. Water and Sediment Diversion Projects

To resolve the uneven distribution of water resources and to ease the high demand of local water resources, China has been conducting water and sediment diversion projects such as the "South-to-North Water Diversion Project" and the "Yellow River-to-Tianjin Water Diversion Project". After implementation of these projects, the streamflow and sediment load decreased in water supply areas and increased in water demand areas. In the "South-to-North Water Diversion Project", the amount of water diversion from the midstream region of the Yangtze River accounted for a very small proportion of the average streamflow, particularly in the wet seasons, and had little impact in abundant water. In the dry seasons, however, the water diversion project reduced the streamflow in the downstream region. Therefore, the saltwater traced back into the estuary of Yangtze River, and the sediment load was reduced downstream.

In the Yellow River, for example, the total amount of water diversion in the lower reaches was 3665.2×10^8 m^3 in 1958–2002, and the annual water diversion was 89.4×10^8 m^3, accounting for 23.1% of the runoff observed at Huayuankou Station during the same period and directly influencing streamflow and sediment changes in the lower reaches [23,55]. The amount of water and sediment diversion in the downstream region of the Yellow River in 2011 is shown in Table 6. These factors accounted for 15.6% and 13.6% of water and sediment discharge, respectively, during the same period.

Table 6. Amount of water and sediment diversion in the downstream reaches of the Yellow River in 2011.

Channel Segment	Xixiayuan–Huayuankou	Huayuankou–Jiahetan	Jiahetan–Gaocun	Gaocun–Sunkou	Sunkou–Aishan	Aishan–luokou	Luokou–Lijin	Lower Reaches of Lijin	Total
Length (km)	109.8	100.8	77.1	118.2	63.9	101.8	167.8	110.0	849.4
Amount of water diversion (10^8 m^3)	4.19	14.57	17.31	11.50	8.5	21.07	22.08	4.66	103.88
Amount of sediment diversion (10^4 t)	62.8	111.3	286.6	242.7	266.3	800.2	464.4	94.5	2328.8

5.2. Other Human Activities

The impacts of human activities on soil and water loss and sediment discharge in rivers include direct and indirect effects. Direct impacts include the erosion-transport-accumulation process induced by human activities such as farmland reclamation, mining, and road construction. The indirect impacts were mainly caused by destruction of vegetation, which accelerated the occurrence and development of soil erosion. With the development of society and the economy, both the strength and breadth of human activities were greater than those in the past. These activities include increases in population, the scale of reclamation, mining, and other infrastructure construction.

6. Impacts of Water and Sediment Discharge Reduction on Utilization of Sediment Resources

6.1. Effects of River Regulation and Flood Control

In the YRB, the sediment load in the lower reaches of Yellow River was greatly reduced in recent decades. Particularly after 2000, the streamflow and sediment load observed at Huayuankou Station were reduced by 52.5% and 61.7%, respectively, compared with those observed in the 1950s to 1960s. Reducing water and sediment discharges has relieved flood pressure and has influenced river regulation to some degree. Dike reinforcement with silt remains a very important project in the lower reaches of the Yellow River [58]. Decreases in sediment concentration by flooding and the threat of beach inundation by flooding have resulted in significant increases in the cost of dike reinforcement with silt.

6.2. Effects of Flood Irrigation and Soil Improvement

The sediment of the Yellow River, particularly flood sediment, is an effective material for soil improvement because it reduces salinity and alkalinity and improves land fertility. Until the early 1990s, the area of soil improvement was 23.2×10^4 hm^2 in the lower reaches of the Yellow River [58]. Soil improvement by silt is usually conducted during the flood season, when the sediment concentration is relatively high. However, this effect will be limited under the reduction of water and sediment resources.

6.3. Effects of Land and Wetland Formation

In the estuary area of the Yellow River, approximately 64% of sediment is precipitated on land and in shallow water, which elevates the riverbed and delta. The Linjin Station observation reported reductions of 88.7% and 67.5% in sediment and water discharge, respectively, compared with values reported in the 1950s and 1960s. Therefore, the land formation rate has decreased. For example, the 23.6 km^2/a land formation rate in 1855–1954 decreased to 8.6 km^2/a in 1992–2001 [58]. In addition, soil, water, and sediment are the main components of wetlands. Therefore, sediment and water discharge reduction will directly affect the quality and formation rate of wetlands.

7. Conclusions

The results of the present study are summarized in the following points:

1. During the past 60 years, the streamflow in northern China, including the HRB, HURB, YRB, LRB, and SRB, showed different decreasing trends. That in the southern rivers, including MRB, PRB and YARB, presented severe fluctuations, although the declining trend did not reach significant levels. For the streamflow in the TRB, HIRB, YZRB, and LCRB, increasing trends were presented. The runoff yield capacity was weakened in the HRB, YRB, SRB, and PRB and enhanced in the LRB and HURB. That in the MRB and YARB remained stable.
2. In the northern rivers, runoff correlated positively with the EAMI and negatively with the WI. In the southern rivers, runoff was mainly influenced by the Tibet–Qinghai monsoon, South Asian monsoon, and westerlies. That in the HIRB and TRB was controlled mainly by the Tibet–Qinghai monsoon and westerlies. Runoff in the YZRB was controlled by the South Asian monsoon and the Tibet–Qinghai monsoon, whereas that in the LCRB was influenced mainly by the East Asian monsoon and westerlies.
3. Sediment loads in the LCRB and YZRB did not present significant change trends. However, sediment loads in other rivers exhibited varying degrees of gradual reduction, the greatest of which was in the 2000s.
4. Underlying surface and precipitation changes jointly influenced the runoff in eastern rivers. Water consumption for industrial and residential purposes, soil and water conservation engineering projects, hydraulic engineering, and underlying surface changes induced by other factors were the main causes of runoff and sediment reduction.

Acknowledgments

This research was jointly funded by a key consulting project from the Chinese Academy of Engineering (2014-XZ-31), the National Natural Science Foundation of China (41301632) and the Chinese Research Academy of Environmental Sciences special funding for basic scientific research (2014-YKY-003).

Author Contributions

Chong Jiang, Fen Li and Linbo Zhang conceived the research idea and designed the methodology, Chong Jiang and Fen Li performed the analyses, and Chong Jiang, Daiqing Li, Fen Li, and Linbo Zhang wrote the paper.

References

1. Alan, D.Z.; Justin, S.; Edwin, P.M.; Bart, N.; Eric, F.W.; Dennis, P.L. Detection of intensification in global and continental scale hydrological cycles: Temporal scale of evaluation. *J. Clim.* **2003**, *16*, 535–547.
2. Easterling, D.R.; Meehl, G.A.; Parmesan, C.; Changnon, S.A.; Karl, T.R.; Mearns, L.O. Climate extremes: Observations, modeling, and impacts. *Science* **2000**, *289*, 2068–2074.
3. Qiu, J. China drought highlights future climate threats. *Nature* **2010**, *465*, 142–143.
4. Zhang, J.Y.; He, R.M.; Qi, J.; Liu, C.S.; Wang, G.Q.; Jin, J.L.; A new perspective on water issues in north China. *Adv. Water Sci.* **2013**, *24*, 303–310.
5. Zhang, Y.J.; Hu, C.H.; Wang, Y.G. Analysis on variation characteristics and influential factors of runoff and sediment of Liaohe River Basin. *Yangtze River* **2014**, *45*, 32–35.
6. World Water Assessment Programme (WWAP). *The United Nations World Water Development Report 4: Managing Water under Uncertainty and Risk*; World Water Assessment Programme: Paris, France, 2012.
7. Vörösmarty, C.J.; McIntyre, P.B.; Gessner, M.O.; Dudgeon, D.; Prusevich, A.; Green, P.; Glidden, S.; Bunn, S.E.; Sullivan, C.A.; Liermann, C.R.; *et al.* Global threats to human water security and river biodiversity. *Nature* **2010**, *467*, 555–561.
8. Intergovernmental Panel on Climate Change (IPCC). *Managing the Risks of Extreme Events and Disasters to Advance Climate Change Adaptation*; Cambridge University Press: Cambridge, UK, 2012.
9. Climate Change 2013: The Physical Science Basis, Summary for Policy Makers, Technical Summary and Frequently Asked Questions. Available online: https://www.ipcc.ch/pdf/assessment-report/ar5/wg1/WG1AR5_SummaryVolume_FINAL.pdf (accessed on 13 October 2015).
10. Matthew, B.C.; Nigel, W.A. Adapting to climate change impacts on water resources in England-An assessment of draft water resources management plans. *Glob. Environ. Chang.* **2011**, *21*, 238–248.
11. Domroes, M.; Peng, G. *The Climate of China*; Springer: Berlin, Germany, 1988.
12. Zhai, P.M.; Zhang, X.B.; Wan, H.; Pan, X.H. Trends in total precipitation and frequency of daily precipitation extremes over China. *J. Clim.* **2005**, *18*, 1096–1108.
13. Ding, Y.; *Monsoons Over China*; Kluwer Academic Publishers: Amsterdam, the Netherland, 1994.
14. Holeman, J.N. The sediment yield of major rivers of the world. *Water Resour. Res.* **1968**, *4*, 737–747.
15. Holland, H.D. River transport to the oceans. In *The Ocean Lithosphere*; John Wiley & Sons: New York, NY, USA, 1981.
16. Walling, D.E.; Fang, D. Recent trends in the suspended sediment loads of the world rivers. *Glob. Planet. Chang.* **2003**, *39*, 111–126.

17. Milliman, J.D. Delivery and fate of fluvial water and sediment to the sea: A marine geologist's view of European rivers. *Sci. Mar.* **2001**, *65*, 121–132.
18. Bobrovitskaya, N.N.; Kokorev, A.V.; Lemeshko, N.A. Regional patterns in recent trends in sediment yields of Eurasian and Siberian rivers. *Glob. Planet. Chang.* **2003**, *39*, 127–146.
19. Liu, C.; Wang, Y.G.; Sui, J.Y. Analysis on variation of seagoing water and sediment load in main rivers of China. *J. Hydraul. Eng.* **2007**, *38*, 1444–1452.
20. Subramainian, V. Sediment load of Indian rivers. *Curr. Sci.* **1993**, *64*, 928–930.
21. Siakeu, J.; Oguchi, T.; Aoki, T.; Esakid, Y.; Jarviee, H.P. Change in riverine suspended sediment concentration in central Japan in response to late 20th century human activities. *Catena* **2004**, *55*, 231–254.
22. Wang, Y.G.; Liu, X.; Shi, H.L. Variation and influence factor of runoff and sediment in the lower and middle Yangtze River. *J. Sed. Res.* **2014**, *29*, 38–47.
23. Miao, C.Y.; Ni, J.R.; Borthwick, A.G.L. Recent changes of water discharge and sediment load in the Yellow River basin. *Prog. Phys. Geogr.* **2010**, *34*, 541–561.
24. Shi, H.L.; Hu, C.H.; Wang, Y.G.; Tian, Q.Q. Variation trend and cause of runoff and sediment load variations in Huaihe River. *J. Hydraul. Eng.* **2012**, *43*, 571–579.
25. Zhang, J.Y.; Zhang, S.L.; Wang, J.X.; Li, Y. Study on runoff trends of the six larger basins in China over the past 50 years. *Adv. Water Sci.* **2007**, *18*, 230–234.
26. Miao, C.Y.; Yang, L.; Liu, B.Y.; Gao, Y.; Li, S.L. Streamflow changes and its influencing factors in the mainstream of the Songhua River basin, Northeast China over the past 50 years. *Environ. Earth Sci.* **2011**, *63*, 489–499.
27. Yang, S.L.; Zhao, Q.Y.; Belkin, I.M. Temporal variation in the sediment load of the Yangtze River and the influences of the human activities. *J. Hydrol.* **2002**, *263*, 56–71.
28. Gao, G.; Chen, D.L.; Xu, C.Y.; Simelton, E. Trend of estimated actual evapotranspiration over China during 1960–2002. *J. Geophys. Res.* **2007**, *112*, D11120.
29. International Research and Training Center on Erosion and Sedimentation (IRTCES) China. *Gazette of River Sedimentation 2011*; China Water Power Press: Beijing, China, 2012.
30. Lu, X.X.; Higgitt, D.L.; Sediment yield variability in the upper Yangtze, China. *Earth Surf. Proc. Land* **1999**, *24*, 1077–1093.
31. Hassan, M.A.; Church, M.C.; Xu, J.X.; Yan, Y.X. Spatial and temporal variation of sediment yield in the landscape: Example of Huanghe (Yellow River). *Geophys. Res. Lett.* **2008**, *35*, L06401.
32. Mann, H.B. Non-parametric tests against trend. *Econometrica* **1945**, *13*, 245–259.
33. Kendall, M.G. *Rank Correlation Methods*; Griffin: London, UK, 1975; pp. 1–200.
34. Chattopadhyay, S.; Jhajharia, D.; Chattopadhyay, G. Trend estimation and univariate forecast of the sunspot numbers: Development and comparison of ARMA, ARIMA and Autoregressive Neural Network models. *C.R. Geosci.* **2011**, *343*, 433–442.
35. Theil, H. A rank invariant method of linear and polynomial regression analysis. *Proc. R. Neth. Acad. Sci.* **1950**, *53*, 1397–1412.
36. Sen, P.K. Estimates of the regression coefficients based on Kendall's tau. *J. Am. Stat. Assoc.* **1968**, *63*, 1379–1389.
37. Jhajharia, D.; Dinpashoh, Y.; Kahya, E.; Choudhary, R.R.; Singh, V.P. Trends in temperature over Godavari River basin in Southern Peninsular India. *Int. J. Climatol.* **2013**, *34*, 1369–1384.

38. Misuses of statistical analysis in climate researchvon; In *Analysis of Climate Variability: Applications of Statistical Techniques*; Storch, H., Navarra, A., Eds.; Springer: Berlin, Germany, 1995; pp. 11–26.
39. Chattopadhyay, S.; Jhajharia, D.; Chattopadhyay, G. Univariate modeling of monthly maximum temperature time series over northeast India: Neural network *versus* Yule-walker equation based approach. *Meteorol. Appl.* **2011**, *18*, 70–82.
40. Kumar, S.; Merwade, V.; Kam, J.; Thurner, K. Streamflow trends in Indiana: Effects of long term persistence, precipitation and subsurface drains. *J. Hydrol.* **2009**, *374*, 171–183.
41. Xu, J.X. Recent variations in water and sediment in relation with reservoir construction in the upper Changjiang River Basin. *J. Mount. Sci.* **2009**, *27*, 385–393.
42. Merriam, C.F. A comprehensive study of the rainfall on the susquehanna valley. *Trans. Amer. Geophys. Union* **1937**, *18*, 471–476.
43. Searcy, J.K.; Hardisoni, C.H.; Langbein, W.B. *Double Mass Curves*; Geological Survey Water Supply Paper 1541-B; US Geological Survey: Washington, DC, USA, 1960.
44. Mu, X.M.; Zhang, X.Q.; Gao, P.; Wang, F. Theory of double curves and its applications in hydrology and meteorology. *J. China Hydrol.* **2010**, *30*, 47–51.
45. Li, Y.Y.; Cao, J.T.; Shen, F.X.; Xia, J. The changes of renewable water resources in China during 1956–2010. *Sci. China Earth Sci.* **2014**, *57*, 1825–1833.
46. Li, J.P.; Wu, Z.W.; Jiang, Z.H.; He, J.H. Can global warming strengthen the East Asian summer monsoon? *J. Clim.* **2010**, *23*, 6696–6705.
47. Li, J.P.; Feng, J.; Li, Y. A possible cause of decreasing summer rainfall in northeast Australia. *Int. J. Climatol.* **2011**, *31*, doi:10.1002/joc.2328.
48. Wang, M.C.; Liang, J.; Shao, M.J.; Shi, G. Preliminary analysis on interannual change of Tibet–Qinghai Plateau monsoon. *Plateau Meteorol.* **1984**, *3*, 76–82.
49. Ma, H.; Yang, D.W.; Tan, S.K.; Gao, B.; Hu, Q.F. Impact of climate variability and human activity on streamflow decrease in the Miyun Reservoir catchment. *J. Hydrol.* **2010**, *389*, 317–324.
50. Chen, Y.N.; Fan, Y.T.; Wang, H.J.; Fang, G.H. Research progress on the impact of climate change on water resources in the arid region of Northwest China. *Acta Geogr. Sin.* **2014**, *69*, 1295–1304.
51. Song, X.M.; Zhang, J.Y.; Zhan, C.S.; Liu, C.Z. Review for impacts of climate change and human activities on water cycle. *J. Hydraul. Eng.* **2013**, *44*, 779–790.
52. He, D.M.; Liu, C.M.; Feng, Y.; Hu, J.M.; Ji, X.; Li, Y.G. Progress and perspective of international river researches in China. *Acta Geogr. Sin.* **2014**, *69*, 1284–1294.
53. Liu, C.; Hu, C.H.; Shi, H.L. Changes of runoff and sediment fluxes of river in mainland of China discharge into Pacific Ocean. *J. Sed. Res.* **2012**, *27*, 70–75.
54. Liu, C.; He, Y.; Zhang, H.Y. Trends analysis of the water and sediment loads of the main rivers in China using water-sediment diagram. *Adv. Water Sci.* **2008**, *19*, 317–324.
55. Liu, C.; Wang, Z.Y.; Sui, J.Y. Variation of flow and sediment of the Yellow River and their influential factors. *Adv. Sci. Technol. Water Resour.* **2008**, *28*, 1–7.
56. Ministry of Water Resources of the People's Republic of China (MWRPRC). *Chinese River Sediment Bulletin 2012*; China Water Power Press: Beijing, China, 2013; pp. 1–60.

57. Ministry of Water Resources of the People's Republic of China (MWRPRC). *Chinese River Sediment Bulletin 2006*; China Water Power Press: Beijing, China, 2007; pp. 1–55.
58. Wang, Y.G.; Hu, C.H.; Shi, H.L. Variation in water and sediment resources and its influence on utilization of sediment resource in the Yellow River Basin. *J. China Inst. Water Resour. Hydropower Res.* **2010**, *8*, 237–245.

Permissions

All chapters in this book were first published in MDPI; hereby published with permission under the Creative Commons Attribution License or equivalent. Every chapter published in this book has been scrutinized by our experts. Their significance has been extensively debated. The topics covered herein carry significant findings which will fuel the growth of the discipline. They may even be implemented as practical applications or may be referred to as a beginning point for another development.

The contributors of this book come from diverse backgrounds, making this book a truly international effort. This book will bring forth new frontiers with its revolutionizing research information and detailed analysis of the nascent developments around the world.

We would like to thank all the contributing authors for lending their expertise to make the book truly unique. They have played a crucial role in the development of this book. Without their invaluable contributions this book wouldn't have been possible. They have made vital efforts to compile up to date information on the varied aspects of this subject to make this book a valuable addition to the collection of many professionals and students.

This book was conceptualized with the vision of imparting up-to-date information and advanced data in this field. To ensure the same, a matchless editorial board was set up. Every individual on the board went through rigorous rounds of assessment to prove their worth. After which they invested a large part of their time researching and compiling the most relevant data for our readers.

The editorial board has been involved in producing this book since its inception. They have spent rigorous hours researching and exploring the diverse topics which have resulted in the successful publishing of this book. They have passed on their knowledge of decades through this book. To expedite this challenging task, the publisher supported the team at every step. A small team of assistant editors was also appointed to further simplify the editing procedure and attain best results for the readers.

Apart from the editorial board, the designing team has also invested a significant amount of their time in understanding the subject and creating the most relevant covers. They scrutinized every image to scout for the most suitable representation of the subject and create an appropriate cover for the book.

The publishing team has been an ardent support to the editorial, designing and production team. Their endless efforts to recruit the best for this project, has resulted in the accomplishment of this book. They are a veteran in the field of academics and their pool of knowledge is as vast as their experience in printing. Their expertise and guidance has proved useful at every step. Their uncompromising quality standards have made this book an exceptional effort. Their encouragement from time to time has been an inspiration for everyone.

The publisher and the editorial board hope that this book will prove to be a valuable piece of knowledge for researchers, students, practitioners and scholars across the globe.

List of Contributors

Pao-Shan Yu, Tao-Chang Yang, Chen-Min Kuo and Hung-Wei Tseng
Department of Hydraulic and Ocean Engineering, National Cheng Kung University, No. 1, University Rd., Tainan 701, Taiwan

Shien-Tsung Chen
Department of Water Resources Engineering and Conservation, Feng Chia University, No. 100, Wenhwa Rd., Seatwen, Taichung 407, Taiwan

Aymar Y. Bossa
West African Science Service Center on Climate Change and Adapted Land Use—WASCAL, Ouagadougou 06, Burkina Faso
Department of Geography, University of Bonn, Meckenheimer Allee 166, Bonn 53115, Germany

Bernd Diekkrüger
Department of Geography, University of Bonn, Meckenheimer Allee 166, Bonn 53115, Germany

Euloge K. Agbossou
Faculty of Agricultural Sciences, University of Abomey-Calavi, Cotonou 01 BP. 526, Benin

Janel Hanrahan
Atmospheric Sciences Department, Lyndon State College, Lyndonville, VT 05851, USA

Paul Roebber and Sergey Kravtsov
Department of Mathematical Sciences, Atmospheric Sciences Group, University of Wisconsin-Milwaukee, Milwaukee, WI 53201, USA

Francis Ndamani
Graduate School of Engineering, Kochi University of Technology, 2-22 Eikokuji, Kochi City, Kochi 780-8515, Japan

Tsunemi Watanabe
School of Economics and Management, Kochi University of Technology, 2-22 Eikokuji, Kochi City, Kochi 780-8515, Japan

Mohammad Kamruzzaman and Simon Beecham
Centre for Water Management and Reuse (CWMR), School of Natural and Built Environments, University of South Australia, Mawson Lakes, SA 5095, Australia

Md Sumon Shahriar
CSIRO Computational Informatics (CCI), Hobart, TAS 7001, Australia

Todd Pagano and Morgan Bida
Laboratory Science Technology Program, Department of Science & Mathematics, Rochester Institute of Technology, Rochester, NY 14623, USA

Jonathan E. Kenny
Department of Chemistry, Tufts University, Medford, MA 02155, USA

Qingting Li, Yue Sui and Huadong Guo
Key Laboratory of Digital Earth Science, Institute of Remote Sensing and Digital Earth, Chinese Academy of Sciences, No. 9 Dengzhuang South Road, Haidian District, Beijing 100094, China

Linlin Lu
Key Laboratory of Digital Earth Science, Institute of Remote Sensing and Digital Earth, Chinese Academy of Sciences, No. 9 Dengzhuang South Road, Haidian District, Beijing 100094, China
State Key Laboratory of Remote Sensing Science, Institute of Remote Sensing and Digital Earth, Chinese Academy of Sciences, Beijing 100101, China

Cuizhen Wang
Department of Geography, University of South Carolina, Columbia, SC 29208, USA

Yingkui Li
Department of Geography, University of Tennessee, Knoxville, TN 37996, USA

Ricardo Arias, M. Luz Rodríguez-Blanco, M. Mercedes Taboada-Castro and M. Teresa Taboada-Castro
Facultad de Ciencias and Centro de Investigaciones Científicas Avanzadas (CICA), University of A Coruña, Campus de A Coruña, A Coruña 15071, Spain

Joao Pedro Nunes and Jan Jacob Keizer
Centre for Environmental and Marine Studies (CESAM) and Department Environment & Planning, University of Aveiro, Aveiro 3810-193, Portugal

Jyun-Long Lee and Wen-Cheng Huang
Department of Harbor and River Engineering, National Taiwan Ocean University, Keelung 20224, Taiwan

Chong Jiang
State Key Laboratory of Environmental Criteria and Risk Assessment, Chinese Research Academy of Environmental Sciences, Beijing 100012, China
College of Global Change and Earth System Science, Beijing Normal University, Beijing 100875, China
Joint Center for Global Change Studies, Beijing 100875, China
Key Laboratory of Regional Eco-Process and Function Assessment and State Environment Protection, Chinese Academy of Environmental Sciences, Beijing 100012, China

Linbo Zhang, Daiqing Li and Fen Li
State Key Laboratory of Environmental Criteria and Risk Assessment, Chinese Research Academy of Environmental Sciences, Beijing 100012, China
Key Laboratory of Regional Eco-Process and Function Assessment and State Environment Protection, Chinese Academy of Environmental Sciences, Beijing 100012, China

Index